SpringerWienNewYork

CISM COURSES AND LECTURES

The series presents lecture notes, monographs, edited works and proceedings in the field of Mechanics, Engineering, Computer Science and Applied Mathematics.
Purpose of the series is to make known in the international scientific and technical community results obtained in some of the activities organized by CISM, the International Centre for Mechanical Sciences.

INTERNATIONAL CENTRE FOR MECHANICAL SCIENCES

COURSES AND LECTURES - No. 530

ELECTROKINETICS AND ELECTROHYDRODYNAMICS IN MICROSYSTEMS

EDITED BY

ANTONIO RAMOS
UNIVERSITY OF SEVILLE, SPAIN

SpringerWienNewYork

This volume contains 143 illustrations

SPIN 80073444

All contributions have been typeset by the authors.

ISBN 978-3-7091-0899-4 SpringerWienNewYork

PREFACE

The manipulation of colloidal particles and fluids in microsystems by using electrical forces has many existing and potential applications. The electrical manipulation at the micrometre scale presents the advantages of voltage-based control and dominance over other forces. The latter is a clear example of the scaling laws of physical systems: in the range above a few millimetres the electrical forces are rather ineffective, but in the micrometre (and submicrometre) scale the electrical forces dominate.

The present book contains the Lecture Notes of a CISM Advanced School on "Electrokinetics and Electrohydrodynamics in Microsystems", held in Udine, Italy, on 22nd-26th June 2009. The aim of the book is to provide a state-of-the-art knowledge on both theoretical and applied aspects of the electrical manipulation of colloidal particles and fluids in microsystems. The book covers the following topics: Dielectrophoresis (DEP), Electrowetting, Electrohydrodynamics (EHD) in microsystems, and Electrokinetics of fluids and particles.

The first three chapters of this book are dedicated to Dielectrophoresis. Chapter I, by Tao Sun and Hywel Morgan, presents an overview of the basis of Dielectrophoresis and its applications for the manipulation and characterization of particles. Chapters II and III, by Nicolas G. Green, examine in depth the theory of Dielectrophoresis. Chapter II provides the basics of Quasi-electrostatics from the perspective of Dielectrophoresis and Chapter III discusses the forces produced by the interaction of electric fields with the induced dipole moments on particles.

Chapter IV, by Pablo García-Sánchez and Frieder Mugele, is dedicated to Electrowetting, where fluid handling is achieved by subdividing the liquid into discrete droplets that are manipulated by electrical forces. The chapter introduces briefly basic concepts of wetting and discusses in detail the fundamental physics behind the electrowetting phenomenon. The second part of the chapter reviews applications of Electrowetting in Lab-on-Chip, Optics, displays and microfluidic devices.

Chapter V, by Antonio Ramos, provides an overview of Electrohydrodynamic pumping in microsystems. The chapter presents first the basic equations of Electrohydrodynamics in the micrometre scale.

Subsequently, five different EHD micropumps are studied and compared: from those that employ forces in the liquid bulk to those that employ forces in the electrical double layer.

Chapter VI, by Chuan-Hua Chen, is devoted to Electrohydrodynamic Stability in the context of the ohmic model. Many EHD systems, such as leaky dielectric liquids or electrolytic solutions, fall into the ohmic regime. The chapter presents first a derivation of the EHD equations in the ohmic regime. Afterwards, basic concepts of EHD stability are illustrated using two model problems: the electrokinetic mixing flow and the EHD cone-jet flow.

The last chapter of the book covers an important part of electromicrofluidics: fluid flows generated by electrical forces in the double layer. Chapter VII, by Martin Z. Bazant, presents an introduction to Induced Charge Electrokinetic (ICEK) Phenomena, where the applied electric field acts on its own induced-charge in an electrolytic solution near a polarizable surface. The chapter discusses the basic physics behind ICEK of colloidal particles (induced charge electrophoresis) and fluids (induced charge electroosmosis).

I would like to take this opportunity to thank all the contributors to this book for their valuable time and the excellence of their work. My sincere thanks are also extended to Prof. Giulio Maier, rector of CISM, to Prof. Paolo Serafini for his help in the editorial work, and to all the CISM staff in Udine. Finally, I would like to acknowledge the help of my colleagues at the University of Seville, Antonio Castellanos, Alberto T. Pérez and Antonio González, in the process of preparing both the CISM course and this book.

Antonio Ramos

CONTENTS

AC Electrokinetic Micro- and Nano-particle Manipulation and Characterization

Tao Sun and Hywel Morgan
School of Electronics and Computer Science
University of Southampton, SO17 1BJ, UK

Abstract Automated or remote manipulation and characterization of particles is a key element in microfluidic devices. Microelectrodes integrated into microfluidic devices can generate large electric fields and field gradients using low voltages. The field gradients can be used to actively drive the motion of particles by dielectrophoresis. In this chapter, the basis of AC electrokinetics is reviewed and example applications for manipulation and characterization of particles are provided.

1 Introduction

Microfluidics involves the manipulation of fluids and particles within a microscale chip. Physical parameters that characterize microfluidics include Reynolds number, diffusion, fluidic resistance, surface area to volume ratio and surface tension (Beebe et al., 2002; Stone et al., 2004). Within microfluidic systems, the manipulation of particles is of fundamental importance in bioanalytical science and biotechnology. Manipulation involves a range of processes including patterning, focusing, sorting, trapping, handling and separation. These operations call on a wide range of techniques such as hydrodynamic focusing (Lee et al., 2001; Rodriguez-Trujillo et al., 2007), electrophoresis (Lacher et al., 2001; Kremser et al., 2004), optical tweezer (Ashkin et al., 1986; Ashkin, 1997), acoustic standing waves (Wiklund et al., 2006; Laurell et al., 2007), magnetism (Pamme, 2006) and AC electrokinetics (Pohl, 1978; Pethig, 1979; Morgan and Green, 2003). Characterization of particles within microfluidic systems provides a quantitative and analytical approach to interrogate the physico-chemical properties of particles, such as size, permittivity and conductivity. In terms of cell handling a number of different methods can be combined to produce a complex technological platform. In this chapter, we describe AC electrokinetic principles and techniques whereby electric fields are used to manipulate and characterize particles. We show how a combination of electric fields and hydrodynamics can be used to manipulate particles. The main focus of the chapter is devoted to explaining the principles and applications of dielectrophoresis (DEP). Examples of applications of various AC electrokinetic techniques are given, including a short overview of single cell impedance analysis methods.

2 AC electrokinetics

AC electrokinetics describes the general behavior of polarisable particles in an AC electric field.

2.1 Polarized Particle

If a homogeneous solid dielectric sphere (for example a solid particle with a radius of R) sits in a homogeneous dielectric medium, charges will accumulate at the interface between the particle and the medium. When an electric field is applied, positive and negative charges are pulled in opposite directions, which gives rise to an effective or induced dipole moment across the particle. The mechanism of the formation of this induced dipole, $\mathbf{p}$, is the first approximation for a polarized particle in an electric field, called the dipole approximation. The net force, $\mathbf{F}$, and torque, Γ on this polarized particle can be calculated from the induced dipole moment, and these are given by:

$$\mathbf{F} = \mathbf{p} \cdot \nabla \mathbf{E} \tag{1}$$

$$\Gamma = \mathbf{p} \times \mathbf{E} \tag{2}$$

The operator $\cdot$ is the dot product; the operator $\times$ the cross product, and ∇ the gradient of (the field). If the electric field is uniform, the net force on the dipole (particle) is zero. For a dipole at some random orientation to the field, the torque will tend to align the dipole (particle) with the electric field.

To quantitatively calculate the effective dipole moment, analysis is performed using spherical coordinates $\{r \geq 0,\ 0 \leq \theta \leq \pi,\ 0 \leq \vartheta \leq 2\pi\}$ with the particle located in a homogeneous and parallel electric field along the negative z-axis, as shown in figure 1.

The potential distributions inside the particle ϕ_p and in the medium ϕ_m are given by (Lorrain et al. 1988):

$$\phi_p = -\left(\frac{3\tilde{\epsilon}_m}{\tilde{\epsilon}_p + 2\tilde{\epsilon}_m}\right) E_0 \cos\theta \tag{3}$$

$$\phi_m = -\left[1 - \frac{R^3}{r^3}\left(\frac{\tilde{\epsilon}_p - \tilde{\epsilon}_m}{\tilde{\epsilon}_p + 2\tilde{\epsilon}_m}\right)\right] E_0 \cos\theta \tag{4}$$

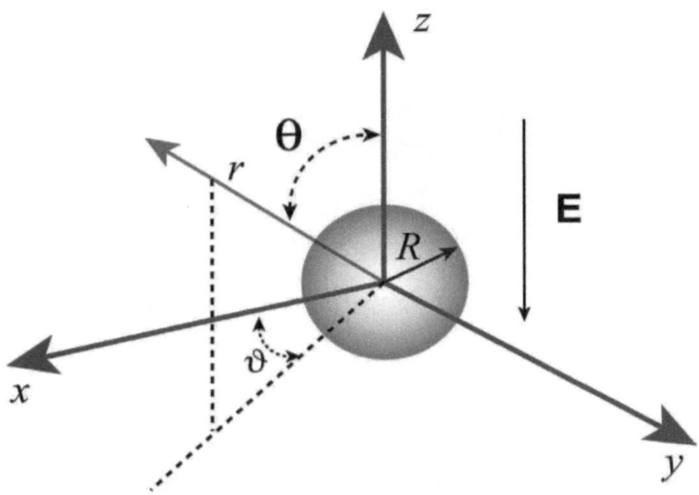

Figure 1. Diagram showing a spherical particle locating in an homogeneous and parallel electric field along with the z-axis.

where $\tilde{\epsilon} = \epsilon - j\sigma/\omega$ is the complex permittivity, $j^2 = -1$ and ω is the angular frequency. The subscripts p and m refer to particle and medium respectively. E_0 is the amplitude of the applied electric field.

The potential of the effective dipole moment $\mathbf{P}_{eff}$ can be considered as an increment in the potential distribution of the applied field, with $\mathbf{P}_{eff}$ given by:

$$\mathbf{P}_{eff} = 4\pi\tilde{\epsilon}_m \left(\frac{\tilde{\epsilon}_p - \tilde{\epsilon}_m}{\tilde{\epsilon}_p + 2\tilde{\epsilon}_m} \right) R^3 \mathbf{E} \tag{5}$$

According to equation 5, the effective dipole moment is frequency-dependent where the dependence is characterized by the Clausius-Mossotti factor $\tilde{f}_{CM}$:

$$\tilde{f}_{CM} = \frac{\tilde{\epsilon}_p - \tilde{\epsilon}_m}{\tilde{\epsilon}_p + 2\tilde{\epsilon}_m} \tag{6}$$

The Clausius-Mossotti factor provides a quantitative evaluation of the polarisability of the particle and the suspending medium. If the particle is not solid, e.g. has a shell like a cell, as shown in figure 2a then the particle complex permittivity $\tilde{\epsilon}_p$ is a function of the dielectric properties of the shell (membrane) and the inside (cytoplasm). This is given by:

$$\tilde{\epsilon}_p = \tilde{\epsilon}_{mem} \frac{\gamma^3 + 2\dfrac{\tilde{\epsilon}_i - \tilde{\epsilon}_{mem}}{\tilde{\epsilon}_i + 2\tilde{\epsilon}_{mem}}}{\gamma^3 - \dfrac{\tilde{\epsilon}_i - \tilde{\epsilon}_{mem}}{\tilde{\epsilon}_i + 2\tilde{\epsilon}_{mem}}} \quad \text{with}\, \gamma = \frac{R_i + d}{R_i} \tag{7}$$

where $\tilde{\epsilon}_{mem}$ is the complex permittivity of the membrane, $\tilde{\epsilon}_i$ the internal properties, and the cell has inner radius R_i and membrane thickness d.

For a cell in suspension, the dielectric properties of the suspending system has two intrinsic relaxation frequencies. The first relaxation occurs at low frequencies and is due to (Maxwell-Wagner) polarization of the cell membrane-suspending medium interface. The second relaxation, occurs at higher frequencies, and is due to polarization between the suspending medium and the cell cytoplasm, where the cell membrane capacitance is in effect short-circuited.

Figure 2b shows these two relaxations as the real and imaginary parts of the Clausius-Mossotti factor of a cell suspended in a medium with different conductivities (see legends for details). It clearly shows that the imaginary part has two peaks, each corresponding to the two relaxations.

2.2 Dielectrophoresis

The phenomenon of DEP originates from the interaction of the induced dipole moment with the applied electric field. In a field gradient there is a net force on the polarized particle that causes it to move towards either the high or low electric field regions depending on the particle (and suspending medium) properties. According to equation 1, two conditions must be satisfied for DEP to occur. First there must be a difference between the polarisability of the particle and medium so that an induced dipole moment is established across the particle. Secondly, the electric field must be non-uniform. Figure 3 shows the field configuration when a particle sits in a non-uniform electric field depending on the particle polarisability. When the particle polarisability is greater than the suspending medium, the particle behaves as a conductor and the electric field vectors bend towards the particle, meeting the surface at right angles. The field inside the particle is nearly zero, as shown in figure 3a. The converse is shown in figure 3b, where the particle polarisability is less than the medium. The field vectors now bend around the particle as if it were an insulator. When the polarisability of the particle and electrolyte are the same it is as if the particle does not exist and the field lines are parallel and continuous everywhere. The imbalance of forces on the induced dipole gives rise to particle movement, i.e. DEP. When the polarisability of the particle is greater than its surrounding

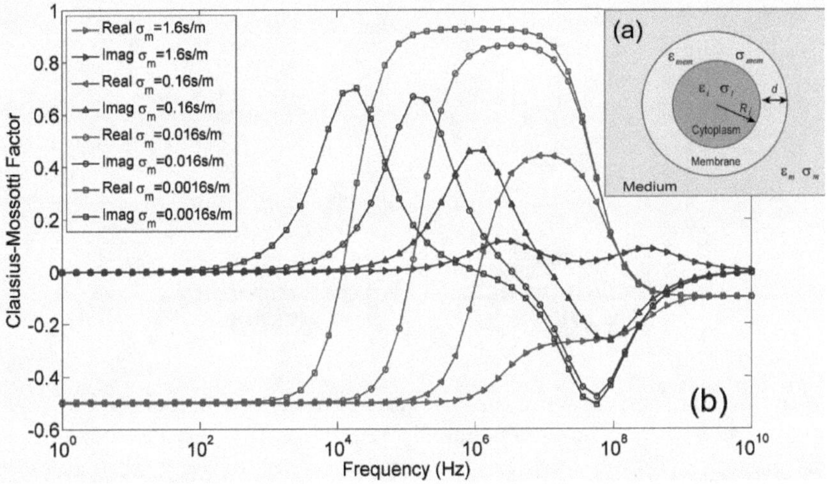

Figure 2. (a) Diagram showing a single shell spherical particle, representing a cell in suspension. (b) Plot showing a spectrum of the real and imaginary parts of the Clausius-Mossotti factor for a particle in an electrolyte, calculated for different suspending medium conductivities. The following parameters were used: $\epsilon_o = 8.854 \times 10^{-12}\,\mathrm{Fm}^{-1}$, $R_i = 3 \times 10^{-6}$ m, $d = 5 \times 10^{-9}$ m, $\epsilon_m = 80\epsilon_o$, $\epsilon_{mem} = 5\epsilon_o$, $\sigma_{mem} = 10^{-8}\,\mathrm{Sm}^{-1}$, $\epsilon_i = 60\epsilon_o$, $\sigma_i = 0.4\,\mathrm{Sm}^{-1}$.

medium, the direction of the dipole is with the field and the particle experiences a positive DEP force (pDEP); the particle moves towards the high field region. The opposite situation gives rise to negative DEP (nDEP); and the particle moves away from regions of high electric fields.

The time-averaged dielectrophoretic force on the dipole is given by:

$$\langle \mathbf{F}_{DEP} \rangle = \frac{1}{2}\mathrm{Re}\left[(\tilde{\mathbf{p}} \cdot \nabla)\tilde{\mathbf{E}}^*\right] = \frac{1}{2}\upsilon\mathrm{Re}\left[\tilde{\alpha}(\mathbf{E} \cdot \nabla)\tilde{\mathbf{E}}^*\right] \tag{8}$$

where $\tilde{\mathbf{p}}$ is the induced dipole moment phasor, υ the volume of the particle, $\tilde{\alpha}$ the effective polarisability and $*$ indicates complex conjugate.

If the non-uniform electric field has no spatially dependent phase, the dielectrophoretic force simplifies to:

$$\langle \mathbf{F}_{DEP} \rangle = \frac{1}{4}\upsilon\mathrm{Re}\left[\tilde{\alpha}\right] \nabla|\tilde{\mathbf{E}}|^2 \tag{9}$$

For a spherical particle, equation 9 becomes:

$$\langle \mathbf{F}_{DEP} \rangle = \pi\epsilon_m R^3 \mathrm{Re}\left[\tilde{f}_{CM}\right] \nabla|\tilde{\mathbf{E}}|^2 \tag{10}$$

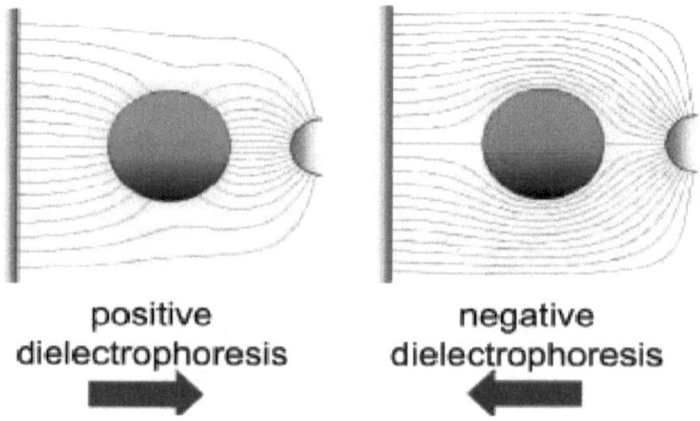

Figure 3. Diagram showing the principle of dielectrophoresis (DEP), which only occurs in a non-homogeneous electric field. (a) particle more polarizable than the medium; positive dielectrophoresis (pDEP) (b) particle less polarizable than the medium; negative dielectrophoresis (nDEP).

According to equation 10, if the electric field is uniform, the gradient of the magnitude of the field is zero ($\nabla|\tilde{\mathbf{E}}|^2 = 0$), which means that there is no DEP force. The frequency dependence and the direction of the DEP force are governed by the real part of the Clausius-Mossotti factor. If the particle is more polarisable than the medium, ($\mathrm{Re}\left[\tilde{f}_{CM}\right] > 0$), the particle is attracted to high intensity electric field regions (pDEP). Conversely, if the particle is less polarisable than the medium, ($\mathrm{Re}\left[\tilde{f}_{CM}\right] < 0$), the particle is repelled from high intensity field regions (nDEP).

2.3 Travelling wave dielectrophoresis

Note that the simplification of equation 8 to equation 9 is based on the assumption that the non-uniform electric field has no spatially dependent phase. In contrast, in electric fields with spatially varying phases, equation 8 can be expanded as:

$$\langle \mathbf{F}_{DEP} \rangle = \frac{1}{4} \upsilon \mathrm{Re}\left[\tilde{\alpha}\right] \nabla|\tilde{\mathbf{E}}|^2 - \frac{1}{2} \upsilon \mathrm{Im}\left(\nabla \times (\mathrm{Re}[\tilde{\mathbf{E}}] \times \mathrm{Im}[\tilde{\mathbf{E}}])\right) \qquad (11)$$

Equation 11 shows that the dielectrophoretic force consists of two components; the first term on the right is the DEP force; the second term is called the travelling wave DEP (twDEP) force. In this case a field with spatially varying phase can be generated with a travelling electric field as shown in figure 4 for an interdigitated electrode array energized with a four phase AC signal. This twDEP force propels a particle along the electrode array. If there is no spatially varying phase, the imaginary part of the electric field is zero ($\text{Im}[\tilde{\mathbf{E}}] = 0$), which means that there is no twDEP. In order for twDEP to be effective, the frequency of the excitation voltage and the conductivity of the medium should be chosen to satisfy two conditions: (a) the particle must experience nDEP so that it can be levitated above the electrode array and (b) the imaginary part of the Clausius-Mossotti factor has to be non-zero.

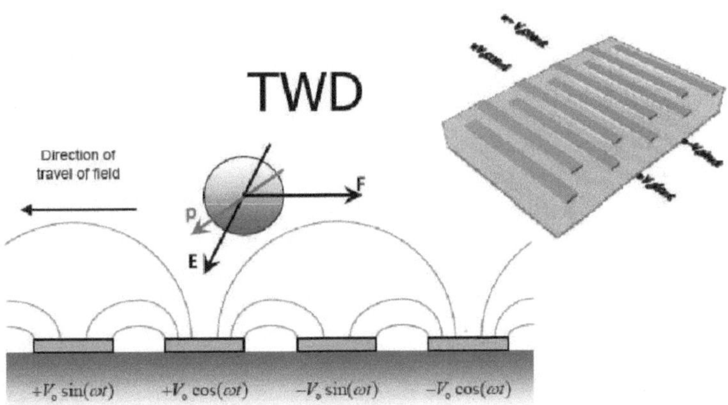

Figure 4. Diagram showing the principle of travelling wave DEP (twDEP). The consecutive phase-shifted signals generate a travelling electric field.

2.4 Electrorotation

When a dipole sits in a field, the interaction between the electric field and this dipole moment leads to a torque on the particle. There is a finite time (or phase delay) between the application of the electric field and the establishment of the dipole moment. If the field vector now changes direction, the vector of the dipole moment will try to follow the changing of field vector. If the field vector rotates then the particle will also rotate. The time-averaged rotating torque is given by:

$$\langle \mathbf{\Gamma}_{ROT} \rangle = \frac{1}{2}\text{Re}\left[\tilde{\mathbf{p}} \times \tilde{\mathbf{E}}^*\right] = -v\text{Im}[\tilde{\alpha}]\left(\text{Re}[\tilde{\mathbf{E}}] \times \text{Im}[\tilde{\mathbf{E}}]\right) \tag{12}$$

For a spherical particle, this becomes:

$$\langle \Gamma_{ROT} \rangle = -4\pi\epsilon_m R^3 \text{Im}[\tilde{f}_{CM}]|\tilde{\mathbf{E}}|^2 \tag{13}$$

Equation 13 shows that the frequency-dependent property of the ROT torque depends on the imaginary part of the Clausius-Mossotti factor. The particle will rotate with or against the electric field, depending on whether the imaginary part of the Clausius-Mossotti factor is negative or positive. If the charge relaxation time constant of the particle is smaller than that of the medium ($\tau_p = \epsilon_p/\sigma_p < \tau_m = \epsilon_m/\sigma_m$), the particle rotates with the changing direction of the field. If $\tau_m < \tau_p$, the particle rotates against the field. In a viscous medium, the particle rotates at a constant angular velocity. The ROT torque can be measured indirectly by analysis of this angular velocity, which is given by (Arnold and Zimmerman 1988)

$$R_{ROT}(\omega) = -\frac{\epsilon_m \mathrm{Im}[\tilde{f}_{CM}]|\tilde{\mathbf{E}}|^2}{2\eta}\xi \tag{14}$$

where $R_{ROT}(\omega)$ is the rotation rate and ξ is a scaling factor that is introduced to consider that neither the viscosity η nor the electric field strength are precisely known. A typical four-electrode configuration used for ROT experiments is shown in figure 5. Here the rotating electric field is generated by four sine waves in phase quadrature.

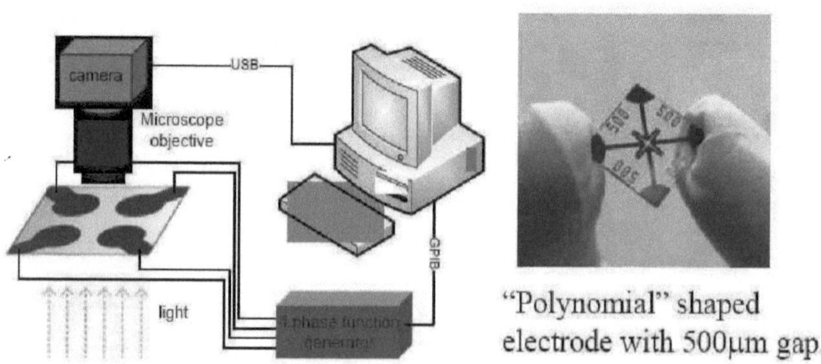

Figure 5. Diagram showing a typical set up of electrorotation (left) and an image of a chip for electrorotation (right).

The frequency spectra of both the DEP force and ROT torque (given by figure 2) can provide information on the dielectric properties of particles in suspension. The relationship between DEP and ROT can be examined using Argand diagrams (Wang et al., 1992, 1993), where the real and imaginary parts of the Clausius-Mossotti factor are mapped onto the complex plane as a function of frequency.

3 Micro-particle Manipulation

Micro-particles in suspension can be manipulated using DEP. The force depends on the magnitude of the field and the gradient, together with the particle volume. The direction of the force depends on the Clausius Mossotti factor, which is a measure of the polarisability of the particle in the suspending medium and importantly varies with the frequency of the applied potential (as shown in figure 2). Particle manipulation and separation have been achieved using a wide range of different electrode configurations. In this section, we review three classical electrode configurations that have been used for DEP manipulation: castellated electrodes, polynomial electrodes and interdigitated electrodes, the later used for field flow fraction (FFF) separation. We then describe recent applications of insulator-based DEP for particle sorting and the use of nDEP for single particle trapping.

3.1 Classical Electrode Configurations

A castellated electrode is shown in figure 6a. The electrodes are designed such that regions of the both positive and negative DEP can be found simultaneously. The array was first used by Pethig's group (Price et al., 1988; Burt et al., 1989) to dielectrophoretically collect particles. Typical electrode dimensions are 10 to 100 μm width and gap. The field is maximum at the electrode tips and minimum in the gaps between electrodes as shown in the figure 6b. This electrode array has been widely used both to separate and to characterize the DEP behavior of particles. Particle experiencing pDEP collect on the tips, and those experiencing nDEP in the gaps.

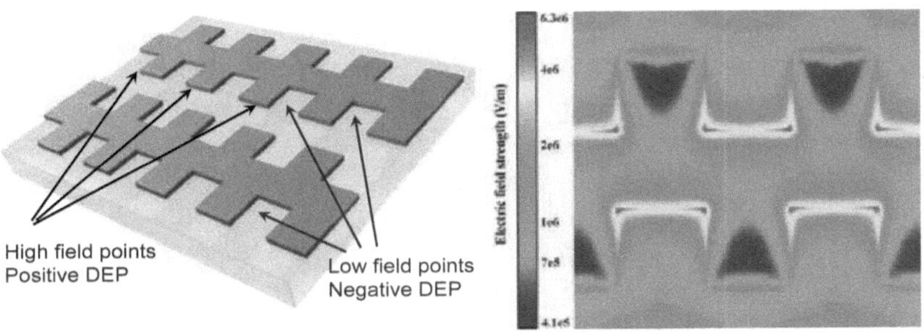

Figure 6. Schematic diagram of a castellated electrode, where typical electrode gap and spacing is 10 μm to 100 μm (left) and (b) an electric field plot showing the high field regions at the electrode tips.

The polynomial electrode design is shown figure 7 and has four electrodes arranged around a centre point, with the edges defined by a hyperbolic function in the centre, with parallel edges out to an arbitrary distance. The principle underlying this electrode design has been described by Huang et al. (1992). In the centre there is a region of almost uniform field gradient; particles can be trapped here by nDEP. At the electrode edges the field gradient is maximum. The polynomial electrode has been used for trapping and characterising a range of biological particles and also nano-particles (Green et al., 2000).

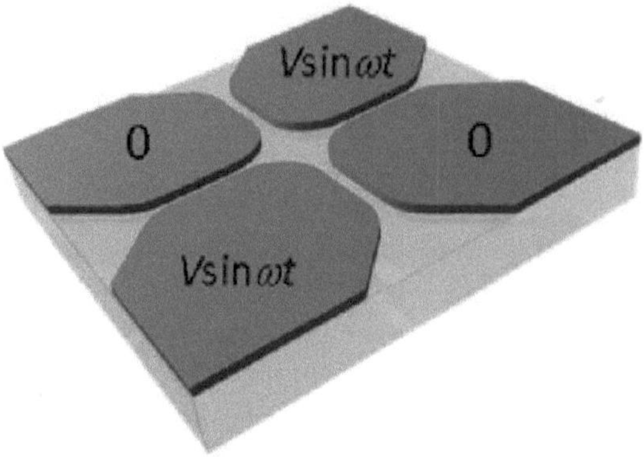

Figure 7. Schematic image of an electrode array defined by a polynomial, used for trapping cells in the centre by negative DEP.

The interdigitated electrode array (figure 8) is often used in DEP separation systems, since it generates a DEP force that decays exponentially from the surface. Depending on the applied frequency, the force either pulls particles towards the electrode edges or pushes particles away under nDEP.

When configured appropriately, this electrode array can also be used for twDEP. Here four AC signals phase shifted by 90° are sequentially applied to the electrodes.

A knowledge of the electric field and field gradient is required to analyse and predict the behavior of particles in these electrode systems. The simplest electrode array is the interdigitated device (figure 8), and the electric field for this system can be determined in a number of ways, including using numerical methods or analytical approaches such as Schwarz-Christoffel Mapping (Sun et al., 2007). In this paper, full analytical solutions were given for the electric field, the dielectrophoretic and travelling-wave dielectrophoretic forces for the interdigitated electrode arrays.

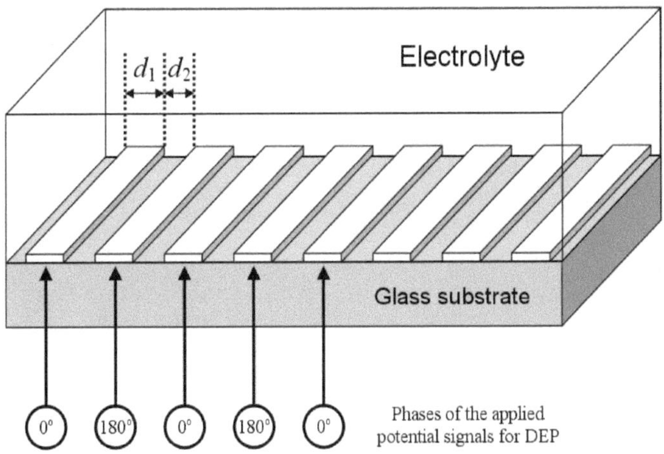

Figure 8. Schematic diagram of an interdigitated electrode array used for DEP (as in the diagram) or also for travelling wave DEP, where the electrodes are addressed by four phase shifted sine waves.

The interdigitated electrode array has been used to develop a DEP based separation system. Using a technique called Field Flow Fraction (FFF) (Rousselet et al., 1998; Gascoyne and Vykoukal, 2004) a deterministic force is combined with a fluid flow to provide a method for particle fractionation. Figure 9 shows a DEP-FFF system. Particles are introduced into the system and when the electric field is switched on, they experience a nDEP which pushes them up and away from the electrodes. This forces is balanced by a downward acting gravitational (buoyancy) force.

Depending on a combination of volume, mass density and polarisability, different particles move to different equilibrium positions in the system. When a laminar flow is applied, the fluid carries particles out of the device at a rate that depends on their original equilibrium position. Since different types of the particles are transported at different rates, a heterogeneous sample can be separated and fractionated along the channel. Recent examples of this technique include device for separating complex mixtures of different blood cells.

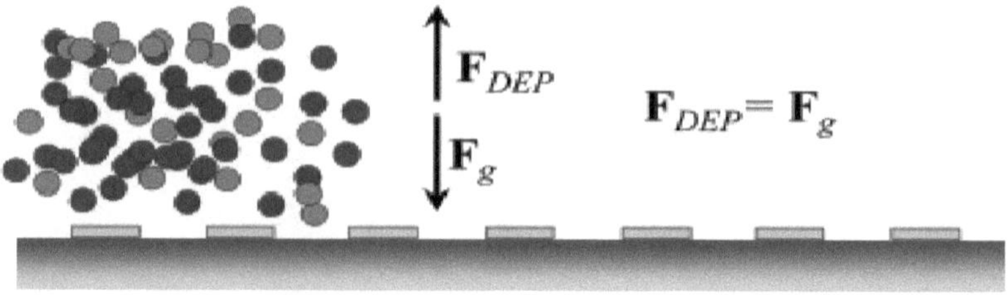

Figure 9. Diagram showing the principle of hyper-layer DEP FFF system. Particles are separated in the vertical direction by a balance of nDEP and gravitational forces. A parabolic fluid flow is applied and particles are fractionated.

3.2 Insulator-based DEP

Dielectrophoresis does not always have to be performed with conducting metal electrodes. The electric field can be modulated using insulating structures placed within a conducting electrolyte fluid. 3D insulating-post arrays (Lapizco-Encinas et al., 2004a,b) have been used to trap and separate live and dead bacteria. In this case large DC voltages are applied across the length of a microchannel. The insulating posts in the channel create obstructions in the pathways of the electric field producing non-uniformities in the electric field distribution in the channel, causing particle DEP. A continuous-flow dielectrophoretic spectrometer system has also been developed based on an insulating DEP technique using 3D geometries (Hawkins et al., 2007). Different field gradients were generated within a structure by fabricating devices with constrictions in a channel as shown in figure 10. These constriction in depth create regions of different field gradient. The device is able to continuously separate particles, as shown in figure 10.

3.3 Single cell trapping

Single cell trapping is important in many applications of biotechnology, such as the study of cell-cell interaction, drug screening and diagnostics. Cells can be trapped at regions of high electric fields (electrode edges or tips) by pDEP, but this requires that the cells are re-suspended in a low conductivity buffer (see figure 3). When cells are suspended in a high conductivity physiological buffer, they only experience nDEP. Electric field

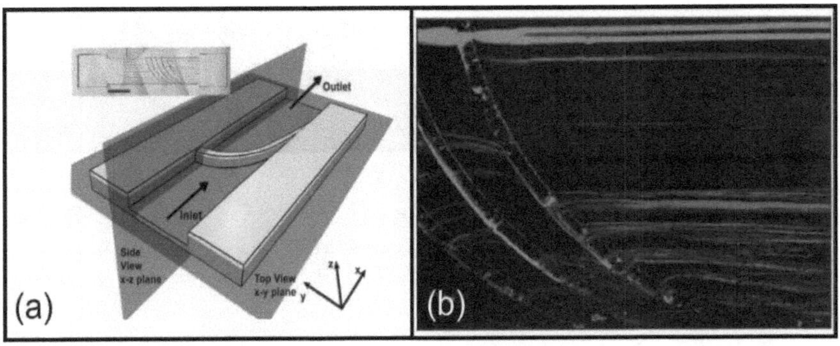

Figure 10. (a) Schematic showing the insulating-post geometry, which incorporates a curved constriction in the channel depth (b) Image showing separation of 2 μm (pseudo-colored green) and 3 μm particles (pseudo-colored red). The 2 μm particles pass the ridge without deflection, while the 3 μm particles are deflected due to higher DEP mobility. Reprinted with permission from (Hawkins et al. 2007), © 2007 American Chemical Society.

cages that generate nDEP forces to trap single cells were first introduced by Fuhr et al. (1992). Since then single cell trapping systems have been widely studied. Müller et al. (1996) used planar quadrupole electrode configuration to trap and concentrate micrometer and sub-micrometer particles. A 3-D microelectrode system (Müller et al., 1999) consisting of two layers of electrode structures was designed to focus, trap and separate cells and latex beads using nDEP. Schnelle et al. (1999, 2000) fabricated an AC cage with octode electrode to trap cells against a fluid stream. Voldman et al. (2001, 2002) and Voldman (2003) developed multiple single cell DEP traps, and a review of cell manipulation technologies based on DEP forces has been published by Voldman (2006). Various electrode geometries such as the quadrupole and octopole electrode, nDEP microwells, point-and-lid geometry and ring-dot geometry were described and evaluated.

A novel design of particle trap that uses nDEP (Thomas et al., 2009) is shown in figure 11. The array of single cell trap consists of a metal ring electrode and a surrounding ground plane - figure 11a. This ring electrodes creates a closed electric field cage in the centre (figure 11) and can be used to trap single cells, using "one wire per trap". Figure 11c shows 15 μm diameter beads trapped against. The behavior of the trap as a function of the fluid flow has been characterized (Thomas et al., 2009).

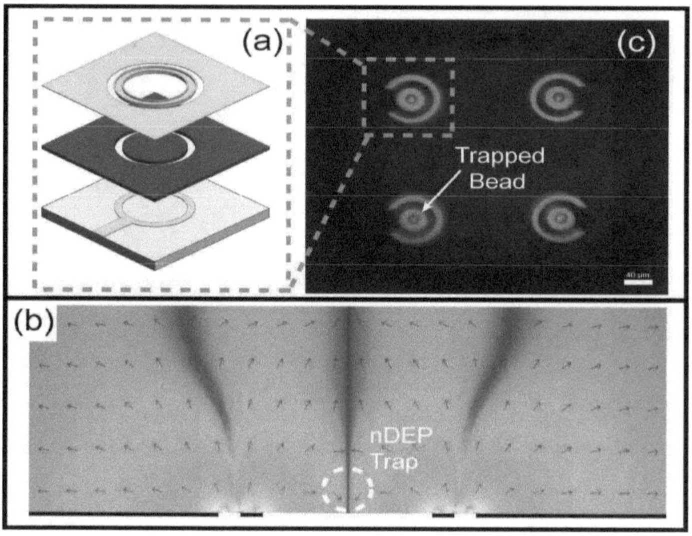

Figure 11. (a) Diagram showing the structure of a ring electrode. (b) Numerical simulation showing DEP force vectors and electric field distribution in the ring trap. (c) Four 40 μm diameter ring traps from an array of 48 traps. Single beads are captured by nDEP. Reprinted with permission from (Thomas et al., 2009), © 2009 Royal Society of Chemistry.

4 Micro-particle Characterization

4.1 Particle characterization

Apart from manipulating particles, DEP has also been used to characterize the dielectric properties of particles. Since the dielectrophoretic force is proportional to the effective polarisability (Clausius-Mossotti factor) of the particle, measurement of the force on a particle can be directly used to determine the permittivity and conductivity of the particle. In practice, it is difficult to measure the dielectrophoretic force on a particle from many reasons, including the effect of electrically induced fluid flow, interaction of a particle with a surface, or Brownian motion for small particles. An alternative is to measure the frequency at which the DEP is zero, or the cross-over frequency (Green and Morgan, 1999; Hughes and Morgan, 1999; Ermolina and Morgan, 2005; Jones, 1995). At this frequency, the real part of the particle polarisability is the same as the suspending medium and the dielectrophoretic force is zero. This point can be measured as a function of medium conductivity and provides sufficient information to determine the dielectric properties of the suspended particles. This type of measurements involves observing the motion of single particles experiencing DEP using a fluorescence microscope. Depending on the applied frequency, particles move either towards an electrode under positive DEP or away from an electrode under negative DEP. At a particular frequency (or frequencies) the DEP force is zero and the particle will remain stationary, i.e. when $\mathrm{Re}[\tilde{f}_{CM}] = 0$. If the properties of the suspending fluid are known, then the effective complex permittivity of the particle can be calculated. According to equation 10, the crossover frequency is defined to be the frequency point that the real part of the Clausius-Mossotti factor equals zero;

$$f_{cross} = \frac{1}{\sqrt{2\pi}} \sqrt{\frac{\sigma_m - \sigma_p}{\epsilon_p - \epsilon_m}} f_{MW} \tag{15}$$

with

$$f_{MW} = \frac{1}{2\pi\tau_{MW}} \tag{16}$$

$$\tau_{MW} = \frac{\epsilon_p + 2\epsilon_m}{\sigma_p + 2\sigma_m} \tag{17}$$

where f_{cross} is the cross-over frequency, f_{MW} is the Maxwell-Wagner relaxation frequency and τ_{MW} is the Maxwell-Wagner time constant of the relaxation.

In a typical experiment, the zero force or cross-over frequency is measured for different suspending medium conductivities spanning two to three decades. For a solid homogeneous spherical particle, a frequency vs. conductivity map can be plotted as shown in figure 12. For frequencies and conductivities corresponding to the shaded area only positive DEP is observed. For all other regions, for example at high conductivities when the particle's effective polarisability is always less than the suspending medium, only negative DEP is observed.

Recently, an evolution of zero force characterization has been developed, called iso-dielectric separation (IDS) technology (Vahey and Voldman, 2009), as shown in figure 13. In this methodology, particles are dielectrophoretically concentrated to a region along an electrical conductivity gradient where the effective polarisability of particles goes to zero. By measuring this isodielectric position and operating with an appropriate frequency and amplitude of the applied voltage, a mixture of particles and cells can be separated and simultaneously characterized.

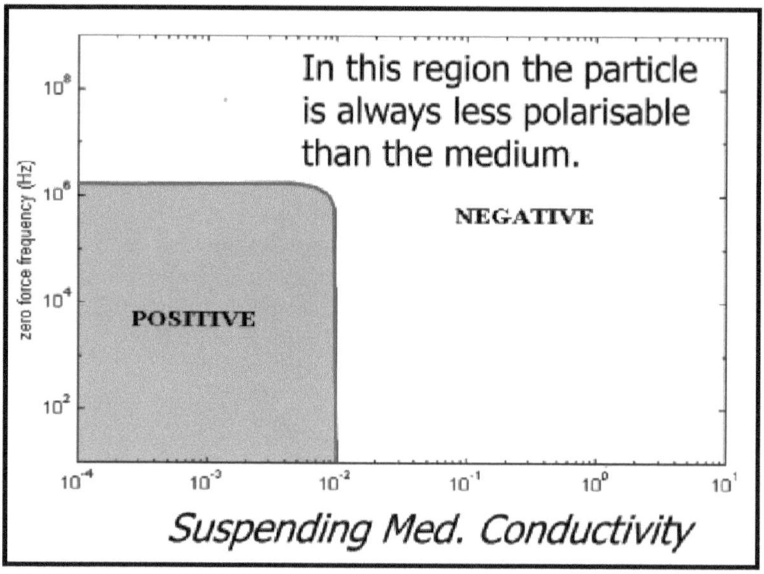

Figure 12. A plot showing the zero force point (cross-over frequency) varies with suspending medium conductivity for a 557 nm diameter particle with relative permittivity 2.55 and conductivity 10 mSm^{-1}.

4.2 DEP for microfluidic systems

Many biomedical and diagnostic applications require fast and accurate analysis of single particles. For example, blood contains many types of cell that can be discriminated on the basis of their optical and/or electrical properties. Morgan et al. (2006) designed and built a device to measure both the optical and electrical-impedance properties of cells at high speed using a microfabricated cytometer. For efficient and high-speed characterization of single particles, the particles need to be focused into the centre of a fluid stream. This can be accomplished using DEP, as shown in figures 14 (a) and (b), where four thin (100 nm) electrodes are fabricated on the base and lid of a microchannel (Holmes et al., 2006). Using a high frequency (20MHz) voltage, the electric field generated by the electrodes produces nDEP force onto the particles pushing particles away from the electrode edges. As the gap between the electrodes decreases, the particles are gradually forced into a narrow focused beam, in which single particles are well aligned, as demonstrated in figure 14 (c).

In this way the properties of the single particles can be measured using

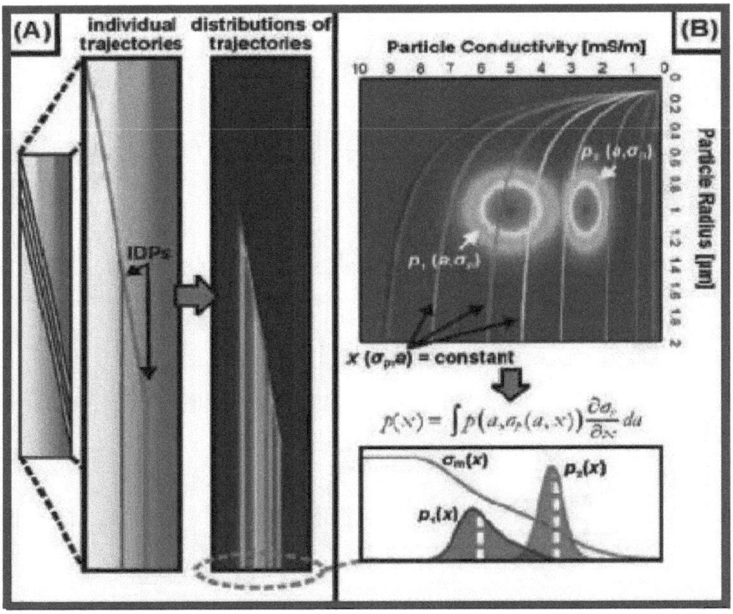

Figure 13. Particles with different electrical properties (i.e. conductivity) go along with different trajectories in an electrical conductivity gradient environment under DEP. Reprinted with permission from (Vahey and Voldman, 2009), © 2009 American Chemical Society.

fluorescence spectroscopy or electrical impedance spectroscopy, as shown in figure 15a. Particles pass through a laser and scatter light: the amount of scatter depends on particle size and shape. Particles can also be labeled with fluorescent antibodies that emit light when excited by the laser. Simultaneously, the electrical impedance of single particles can be measured. This is done by fabricating chips with microelectrodes precisely positioned in the microchannel. The electrodes are a similar size to a particle (typically 10 μm to 20 μm wide, with similar gaps) and are energized with an AC signal of a few hundred millivolts. Particles flow through the impedance detection region, confined by the microelectrodes, one at a time, so that sensitive circuitry can determine the electrical properties of single cells (Morgan et al., 2007). This type of impedance spectroscopy can discriminate between cells without resorting to labeling. Figure 15b shows the simultaneously measured fluorescent signal and impedance signal of a 5.49 μm bead, while a higher impedance amplitude signal of a 7.18 μm bead is also captured.

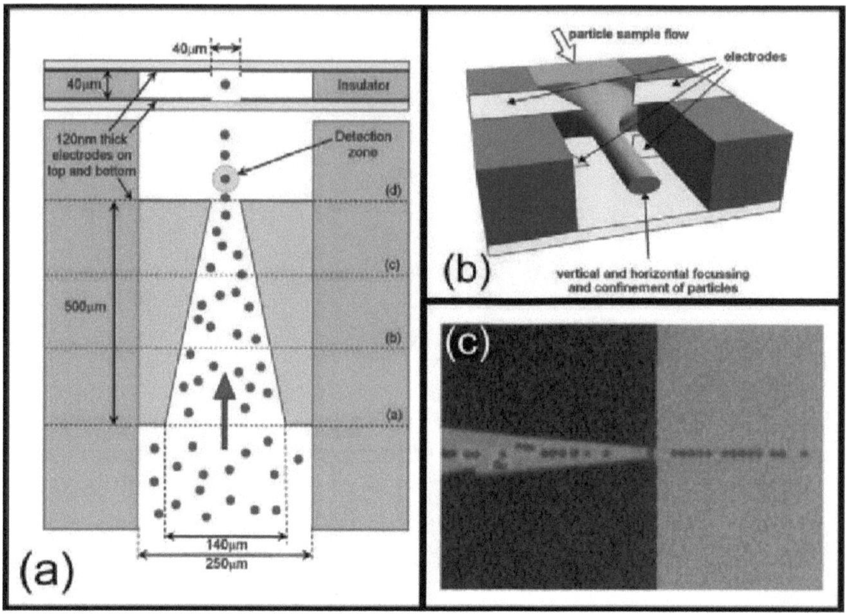

Figure 14. (a) Schematic showing the four focusing electrodes in top-down and cross-sectional view. (b)Diagram showing the principle of 3D dielectrophoretic focusing. (c) Photograph showing the focusing of 6 μm diameter latex beads. Reprinted with permission from (Holmes et al., 2006), © 2006 Elsevier.

A challenge in single particle impedance analysis was to develop a way of measuring a broad band of frequencies at once, in the time it takes a particle to pass through the detection zone. This has been resolved by exciting the cells using a pseudo-random binary signal, maximum length sequences. Impedance data of single particle distributed at 512 frequencies can be measured within 1 ms (Sun et al., 2007a,b).

5 Nanoscale DEP

The forces acting on particles in suspension can be categorized as stochastic, i.e. Brownian motion, and deterministic, such as viscous drag, gravitational or dielectrophoretic. When an electric field is applied to a fluid there are other electric-field induced forces (electrohydrodynamic) in the systems. The Brownian motion of a particle arises from the thermal energy of the

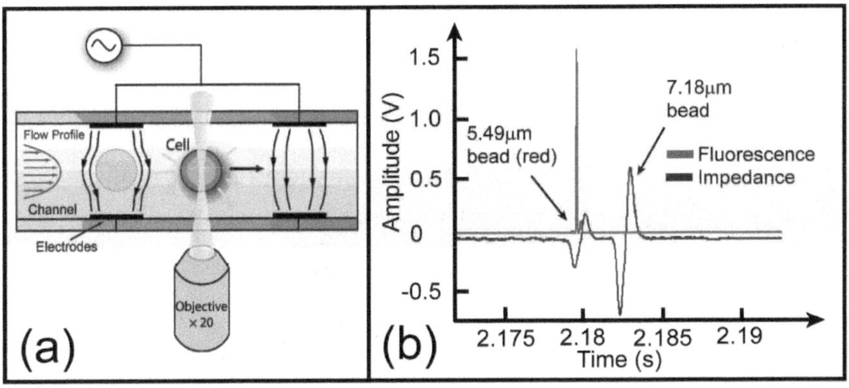

Figure 15. (a) Diagram showing the simultaneously optical and electrical characterization for single particle analysis in a microchip. (b) Plot showing the fluorescent and impedance signal of a single 5.49 μm bead. The 7.18 μm has no fluorescent signal but displays higher impedance amplitude.

system, which causes the particles to move in a random way. There is very little that can be done to control Brownian motion, other than cooling or increasing the viscosity of the liquid. In contrast to stochastic force, the deterministic forces are not random and can be controlled.

5.1 Navier-Stoke's equation and the Stoke's drag force

The Navier-Stokes equation is the equation of motion for the fluid and is derived from conservation of momentum arguments. For an incompressible Newtonian fluid, the Navier-Stokes equation is:

$$\rho_m \frac{\partial \mathbf{u}}{\partial t} + \rho_m(\mathbf{u} \cdot \nabla)\mathbf{u} = -\nabla p + \eta \nabla^2 \mathbf{u} + \mathbf{f} \tag{18}$$

where ρ_m is the mass density, $\mathbf{u}$ is the velocity of the fluid, t is time, p is the pressure, η is the viscosity and $\mathbf{f}$ is the total applied force (force per unit volume) acting on the fluid.

The ratio of the inertial term, $\rho_m(\mathbf{u} \cdot \nabla)\mathbf{u}$, to the viscous term, $\eta \nabla^2 \mathbf{u}$ is the Reynolds number (Re), a parameter that is used to characterize microfluidic systems:

$$\mathrm{Re} = \frac{\rho_m u_0 l_0}{\eta} \tag{19}$$

with l_0 a length scale and u_0 a typical velocity. For low values of Reynolds number (Re$\ll 1$) the viscous term dominates, whilst for high values (Re$\gg 1$)

the inertial term dominates. In microfluidic systems, microfabricated channels and chambers are used to guide the fluid through the device. Typical dimensions for these channels vary between $10\,\mu$m and 1 mm. The flow in such channels is generally laminar i.e. the fluid flow follows streamlines and is free of turbulence.

For a particle moving in the fluid, the fluid exerts a drag force on the particle that affects the velocity of the particle. The force is known as the Stokes force

$$\mathbf{F}_\eta = -f\mathbf{u} \tag{20}$$

where the constant f is the friction factor, a factor that depends on particle parameters such as size, shape and surface characteristics. For a spherical particle of radius R, the Stokes force is given by:

$$\mathbf{F}_\eta = -6\pi\eta R\mathbf{u} \tag{21}$$

5.2 Gravitational force

For a particle with a mass density of ρ_p, suspended in a medium with density of ρ_m, the effective mass of the particle is equal to the volume of the particle times the difference in mass densities between the particle and the suspending medium. In a gravitational field, the magnitude of the gravitational force, $\mathbf{F}_g$ is given by:

$$\mathbf{F}_g = \frac{4}{3}\pi R^3(\rho_p - \rho_m)\mathbf{g} \tag{22}$$

where $\mathbf{g}$ is the gravitational acceleration.

The gravitational force is counteracted by the Stoke's drag force, therefore substitute equation 21 into 22 to give the sedimentation velocity of the particle:

$$u_g = \frac{2}{9}\frac{R^2(\rho_p - \rho_m)g}{\eta} \tag{23}$$

The magnitude of this velocity is variable, for example a 500 nm latex sphere has a mass density of 1050 kg/m^3 giving a sedimentation velocity of only 7 nm/s, indicating the particle moves a distance of the order of its diameter in a minute. However, for a cell, which is much greater, the sedimentation velocity can approach 1 to 10 μm/s in water. Although the gravitational forces acting on sub-micron particles are small, they are not insignificant and cannot always be ignored.

5.3 Electric field induced forces (electrohydrodynamics)

There are two major forces that can be induced through the interaction of the applied electric field with the fluid. The first is the electrothermal force, which is due to the interaction between the electric field and the gradients of the dielectric properties (permittivity and conductivity) of the fluid. These gradients in dielectric properties are generally caused by heating of the fluid, either directly from the electric field (Joule heating) or from some other external sources. This interaction gives rise to an electrothermal body force (Ramos et al., 1998). A second force is due to the interaction between the applied electric field and the induced free charges in the electrical double layer between the electrode and the electrolyte, which causes the fluid to move, a flow called AC electroosmosis (Ramos et al., 1998). At low frequencies, the potential applied to a metal electrode causes charge accumulation in the electrode-electrolyte double layer. In AC electrokinetics, the difference between electrothermal flow and AC electroosmosis is generally observed in the frequency domain. At low frequencies ($<$ 100 kHz), AC electroosmosis dominates the fluid flow. As the frequency increases, this flow diminishes so that above 100 kHz, the behavior of the fluid is dominated by electrothermal force. However, the latter force is only substantial if there is significant temperature gradients in the fluid, for example due to Joule heating in a fluid of high electrical conductivity. Detailed descriptions on electrothermal and electroosmotic effects can be found elsewhere (Ramos et al., 1998; Castellanos et al., 2003; Morgan and Green, 2003).

5.4 Scaling laws

In order to move particles in a deterministic manner a knowledge of the forces acting on the particles is required. Importantly, the displacement due to the deterministic force should dominate over the random or stochastic force. The two main forces that act on particles in addition to DEP are gravitational and Brownian motion. The relative scale of these forces is often difficult to determine, but can be evaluated analytically for some simple geometries. In particular one of the easiest geometries is a semi-infinite parallel plate structure, consisting of two coplanar rectangular electrodes with an infinitely small gap, as shown in figure 16a (Castellanos et al., 2003). In this geometry, the field as a function of radial distance (r) is $E = V/r$, where V is the amplitude of the applied voltage and r is the distance to the centre of the gap. The influence of Brownian motion, gravity and DEP on a single particle can be calculated for this electrode structure, and the results are summarized in figure 16b (Castellanos et al., 2003). This plot shows the displacement of a particle during a time interval of one second as a function of particle radius, R. For an electrode gap of 25 μm, and with an applied potential of 5 volts, it can be seen that the displacement due to Brownian motion is greater than that due to DEP for a particle less than 0.4 μm diameter. Also, gravity is less important than DEP for any particle sizes, since both scale as R^2. The deterministic manipulation of particles smaller than 0.4 μm can be achieved if the magnitude of the applied voltage is increased, or the characteristic length of the system r is reduced. One simple scaling law comes from the relationship between field gradient and electrode gap. For example, reducing the electrode gap by one order of magnitude increases the DEP force by three orders of magnitude. Although figure 16b shows that it is relatively easy to move small particles simply by increasing the electric field, this assertion assumes that there are no other forces. This is generally not the case, since depending on the frequency and the conductivity of the suspending medium there is often an electrohydrodynamic force on the fluid.

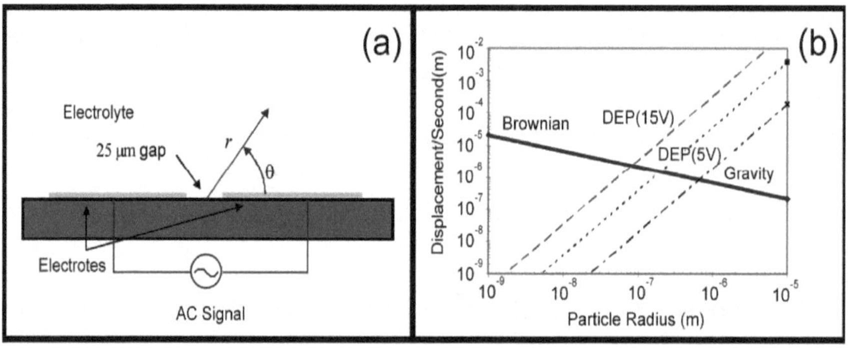

Figure 16. (a) Simple electrode geometry consisting of two parallel plate electrodes with a small gap used to calculate the typical particle displacements. (b) Particle displacement in one second versus particle radius for a particle of mass density 1050 kgm^{-3}. The characteristic length used in this figure is $r = 25$ μm. Reprinted with permission from (Castellanos et al., 2003), © Institute of Physics.

5.5 Nano-particle separation

Polymer particles do not behave as perfect insulators. They have a net surface charge density and are characterized by a surface conductance. Charges move along the surface, giving rise to a surface conductance, K_s, which is related to an equivalent bulk conductivity of the particle through:

$$\sigma_p = 2K_s/R \tag{24}$$

According to this equation, surface conductance plays a significant role in determining the conductivity in small sized ($< 1\mu$m) particles since it is inversely proportional to particle radius. At low frequencies, small polystyrene particles are generally more polarisable than the suspending medium and experience pDEP. At high frequencies, the properties of the system are dominated by permittivity, the particles have a much lower permittivity that water, they are less polarisable and therefore experience nDEP. Separation of sub-micrometre particles can be achieved in a number of ways, and was demonstrated using electrodes such as castellated or polynomial (Green et al., 2000). Apart from size-based separation, a mixture of the same size particles can be separated because of differences in surface charge densities, as given by equation 24.

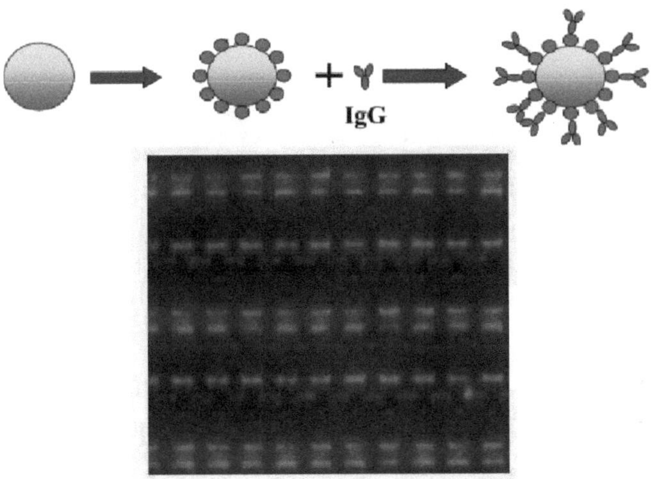

Figure 17. Schematic of surface-modified latex particles (above) and photograph (below) showing separation of unmodified and protein conjugated 216 nm diameter latex particles due to positive DEP forces of different magnitude. The protein-conjugated beads form pearl chains between the tips of the castellated electrode array. The unmodified particles (brighter in the picture) experience a significantly stronger positive DEP force and are attracted to the electrode tips. Reprinted with permission from (Morgan et al., 1999), © 1999 Elsevier.

One method of differentiating the same sized particles is to modify their surface conductance by functionalizing the surface of the particles. Binding of molecules such as proteins to small (100n m) beads changes the surface conductance such that there is a measurable change in DEP behavior. Figure 17a shows a schematic diagram of binding antibodies to the surface to nano-sized latex beads, figure 17b shows the separation of a mixture of 200 nm beads. The red beads are conjugated with antibody, and form pearl chains between the electrodes, while green beads are un-conjugated, experience a stronger DEP force and collect on the tips of the electrodes.

More recently, DEP has been used to manipulate and trap a range of micro and nano-particles, including DNA molecules of various length, for further details see for example (Tuukkanen et al., 2007; Wälti et al., 2007), and combinations of DEP and electrohydrodyanmic forces have also been used to trap and concentrate nano-particles, e.g. (Bown and Meinhart, 2006).

6 Conclusions

In AC electric fields, the electrokinetic forces act on the induced dipole in a particle causing particle movement or rotation. When the electric field acts on the fluid, electrohydrodynamic forces occur, and these are either AC electroosmosis or electrothermal forces that arise from gradients in conductivity and/or permittivity. The DEP force depends on the gradient of the energy density, which changes on the length scale of the electrodes and is a short range effect. The DEP force can be modulated by changing the frequency and electrical properties of the suspending medium. The electrokinetic forces scale in a complex manner with system dimensions, frequency, field, etc. In separation systems, the buoyancy force can be significant (as in FFF) but often the magnitude of this force is much lower than the other forces for micron sized particles. AC electrokinetic manipulation of particles and fluids is a powerful enabling technology which will continue to find many applications in microfluidic systems.

Bibliography

A. Ashkin. Optical trapping and manipulation of neutral particles using lasers. *Proc. Natl. Acad. Sci. USA*, 94:4853–4860, 1997.

A. Ashkin, J.M. Dziedzic, J.E. Bjorkholm, and Steven Chu. Observation of a single-beam gradient force optical trap for dielectric particles. *Opt. Lett.*, 11:288–290, 1986.

D.J. Beebe, G.A. Mensing, and G.M. Walker. Physics and applications of microfluidics in Biology. *Annu. Rev. Biomed. Eng.*, 4:261–286, 2002.

M.R. Bown and C.D. Meinhart. AC electroosmotic flow in a DNA concentrator. *Microfluidics Nanofluidics*, 2:513–523, 2006.

J.P.H. Burt, T.A.K. Al-Ameen, and R. Pethig. An optical dielectrophoresis spectrometer for low-frequency measurements on colloidal suspensions. *J. Phys. E: Sci. Instrum.*, 22:952–957, 1989.

A. Castellanos, A. Ramos, A. González, N.G. Green, and H. Morgan. Electrohydrodynamics and dielectrophoresis in microsystems: Scaling laws. *J. Phys. D: Appl. Phys.*, 36:2584–2597, 2003.

I. Ermolina and H. Morgan. The electrokinetic properties of latex particles: comparison of electrophoresis and dielectrophoresis. *J. Colloid Interface Science*, 285:419–428, 2005.

G. Fuhr, W.M. Arnold, R. Hagedorn, T. Müller, W. Benecke, B. Wagner, and U. Zimmermann. Levitation, holding, and rotation of cells within traps made by high-frequency fields. *Biochimica et Biophysica Acta*, 1108:215–223, 1992.

P.R.C. Gascoyne and J. Vykoukal. Dielectrophoresis-based sample handling in general-purpose programmable diagnostic instruments. *Proc. IEEE Inst. Electr. Electron. Eng.*, 92:22–42, 2004.

N.G. Green and H. Morgan. Dielectrophoresis of submicrometer latex spheres. 1. Experimental results. *J. Phys. Chem. B*, 103:41–50, 1999.

N.G. Green, A. Ramos, and H. Morgan. Ac electrokinetics: a survey of sub-micrometer particle dynamics. *J. Phys. D: Appl. Phys.*, 33:632–641, 2000.

B.G. Hawkins, A.E. Smith, Y.A. Syed, and B.J. Kirby. Continuous-Flow Particle Separation by 3D Insulative Dielectrophoresis Using Coherently Shaped, dc-Biased, ac Electric Fields. *Analytical Chemistry*, 79:7291–7300, 2007.

D. Holmes, H. Morgan, and N.G. Green. High throughput particle analysis: Combining dielectrophoretic particle focussing with confocal optical detection. *Biosensors Bioelectronics*, 21:1621–1630, 2006.

Y. Huang, R. Holzel, R. Pethig, and X-B Wang. Differences in the ac electrodynamics of viable and non-viable yeast cells determined through combined dielectrophoresis and electrorotation studies. *Phys. Med. Biol.*, 37:1499–1517, 1992.

M.P. Hughes and H. Morgan. Dielectrophoretic characterization and separation of antibody coated submicrometer latex spheres. *Analytical Chemistry*, 71:3441–3445, 1999.

T.B. Jones. *Electromechanics of particles.* Cambridge University Press, 1995.

L. Kremser, D. Blaas, and E. Kenndler. Capillary electrophoresis of biological particles: Viruses, bacteria, and eukaryotic cells. *Electrophoresis*, 25:2282–2291, 2004.

N.A. Lacher, K.E. Garrison, R.S. Martin, and S.M. Lunte. Microchip capillary electrophoresis/ electrochemistry. *Electrophoresis*, 22:2526–2536, 2001.

B.H. Lapizco-Encinas, B.A. Simmons, E.B. Cummings, and Y. Fintschenko. Insulator-based dielectrophoresis for the selective concentration and separation of live bacteria in water. *Electrophoresis*, 25:1695–1704, 2004a.

B.H. Lapizco-Encinas, B.A. Simmons, E.B. Cummings, and Y. Fintschenko. Dielectrophoretic Concentration and Separation of Live and Dead Bacteria in an Array of Insulators. *Analytical Chemistry*, 76:1571–1579, 2004b.

T. Laurell, F. Petersson, and A. Nilsson. Chip integrated strategies for acoustic separation and manipulation of cells and particles. *Chem. Soc. Rev.*,, 36:492–506, 2007.

G.B. Lee, C.I. Hung, B.J. Ke, G.R. Huang, B.H. Hwei, and H.F. Lai. Hydrodynamic Focusing for a Micromachined Flow Cytometer. *J. Fluids Eng.*, 123:672–679, 2001.

H. Morgan and N.G. Green. *AC Electrokinetics: colloids and nanoparticles.* Research Studies Press, 2003.

H. Morgan, M.P. Hughes, and N.G. Green. Separation of submicron bioparticles by dielectrophoresis. *Biophys. J.*, 77:516–525, 1999.

H. Morgan, D. Holmes, and N.G. Green. High speed simultaneous single particle impedance and fluorescence analysis on a chip. *Current Applied Physics*, 6:367–370, 2006.

H. Morgan, T. Sun, D. Holmes, S. Gawad, and N.G. Green. Single cell dielectric spectroscopy. *J. Phys. D: Appl. Phys.*, 40:61–70, 2007.

T. Müller, A. Gerardino, T. Schnelle, S.G. Shirley, F. Bordoni, G. De Gasperis, R. Leoni, and G. Fuhr. Trapping of micrometre and submicrometre particles by high-frequency electric fields and hydrodynamic forces. *J. Phys. D: Appl. Phys.*, 29:340–349, 1996.

T. Müller, G. Gradl, S. Howitz, S. Shirley, T. Schnelle, and G. Fuhr. A 3-d microelectrode system for handling and caging single cells and particles. *Biosensors and Bioelectronics*, 14:247–256, 1999.

N. Pamme. Magnetism and microfluidics. *Lab Chip*, 6:24–38, 2006.

R. Pethig. *Dielectric and electronic properties of biological materials.* John Wiley & Sons, 1979.

H.A. Pohl. *Dielectrophoresis.* Cambridge University Press, 1978.

J.A.R. Price, J.P.H. Burt, and R. Pethig. Applications of a new optical technique for measuring the dielectrophoretic behaviour of micro-organisms. *Biochimica et Biophysica Acta*, 964:221–230, 1988.

A. Ramos, H. Morgan, N.G. Green, and A. Castellanos. Ac electrokinetics: a review of forces in microelectrode structures. *J. Phys. D: Appl. Phys.*, 31:2338–2353, 1998.

R. Rodriguez-Trujillo, C.A. Mills, J. Samitier, and G. Gomila. Low cost micro-Coulter counter with hydrodynamic focusing. *Microfluid. Nanofluid.*, 3:171–176, 2007.

J. Rousselet, G.H. Markx, and R. Pethig. Separation of erythrocytes and latex beads by dielectrophoretic levitation and hyperlayer field-flow fractionation. *Colloids and Surfaces A: Physicochemical and Engineering Aspects*, 140:209–216, 1998.

T. Schnelle, T. Müller, S. Fiedler, and G. Fuhr. The influence of higher moments on particle behaviour in dielectrophoretic field cages. *J. Electrostatics*, 46:13–28, 1999.

T. Schnelle, T. Müller, and G. Fuhr. Trapping in AC octode field cages. *J. Electrostatics*, 50:17–29, 2000.

H.A. Stone, A.D. Stroock, and A. Ajdari. Engineering flows in small devices: Microfluidics toward a Lab-on-a-Chip. *Annu. Rev. Fluid Mech.*, 36:381–411, 2004.

T. Sun, H. Morgan, and N.G. Green. Analytical solutions of ac electrokinetics in interdigitated electrode arrays: Electric field, dielectrophoretic and traveling-wave dielectrophoretic forces. *Phys. Rev. E*, 76:046610, 2007.

T. Sun, S. Gawad, C. Bernabini, N.G. Green, and H. Morgan. Broadband single cell impedance spectroscopy using maximum length sequences: theoretical analysis and practical considerations. *Measurement Science & Technology*, 18:2859–2868, 2007a.

T. Sun, D. Holmes, S. Gawad, N.G. Green, and H. Morgan. High speed multi-frequency impedance analysis of single particles in a microfluidic cytometer using maximum length sequences. *Lab Chip*, 7:1034–1040, 2007b.

R.S. Thomas, H. Morgan, and N.G. Green. Negative DEP traps for single cell immobilisation. *Lab Chip*, 9:1534–1540, 2009.

S. Tuukkanen, A. Kuzyk, J.J. Toppari, H. Hakkinen, V.P. Hytonen, E. Niskanen, M. Rinkio, and P. Torma. Trapping of 27 bp-8 kbp DNA and immobilization of thiol-modified DNA using dielectrophoresis. *Nanotechnology*, 18:295204, 2007.

M.D. Vahey and J. Voldman. High-Throughput Cell and Particle Characterization Using Isodielectric Separation. *Analytical Chemistry*, 81: 2446–2455, 2009.

J. Voldman. Biomems: Building with cells. *Nature Materials*, 2:433–434, 2003.

J. Voldman. Electrical forces for microscale cell manipulation. *Annu. Rev. Biomed. Eng.*, 8:425–454, 2006.

J. Voldman, R.A. Braff, M. Toner, M.L. Gray, and Martin A. Schmidt. Holding forces of single-particle dielectrophoretic traps. *Biophysical J.*, 80:531–542, 2001.

J. Voldman, M.L. Gray, M. Toner, and Martin A. Schmidt. A microfabrication-based dynamic array cytometer. *Analytical Chemistry*, 74:3984–3990, 2002.

C. Wälti, W.A. Germishuizen, P. Tosch, C.F. Kaminski, and A.G. Davies. AC electrokinetic manipulation of DNA. *J. Phys. D: Appl. Phys.*, 40: 114–118, 2007.

X-B Wang, R. Pethig, and T.B. Jones. Relationship of dielectrophoretic and electrorotational behaviour exhibited by polarized particles. *J. Phys. D: Appl. Phys.*, 25:905–912, 1992.

X-B Wang, Y. Huang, R. Holzel, J.P.H. Burt, and R. Pethig. Theoretical and experimental investigations of the interdependence of the dielectric, dielectrophoretic and electrorotational behaviour of colloidal particles. *J. Phys. D: Appl. Phys.*, 26:312–322, 1993.

M. Wiklund, C. Günther, R. Lemor, M. Jäger, G. Fuhr, and H.M. Hertz. Ultrasonic standing wave manipulation technology integrated into a dielectrophoretic chip. *Lab Chip*, 6:1537–1544, 2006.

Electrostatics and Quasielectrostatics

Nicolas G. Green
School of Electronics and Computer Science University of Southampton
Highfield, Southampton, SO17 1BJ, United Kingdom

Abstract This chapter discusses the basics of electrostatic and quasielectrostatic systems from the perspective of charges and forces on charges. The fundamental equations and physical concepts are discussed along with relevant constructs such as the electric field and dipole moments. Polarisation and dielectrics are discussed in detail, outlining the mechanisms and special cases relevant to electrokinetic phenomena.

1 Introduction

An introduction to dielectrophoresis and electromechanics in general requires a discussion of the background theory, in this case the static approximation to Maxwell's equations. There are a large number of textbooks in this area which can be consulted for further information: a few suggested examples are given in the bibliography at the end of the chapter (Stratton, 1941; Maxwell, 1954; Bleaney and Bleaney, 1962; Kraus, 1984; Lorrain et al., 2000). This chapter will discuss electrostatics and the small current approximation termed quasi-electrostatics. It will cover the basics of electric fields, dielectrics and polarisation and, more importantly for electromechanics, the additional polarisation arising at the interface between a particle and a suspending fluid. An applied electric field both generates this polarisation charge and then acts on it to produce a force, which produces a number of effects including dielectrophoresis. The chapter begins with the basic equations required in order to describe polarisation and dielectrics, then adds successive layers of complexity to the behaviour of real systems and non-homogeneous materials.

1.1 Charge and forces on charges

In this discussion, it is assumed that charges are either stationary (electrostatics) or have small velocities and accelerations. The movement of the charge does not change the electric field and magnetic fields are small,

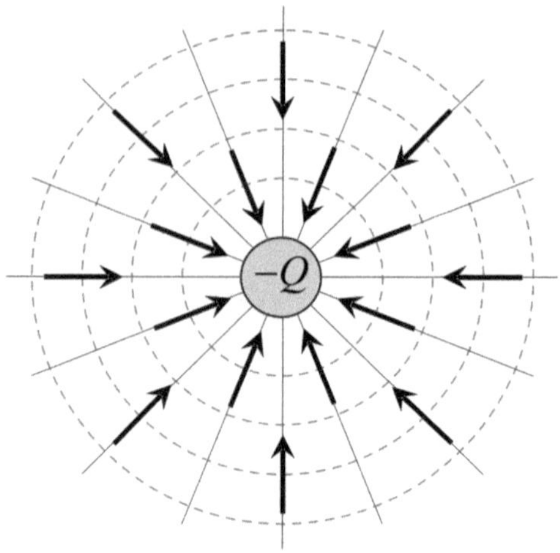

Figure 1. The electric field lines (solid), direction of the electric field (arrows) and equipotential lines (dashed) for a negative point charge.

giving the quasi-electrostatic approximation. The force exerted by a stationary point charge Q_1 on a second stationary point charge Q_2 is described by Coulomb's law

$$\mathbf{F_{12}} = \frac{Q_1 Q_2}{4\pi\varepsilon_o r^2}\mathbf{\hat{r}_{12}} \tag{1}$$

where r is the separation of the charges and the unit vector $\mathbf{\hat{r}}_{12}$ points from Q_1 to Q_2. The constant ε_o is referred to as the *permittivity of free space* and has the value 8.854×10^{-12} Farads/metre (F/m). The force is attractive if the two charges are opposite in sign and repulsive if they have the same sign. The unit of charge is the Coulomb(C).

1.2 Electric field and electrical potential

The *electric field strength* $\mathbf{E}$ from a charge Q, is the force per unit charge that produces the Coulomb force on an arbitrary unit positive test charge, so that:

$$\mathbf{E} = \frac{Q}{4\pi\varepsilon_o r^2}\mathbf{\hat{r}_{12}} \tag{2}$$

The direction and magnitude of the field for a negative point charge is shown by the dotted lines and vectors in figure 1.

The *electric potential* is defined to be the scalar ϕ such that the electric field is given by $\mathbf{E} = -\nabla\phi$. The unit of electric potential is Volts (V) defined by the relationship Volts = Joules/Coulomb (V=J/C) and the unit of **E** is therefore V/m. The lines of constant potential for a point charge are also shown in figure (1).

1.3 Basic equations: Gauss, Poisson, Laplace

The basic electrostatic equations which underpin electrical and electrokinetic systems are based on distributions of charge in a material, of which there are two basic types: free and bound. Those charges sufficiently free to move through the material from atom to atom give rise to currents. Those that are bound to atoms and molecules are responsible for the polarisation of the material. The fundamental electrostatic equations relate these different types of charge to the electric field and electrical potential.

Gauss's Law relates **E** to the volume electric charge density ρ and in differential form this is:

$$\nabla \cdot \mathbf{E} = \frac{\rho}{\varepsilon_o} \tag{3}$$

Substituting for the potential gives Poisson's equation

$$\nabla^2 \phi = -\frac{\rho}{\varepsilon_o} \tag{4}$$

In most cases, the charge density is zero, giving Laplace's equation

$$\nabla^2 \phi = 0 \tag{5}$$

We will now examine the physical origin and behaviour of the two types of charge, starting with bound charges.

2 Bound charges, polarisation and dielectrics

A dielectric material is a material that contains charges which *polarise* under the influence of an applied electric field. These charges are bound within the material and can only move short distances when the field is applied, the negative and positive charges moving in opposite directions to form induced dipoles. Some materials also consist of molecules with permanent dipoles which polarise by orienting with the field, an example of which is pure water. A fundamental discussion of dipoles and polarisation is necessary to derive the equations for dielectrics but this is also the physical mechanism responsible for electrokinetic and electromechanical forces (Jones, 1995; Morgan and Green, 2003).

2.1 Dipoles

The electrical dipole is formed from a simple distribution of charges: two charges of the same magnitude Q and opposite sign, separated by a distance d. The dipole moment is the vector $\mathbf{p}$ where

$$\mathbf{p} = Q\mathbf{d} \tag{6}$$

and the vector $\mathbf{d}$ is the displacement vector from the negative to the positive charge. The dipole moment has units of Coulomb-metre or Debye (where 1 Debye = 3.33×10^{-30} C m). In spherical polar co-ordinates, the potential of a dipole is

$$\phi = \frac{\mathbf{p} \cdot \hat{\mathbf{r}}}{4\pi\varepsilon_o r^2} = \frac{|\mathbf{p}|\cos\theta}{4\pi\varepsilon_o r^2} \tag{7}$$

and the electric field is

$$\mathbf{E} = \frac{|\mathbf{p}|}{4\pi\varepsilon_o r^3}(2\cos\theta\hat{\mathbf{r}} + \sin\theta\hat{\theta}) \tag{8}$$

Both expressions are only valid for large distances ($r \gg d$) and the dipole field falls off as the inverse third power of r. Close to the dipole, the field expression contains terms with a higher order power of r. A plot of the field and potential from a simple dipole are shown in figure (2).

2.2 Dielectrics

A dielectric material is a material that contains charges which *polarise* under the influence of an applied electric field. These charges are bound within the material and can only move short distances when the field is applied, the negative and positive charges moving in opposite directions to form *induced* dipoles. This type of material is referred to as *non-polar* (Grant et al., 1978; Pethig, 1979).

The molecules of some materials can also possess permanent dipole moments. Water for example has a dipole moment of 1.8 Debye, arising from the structure shown in figure (3). The two hydrogen atoms in the molecule are on the same side of the oxygen atom as shown, making a bond angle of 106°. In this molecule, the oxygen is slightly electronegative and the two protons slightly electro-positive so that the molecule has a dipole moment pointing from the oxygen atom to the point midway between the two hydrogen atoms. Permanent dipoles polarise by orienting with the field and materials containing this type of molecule are termed *polar* materials.

The average dipole moment of the molecules of a material is proportional to the magnitude of the field, i.e.

$$\mathbf{p_{av}} = \alpha\mathbf{E}' \tag{9}$$

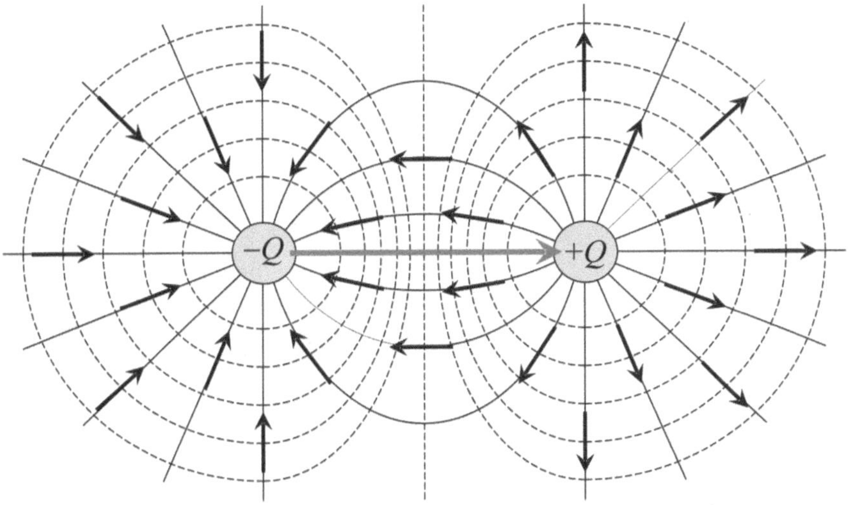

Figure 2. The electric field lines (solid), direction of the electric field (arrows) and equipotential lines (dashed) for a simple dipole consisting of two opposite charges. The large grey arrow indicates the direction of the dipole moment.

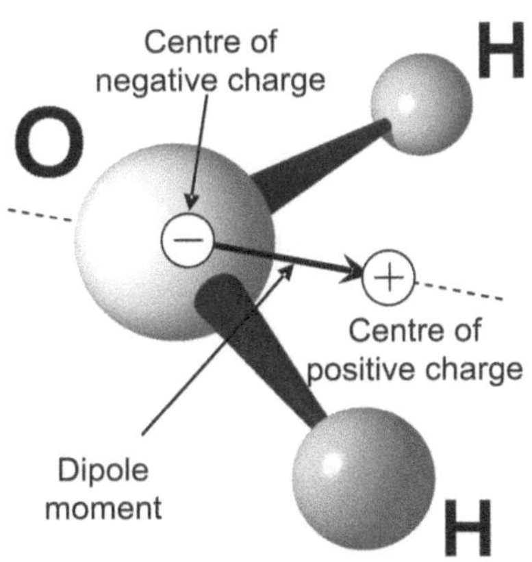

Figure 3. The electric field lines (solid), direction of the electric field (arrows) and equipotential lines (dashed) for a simple dipole consisting of two opposite charges. The large grey arrow indicates the direction of the dipole moment.

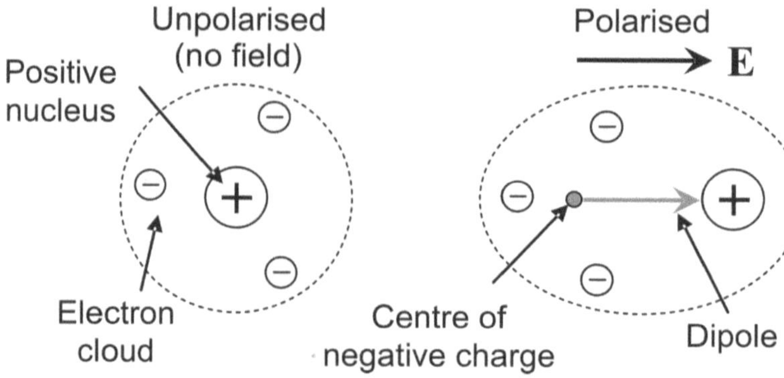

Figure 4. Schematic diagram of the displacement of the electron cloud around a positive nucleus under and applied field. The separation of the centres of negative charge and the positive nucleus produces a dipole.

where $\mathbf{E}'$ is the local electric field in the vicinity of the dipole. The proportionality α is called the polarisability (average dipole moment per unit field strength) and is a measure of the response of the dielectric to the electric field. It has units of $(\mathrm{CV^{-1}m^2})$ or $(\mathrm{Fm^2})$.

In a dielectric consisting of n molecules per cubic metre, the polarisation $\mathbf{P}$ (the dipole moment per unit volume) is

$$\mathbf{P} = n\mathbf{p_{av}} = n\alpha\mathbf{E}' \tag{10}$$

The polarisation and the displacement of the charge gives rise to a net charge at points in the dielectric or at the surface. These charges are referred to as bound or polarisation charges. The bound volume charge density ρ_b is given by

$$\rho_b = -\nabla \cdot \mathbf{P} \tag{11}$$

2.3 Polarisation mechanisms

There are several molecular polarisation mechanisms that can occur when an electric field is applied to a dielectric:

- *Electronic polarisation* α_e: In an electric field, the centre of charge of the electron cloud in an atom moves slightly with respect to the centre of charge of the nucleic charges. For an applied field of $10^6 - 10^7$ V/m (compared to the internal field of the atom which is of the order of 10^{11} V/m), the displacement is of the order of 10^8 times the diameter of an atom.

- *Atomic* α_a: In a crystalline solid such as potassium chloride, the ions of different sign move in different directions when subjected to an electric field. The small displacement of the charges in the lattice causes atomic polarisability.
- *Orientational* α_d: As described earlier in this chapter, many molecules possess permanent dipole moments. Orientational polarisation is the polarisation arising from the alignment of these permanent dipoles in polar dielectrics.
- *Interfacial polarisation* α_i: In addition to local molecular processes, cumulative long-range charge transport causes polarisation of dielectrics. At interfaces between different materials or locations where the dielectric is inhomogeneous (internal interfaces) or at the surface, causing macroscopic distortion of the field. Similarly, relatively long-range transport of ions occurs along and around the surface of polyelectrolytes. This manifests itself as an increase in the charge storage capacity of the dielectric or an increase in the permittivity of the dielectric. This is the basis of interfacial polarisation. It has an important role to play in AC electrokinetics, since it is the origin of the induced dipole on particles. This will be discussed in greater detail.

Assuming that the polarisability mechanisms act independently, then the total polarisability of a dielectric is the sum of the polarisabilities.

$$\alpha_T = \alpha_e + \alpha_a + \alpha_o + \alpha_i \tag{12}$$

Each polarisability has its own characteristic frequency response, which will be discussed later.

2.4 The electric flux density

Considering the free ρ_f and bound ρ_b charge densities separately, Gauss's law can be written as

$$\nabla \cdot \mathbf{E} = \frac{\rho_f + \rho_b}{\varepsilon_o} \tag{13}$$

Using equation (11) for the bound charge density, this can be re-written as

$$\nabla \cdot (\varepsilon_o \mathbf{E} + \mathbf{P}) = \rho_f \tag{14}$$

The vector $\mathbf{D} = \varepsilon_o \mathbf{E} + \mathbf{P}$ is called the electric flux density (in C/m^2) and its divergence is equal to the free volume charge density

$$\nabla \cdot \mathbf{D} = \rho_f \tag{15}$$

2.5 The electric susceptibility and the relative permittivity

Most dielectrics are linear, with **P** proportional to **E** such that

$$\mathbf{P} = \varepsilon_o \chi_{ae} \mathbf{E} \tag{16}$$

where χ_{ae} is the electric susceptibility of the dielectric. Therefore,

$$\mathbf{D} = \varepsilon_o (1 + \chi_{ae}) \mathbf{E} = \varepsilon_o \varepsilon_r \mathbf{E} \tag{17}$$

where $\varepsilon_r = 1 + \chi_{ae}$ is a dimensionless number referred to as the relative permittivity of the dielectric. The permittivity of the dielectric $\varepsilon = \varepsilon_o \varepsilon_r$ is the constant of proportionality between **D** and **E**.

3 Charge, currents and Ohm's law

So far, the dielectrics have been assumed to be ideal with only bound charge responding to the applied field. In general, a real dielectric will contain some free charge which will also respond. The free charge is not bound to a particular location in the material and moves through it in response to an applied electric field. The electric current density, **J** (charge per unit time: A/m^2) contribution from a charge carrier is given by

$$\mathbf{J} = \rho \mathbf{v}_c \tag{18}$$

where ρ is the density and $\mathbf{v}_c$ the mean drift velocity of the charge carrier. The mean drift velocity is normally proportional to the electric field:

$$\mathbf{v}_c = \mu \mathbf{E} \tag{19}$$

which defines μ, the mobility of the charge in the electric field (in units of m^2/Vs). Substituting this into equation (18) gives

$$\mathbf{J} = n q_c \mu \mathbf{E} \tag{20}$$

where n is the number density of the carrier (m^{-3}) and q_c is the charge of the carrier.

The electrical conductivity, σ (S/m), of the material is defined by Ohm's Law:

$$\mathbf{J} = \sigma \mathbf{E} \tag{21}$$

If there is only one charge carrier in the material, for example residual charge in a dielectric where the electrons separate from the material leaving bound positive charges with zero mobility, equation 20 gives the total current density for the material and the conductivity is given by $\sigma = n q_c \mu$.

3.1 Water electrolytes and conductivity

The relationship between the current density and the electrical conductivity of an aqueous solution of ions is more complicated since the current can be carried by many different types of ion, each with a different mobility. The current density for each ion of type j, with valence z_j is

$$\mathbf{J}_j = \begin{cases} z_j F c_j \mu_j \mathbf{E} \\ z_j n_j q \mu_j \mathbf{E} \end{cases} \tag{22}$$

This expression includes the normal expression in electrochemistry with F the Faraday (9.6487×10^4C/mol), c_j the molar concentration (mol/l), μ_j the mobility of the ion, and the equivalent expression to equation (20), with n_j the number density of the ion). Comparison of the two expressions for in equation (22) demonstrates that the Faraday is $N_A q$, where N_A is Avogadro's number (6.02252×10^{23} mol^{-1}).

The ionic conductivity λ_j (Sm^2/mol) is defined as

$$\lambda_j = z_j F \mu_j \tag{23}$$

and the conductivity of the solution is given by the sum of the contributions of each ion

$$\sigma = \sum_j \lambda_j c_j \tag{24}$$

For a symmetrical electrolyte (such as KCl) with ionic conductivities λ_+ and λ_-, the conductivity is given by

$$\sigma = (\lambda_+ + \lambda_-)c = \Lambda c \tag{25}$$

where Λ is the molar conductivity (Sm^2/mol). At low concentrations ($c \to 0$), the ions can move freely without interaction and the molar conductivity is a constant value, referred to as the limiting conductivity $\Lambda_{c=0}$

$$\Lambda_{c=0} = F(\mu_+ + \mu_-) \tag{26}$$

The molar conductivity is constant for molarities up to approximately 10^{-3} Molar (M). For higher concentrations, the ions interact with each other and the molar conductivity is influenced by higher order effects. This means that at low concentrations, the conductivity is directly proportional to the concentration

Using potassium chloride (KCl) as an example, the mobilities of the potassium K^+ and the chloride Cl^- ions are almost the same, with $\mu_{K+} = 7.62 \times 10^{-8}$ m^2/Vs, and $\mu_{Cl^-} = 7.91 \times 10^{-8}$ m^2/Vs (CRC handbook). Therefore, for KCl, $\Lambda_{c=0} = F(\mu_+ + \mu_-) = 9.6485 \times 10^4(7.62 \times 10^{-8} + 7.91 \times 10^{-8}) = 0.01498$ (Sm^2/mol) and 1 mM KCl has a conductivity of 14.98 $\times 10^{-8}$S/m at 298K.

3.2 Mobility and hydration radius

The mobility of the ion is related to its diffusion constant D_j (m^2/s), by the Einstein equation:

$$\mu_j = D_j \frac{q}{k_B T} = D_j \frac{F}{RT} \tag{27}$$

where $R = N_A k_B$ is the (molar) gas constant (8.3143 $JK^{-1}mol^{-1}$) and k_B is the Boltzmann constant. The diffusion constant of each ion is given by

$$D_j = \frac{k_B T}{6\pi\eta a_j} = \frac{qRT}{6\pi F\eta a_j} \tag{28}$$

where a_j is the radius of the ion. However, when an ion is placed in water, ion-water interactions produce a change in the properties of the medium close to the ions. This interaction is responsible for the ion remaining in solution, and when the suspending medium is water, the process is known as solvation or hydration. The water molecules have a large dipole moment and are highly polarisable: this high polarisability that water is such a good solvent for ions. When an ion is placed in water, the charge produces a local field which polarises the water molecules around it. This region of polarised water is referred to as the ionic atmosphere: a region that screens the ion from the other ions in the medium, ensuring electroneutrality on the global scale. Locally, the electrical potential resulting from the ion falls off exponentially with distance, and the length scale is given by the Debye length (the distance at which the potential falls to 1/e of its maximum value), written as λ_D or κ^{-1} where, for a monovalent ion

$$\kappa = \sqrt{\frac{q^2 n_o}{\varepsilon k_B T}} \equiv \sqrt{\frac{1}{D}\frac{\sigma}{\varepsilon}} \tag{29}$$

In this expression $\varepsilon = \varepsilon_o \varepsilon_r$, and n_o is the number density of ions in the bulk.

The creation of the polarised ionic atmosphere produces a region around the ion that has very different properties from the bulk fluid. The polar molecules in this region are prevented from responding to an applied field and, as the ion moves, its atmosphere moves with it, so that the moving object is bigger than the naked ion. The radius of the solvated or hydrated ion is called the hydration radius and is the one that should be used for determining diffusion constants and mobilities.

3.3 The charge relaxation time

An important characteristic of an electrolyte is the time required for an ion to move a distance of the order of the Debye length by diffusion. This is called the charge relaxation time and is given by

$$\tau_q = \frac{1}{D\kappa^2} \tag{30}$$

Substituting for the Debye length and the conductivity in this expression, the charge relaxation time is also

$$\tau_q = \frac{\varepsilon}{\sigma} \tag{31}$$

The angular frequency associated with this time $\omega_q = 1/\tau_q$, is referred to as the charge relaxation frequency.

3.4 The charge conservation equation

The charge conservation equation relates the rate of change with respect to time of the volume charge density ρ to the current density $\mathbf{J}$

$$\nabla \cdot \mathbf{J} = -\frac{\partial \rho}{\partial t} \tag{32}$$

In the steady state, the right hand side of this equation is zero and therefore $\nabla \cdot \mathbf{J} = 0$. This equation is used in conjunction with Gauss's Law to include the effect of conductivity on charging processes in real dielectrics.

4 Quasi-electrostatics: AC fields and complex permittivity

This section examines the polarisation of a dielectric with a permittivity ε and conductivity σ in response to an applied AC electric field of frequency f and angular frequency $\omega = 2\pi f$. It introduces the concept of complex permittivity, which describes the frequency dependent response of the dielectric to the field. Also introduced is the idea of using equivalent circuits, consisting of electrical components, to represent a physical system.

As discussed, Gauss's Law is used to calculate electric fields and forces. A harmonic AC field, will be defined using phasor notation as:

$$\mathbf{E} = \text{Re}\left[\tilde{\mathbf{E}} e^{i\omega t}\right] \tag{33}$$

This is related to an equivalent complex potential defined by

$$\phi = \text{Re}\left[\tilde{\phi} e^{i\omega t}\right] \tag{34}$$

by the expression $\tilde{\mathbf{E}} = -\nabla\tilde{\phi}$. Gauss's Law can then be expanded for real dielectrics using the charge conservation equation for the free charge. In a harmonic field of angular frequency ω, the operator $\partial/\partial t = i\omega$ and the charge conservation equation is

$$\nabla \cdot \mathbf{J}_f = -\frac{\partial \rho_f}{\partial t} \rightarrow \nabla \cdot \sigma \mathbf{E} = -i\omega\rho_f \tag{35}$$

assuming that the free charge is solely the responsible for the conductivity. This can be substituted into Gauss's Law as follows

$$\nabla \cdot \mathbf{D} = \rho_f \Rightarrow \nabla \cdot (\varepsilon \mathbf{E}) = -\nabla \cdot \left(\frac{\sigma}{i\omega}\mathbf{E}\right) \Rightarrow \nabla \cdot \left(\varepsilon + \frac{\sigma}{i\omega}\mathbf{E}\right) = 0$$

to give

$$\nabla \cdot (\tilde{\varepsilon}\mathbf{E}) = 0 \tag{36}$$

where $\tilde{\varepsilon}$ is the complex permittivity given by:

$$\tilde{\varepsilon} = \varepsilon_o \varepsilon_r - i\frac{\sigma}{\omega} \tag{37}$$

Another way of considering this problem is to calculate the total current in the dielectric. There are two current densities, the conduction $\mathbf{J}_d$ (equation 21) and the displacement $\mathbf{J}_d = \partial\mathbf{D}/\partial t = i\omega\varepsilon\mathbf{E}$. The total current density is given by the sum of the two

$$\mathbf{J} = (\sigma + i\omega\varepsilon)\mathbf{E} = \tilde{\sigma}\mathbf{E} \tag{38}$$

where $\tilde{\sigma}$ is the complex conductivity.

A simple circuit model can be used to illustrate the relationship between permittivity, conductivity and polarisation. Considering a simple parallel plate capacitor, shown in figure (5), with area A and separation d, which contains a homogeneous dielectric. The plates of the capacitor have a potential ϕ of angular frequency ω applied between them. With a loss-free dielectric of permittivity ε filling the gap, the impedance is

$$Z = \frac{1}{i\omega C} \tag{39}$$

where $i^2 = -1$ and $C = \varepsilon(A/d)$ is the capacitance (F). If the dielectric is lossy (figure 6) and has both a permittivity and conductivity, the current in the circuit is the same as if the circuit were replaced with a loss-free capacitor (capacitance C), in parallel with a resistor of resistance $R = (1/\sigma)(d/A)$. The total impedance of the circuit is then

$$Z = \frac{1}{1/R + i\omega C} = \frac{R}{1 + i\omega RC} = \frac{R}{1 + i\omega\tau_c} \tag{40}$$

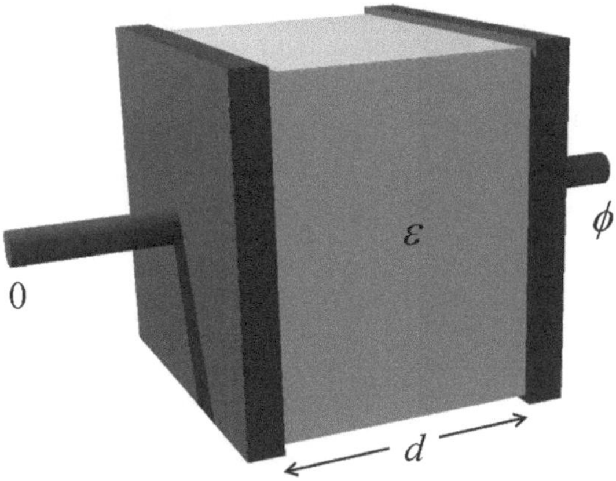

Figure 5. Schematic diagram of a simple parallel plate capacitor containing an ideal dielectric.

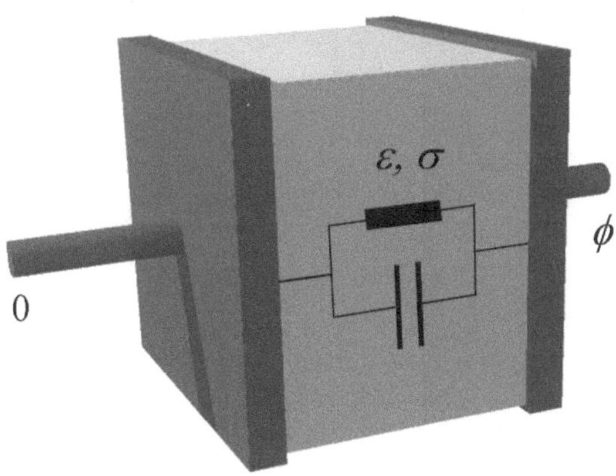

Figure 6. Schematic diagram of a simple parallel plate capacitor containing an dielectric which has a permittivity and a conductivity. Also shown is the equivalent circuit for this capacitor consisting of a resistor and capacitor in parallel.

where τ_c is the charge relaxation time. This demonstrates the significance of the charge relaxation frequency: below it, free charge and conduction are largely responsible for polarisation; and above it, the permittivity dominates. Putting the expressions for R and C into this equation, it can be re-written

$$Z = \frac{1}{\frac{\sigma A}{d} + i\omega\frac{\varepsilon A}{d}} = \frac{1}{i\omega}\frac{1}{\frac{\varepsilon A}{d} + \frac{\sigma A}{i\omega d}} = \frac{1}{i\omega}\frac{1}{\left(\varepsilon - \frac{i\sigma}{\omega}\right)\frac{A}{d}} = \frac{1}{i\omega C_1} \tag{41}$$

where

$$C_1 = \tilde{\varepsilon}\frac{A}{d} \tag{42}$$

4.1 Energy in a dielectric

In a lossy dielectric the total current is given by the sum of $\mathbf{J}_f = \sigma\mathbf{E}$ and $\mathbf{J}_d = i\omega\varepsilon\mathbf{E}$. A useful expression is the loss tangent given by the ratio of the two:

$$\tan\delta = \frac{\sigma}{\omega\varepsilon} \tag{43}$$

where δ is the angle between the conduction and displacement currents, referred to as the loss angle, which for a perfectly insulating (ideal) dielectric, is zero. If the complex permittivity is written as $\tilde{\varepsilon} = \varepsilon' - i\varepsilon''$, the loss angle can also be written as $\tan\delta = \varepsilon''/\varepsilon'$. In general, the imaginary part ε'' pertains to the dissipation of energy as heat and the real part ε' to the storage of energy. The energy stored in a non-ideal dielectric is given by

$$U = \frac{1}{2}\int_v \varepsilon'|\mathbf{E}|^2 dv = \frac{1}{2}\int_v \mathrm{Re}\left[\tilde{\mathbf{D}}\cdot\tilde{\mathbf{E}}^*\right] dv \tag{44}$$

4.2 Summary

Looking again at the polarisation $\mathbf{P}$, for a real dielectric, both conduction and displacement currents flow, and both free and bound charges separate and result in polarisation. This results in a phase lag between the driving field and the induced polarisation, with the polarisability now a complex quantity:

$$\mathbf{P} = \varepsilon_o\left(\frac{\tilde{\varepsilon}}{\varepsilon_o} - 1\right)\mathbf{E} = n\tilde{\alpha}\mathbf{E} \tag{45}$$

The complex permittivity is often written as $\tilde{\varepsilon} = \varepsilon' - i\varepsilon''$ to include further complex terms arising from relaxation mechanisms.

5 Dielectric relaxations

The polarisation of a dielectric involves the movement of charge to create dipoles. The rate of this movement is finite and as a result each polarisation mechanism has a different characteristic time associated with it: the period required to achieve maximum polarisation. At the frequency for which the period of the field and the characteristic time are the same, maximum energy is dissipated by the system. Below this frequency maximum polarisation and energy storage occurs. Above this frequency the polarisation no longer reaches its maximum and at very high frequencies, the mechanism does not respond to the field and no polarisation occurs. For example water, which has a relative permittivity of ~80 below 10^8 Hz, has a relaxation frequency of 2×10^{10} Hz. At frequencies greater than 10^{11} Hz, the permanent dipoles are not able to orient with the field and the permittivity of water drops to ~2. The fall in polarisability results in a decrease in energy storage (and permittivity) and is referred to as a dielectric relaxation.

5.1 Orientational relaxation

Of the basic polarisation mechanisms outlined previously, orientational polarisation has the longest relaxation time. Atomic and electronic polarisation will align with the field up to frequencies of the order of 10^{14} Hz and we can consider them to be constant. The polarisation due to these mechanisms is of the form of equation (16)

$$\mathbf{P}_{ae} = \varepsilon_o \chi_{ae} \mathbf{E} \tag{46}$$

The orientational polarisation has a characteristic relaxation time τ_{or} associated with the time taken for the permanent dipoles to re-orient with the field and is given by

$$\mathbf{P}_{or} = \frac{\varepsilon_o \chi_{or}}{1 + i\omega\tau_{or}} \mathbf{E}$$

where χ_{or} is the low frequency limit for the orientational susceptibility. The total frequency dependent polarisation is then

$$\mathbf{P}_{tot} = \varepsilon_o \left(\chi_{ae} + \frac{\chi_{or}}{1 + i\omega\tau_{or}} \right) \mathbf{E} \tag{47}$$

At the low frequency limit $\chi = \chi_{ae} + \chi_{or} = \varepsilon_s - 1$, where ε_s is the relative permittivity measured in a static electric field. At the high frequency limit $\chi = \chi_{ae} = \varepsilon_\infty - 1$, where ε_∞ is the relative permittivity at sufficiently high frequency that no orientational polarisation occurs. As a result, $\chi_{or} =$

$\varepsilon_s - \varepsilon_\infty$ and equation (47) can be written as

$$\mathbf{P}_{tot} = \varepsilon_o(\tilde{\varepsilon}_d - 1)\mathbf{E} \tag{48}$$

where $\tilde{\varepsilon}_d$ is a complex term given by

$$\tilde{\varepsilon}_d = \varepsilon_\infty + \frac{\varepsilon_s - \varepsilon_\infty}{1 + i\omega\tau_{or}} \tag{49}$$

The total complex permittivity of the dielectric is therefore

$$\tilde{\varepsilon} = \varepsilon_o\left(\varepsilon_\infty + \frac{\varepsilon_s - \varepsilon_\infty}{1 + i\omega\tau_{or}}\right) - i\frac{\sigma}{\omega} \tag{50}$$

Writing this in the form $\tilde{\varepsilon} = \varepsilon' - i\varepsilon''$, ε' and ε'' are given by the Debye relations

$$\varepsilon' = \varepsilon_o\left(\varepsilon_\infty + \frac{\varepsilon_s - \varepsilon_\infty}{1 + \omega^2\tau_{or}^2}\right) \tag{51}$$

$$\varepsilon'' = \varepsilon_o\left(\frac{(\varepsilon_s - \varepsilon_\infty)\omega\tau_{or}}{1 + \omega^2\tau_{or}^2}\right) + \frac{\sigma}{\omega} \tag{52}$$

As discussed previously, a non-zero ε'' implies a phase lag between D and E. This has important ramifications for energy dissipation in the system. When the loss part of the complex permittivity is zero, D and E are in phase and any energy required to produce polarisation of the material in one half-cycle of the field is given back to the driving source in the second half-cycle. In other words, the amount of energy lost through the loss mechanism per cycle is negligible.

When ε'' is non-zero, energy is dissipated in the system by two mechanisms. Firstly, if the dielectric is non-ideal and has a finite conductivity, then energy is lost through Joule heating, the second term on the r.h.s. of equation (52). Secondly, energy is lost due to the electrical conductivity that arises from the relaxation mechanism: the first term on the r.h.s. of equation (52). The electric field orients the dipoles against the randomising effects of Brownian motion. At high frequencies no energy is lost since there is insufficient time for the dipoles to orient with the field and no energy is stored. At low frequencies the dipoles are oriented and work is done in moving the dipoles in a viscous medium. The power lost per cycle of AC field is low but this increases with frequency, since the number of cycles per second increases. It reaches a maximum at the characteristic angular frequency $\omega_{or} = 1/\tau_{or}$, when the time required for maximum orientation of the dipoles is exactly equal to one half-cycle of the field. Maximum energy dissipation from this mechanism occurs at this frequency.

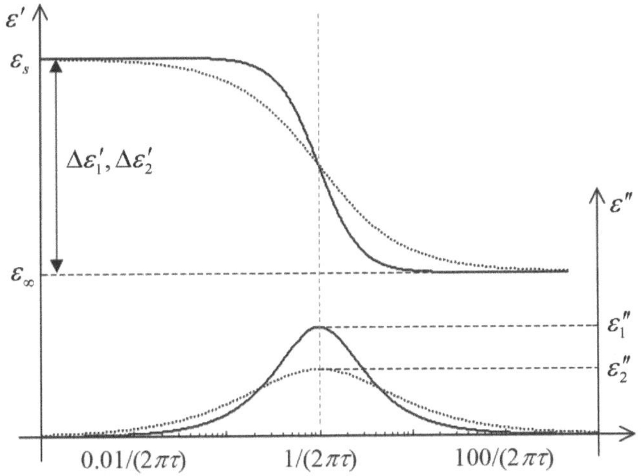

Figure 7. Plot of the variation of the real ε' and imaginary ε'' parts of the complex permittivity as a function of dimensionless frequency. The solid line indicates the behaviour of an ideal relaxation with single frequency τ and the dotted line, the behaviour of a spread of relaxation times.

The variation of ε' and ε'' with ω gives a dispersion in the relative permittivity as shown by the solid line in figure (7). It can be seen that the width of the loss peak extends over approximately four decades in frequency and that for a single Debye-type relaxation the width of the peak at half height is 1.14 decades in frequency.

Such dielectric loss peaks can be analysed to obtain a great deal of information about the molecular material under investigation. For example the width of the loss peak is often wider than the 1.14 decades characteristic of the ideal Debye behaviour, where there is only a single relaxation time. Cole and Cole modified equation (50) to account for a distribution in relaxation times more representative of experimental samples as follows

$$\tilde{\varepsilon} = \varepsilon_o \left(\varepsilon_\infty + \frac{\varepsilon_s - \varepsilon_\infty}{1 + i(\omega\tau_{or})^{(1-\beta)}} \right) - i\frac{\sigma}{\omega} \tag{53}$$

where β equals 0 for a single relaxation time and tends to 1 for an infinite number of relaxation times. A distribution in relaxation times means that the dispersion occurs at a frequency equal to the average of the relaxation times, but is broader than for a single relaxation time. This is shown by the dotted line in figure (7).

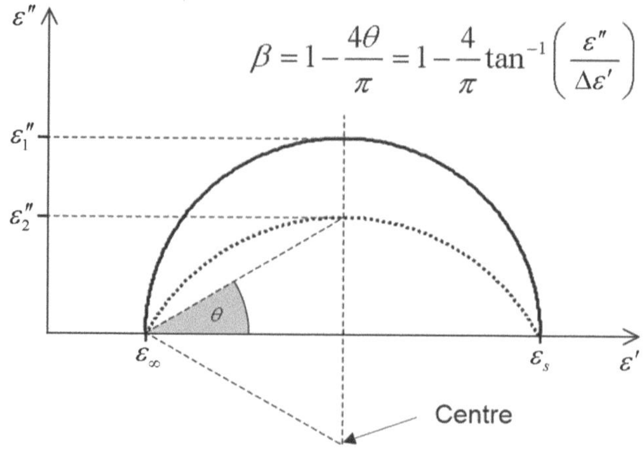

Figure 8. Plot of the variation of the imaginary ε'' versus tje real ε' parts of the complex permittivity on a Cole-Cole plot. Again, the solid line indicates the behaviour of an ideal relaxation with single frequency τ and the dotted line, the behaviour of a spread of relaxation times. Also shown is the relationship between the spread parameter β and the angle θ measured on the plot, which is used in characterisation of the relaxation.

A useful method of analysing this type of data is by using a Cole-Cole plot (Cole and Cole, 1941), where the imaginary part ε'' is plotted against the real part ε'. For a single relaxation, the plot is a semicircle with its centre on the horizontal axis, as shown by the solid line in figure (8). If the plot forms an arc of a circle with the centre below the axis, then there is a distribution of relaxation times as shown by the dotted line in the figure for $\beta = 0.1$. The relationship between the angle θ and the parameter β is shown in the figure. The interrelationship between the real and imaginary components of the complex permittivity can be quantitatively described by the Kramers-Krönig relationships

$$\varepsilon'(\omega) - \varepsilon_\infty = \frac{2}{\pi}\int_0^\infty \frac{\varpi\varepsilon''(\varpi)}{\varpi^2 - \omega^2} d\varpi \tag{54}$$

$$\varepsilon''(\omega) = \frac{2\omega}{\pi}\int_0^\infty \frac{\varepsilon'(\varpi) - \varepsilon_\infty}{\varpi^2 - \omega^2} d\varpi \tag{55}$$

where ϖ is a dummy frequency variable over which ε'' has to be integrated to find ε' or vice versa. The Kramers-Krönig relations show that if either the real or the imaginary component of the complex permittivity is known, then the other can be derived from the above equations. Practically, the Kramers-Krönig relationships can be used to equate the real and imaginary components of the polarisability of a system.

6 Interfacial Polarisation

Real systems often consist of a number of different dielectrics each with different electrical properties: in dielectrophoresis for example, the system is a suspension of dielectric particles in a dielectric fluid. When an electric field is applied to the system, surface charge accumulates at discontinuities (or interfaces) between the dielectrics due to the differences in electrical properties. Since the polarisabilities of each dielectric are frequency dependent, the magnitude of the surface charge is also frequency dependent and the total complex permittivity of the system exhibits dispersions solely due to the polarisation of the interfaces. This is referred to as *Maxwell-Wagner interfacial polarisation.*

If the system consists of a single dielectric particle and a suspending fluid, the distribution of the surface charge density around the closed surface of the particle gives rise to an induced dipole moment. The action of an electric field on this dipole gives rise to ponderomotive forces and torques; the basis of AC electrokinetics. This will be discussed in a subsequent chapter, with the remainder of this chapter discussing the phenomena of interfacial polarisation and the implications for modelling and analysing materials.

6.1 Interfacial polarisation

The mechanism of interfacial polarisation using the parallel plate capacitor is shown in figure (9), which contains two lossy dielectrics with different electrical properties. Extending the simple model presented previously, this capacitor can be modelled as the equivalent circuit shown. The two dielectrics have permittivities ε_1 and ε_2 and conductivities σ_1 and σ_2, respectively. The impedance of the whole circuit is

$$Z = Z_1 + Z_2 = \frac{R_1}{1 + i\omega R_1 C_1} + \frac{R_1}{1 + i\omega R_2 C_2} \tag{56}$$

This can again be re-written as a single capacitance with a complex permittivity. The resulting complex permittivity contains a dispersion arising solely from the difference in the permittivities and conductivities of the two dielectrics. Using the Debye formulations (equations (51) and (52)), this

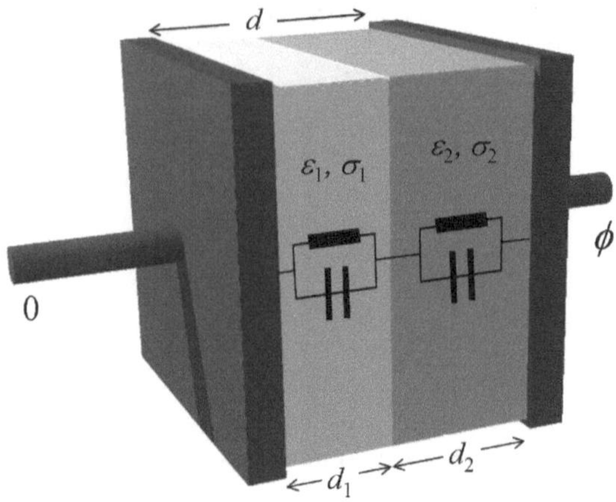

Figure 9. Schematic diagram of a simple parallel plate capacitor containing two dielectric slabs both of which have a different permittivity and a conductivity. Also shown is the equivalent circuit for this capacitor consisting of two parallel resistors/capacitor circuits in series.

complex permittivity is

$$\tilde{\varepsilon} = \varepsilon' - i\varepsilon'' = \varepsilon_o \left[\varepsilon_{hf} + \frac{\varepsilon_{lf} - \varepsilon_{hf}}{1+\omega^2\tau^2}\right] - i\varepsilon_o \left[\frac{(\varepsilon_{lf} - \varepsilon_{hf})\omega\tau_{or}}{1+\omega^2\tau^2} + \frac{\sigma}{\varepsilon_o \omega}\right] \quad (57)$$

where ε_{hf} is the high frequency permittivity, ε_{lf} the low frequency permittivity, τ the relaxation time and σ the system conductivity, each given by

$$\varepsilon_{hf} = \frac{d\varepsilon_1\varepsilon_2}{d_1\varepsilon_2 + d_2\varepsilon_1} \quad (58)$$

$$\varepsilon_{lf} = \frac{d(d_1\varepsilon_1\sigma_2^2 + d_2\varepsilon_2\sigma_1^2)}{(d_1\sigma_2 + d_2\sigma_1)^2} \quad (59)$$

$$\tau = \varepsilon_o \frac{d_1\varepsilon_2 + d_2\varepsilon_1}{d_1\sigma_2 + d_2\sigma_1} \quad (60)$$

$$\sigma = \frac{d\sigma_1\sigma_2}{d_1\sigma_2 + d_2\sigma_1} \quad (61)$$

6.2 The induced effective dipole moment of a particle

An applied field causes charge to build up at the interface between two dielectrics of different permittivity and/or conductivity. This is equally

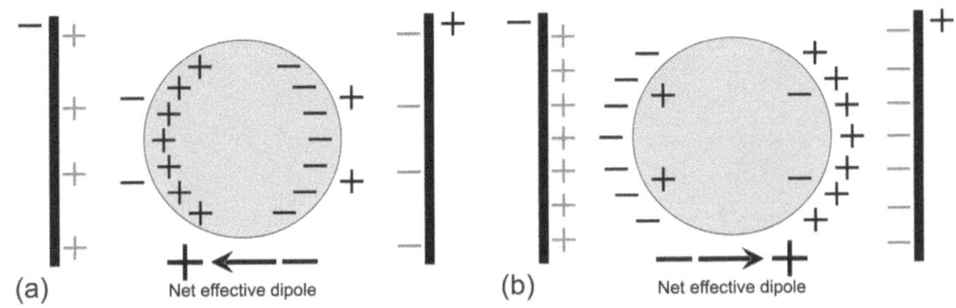

Figure 10. Schematic diagram of the polarisation of a particle in a uniform electric field for when the particle is (a) more polarisable and (b) less polarisable than the medium. The dipole is parallel to the field if the particle is more polarisable and anti-parallel if less polarisable.

true irrespective of the geometry of the system. For a dielectric particle suspended in a dielectric fluid, when the electric field is applied, charge of opposite sign accumulates at either side of the particle as shown in figure (10). When the polarisabilities of the two dielectrics are different, the net charge at the interface is non zero and the sign of the net charge is opposite on either side of the particle. This induced or effective dipole moment depends on the properties of both the particle and the suspending medium (or electrolyte), and on the frequency of the applied field. This section will discuss the polarisation and induced dipole moments of solid spherical particles, ellipsoids and more complicated shelled particles.

6.3 The effective dipole moment of a spherical particle

The simplest case is that of a homogeneous solid dielectric sphere of radius a suspended in a homogenous dielectric medium, shown schematically in figure (11). The applied electric field far from the origin is taken to be uniform and anti-parallel to the z-axis i.e. $\mathbf{E} = -E\hat{\mathbf{z}}$. Without loss of generality, the sphere can be assumed to have its centre at the origin, making the problem axially symmetric and two-dimensional in spherical polar co-ordinates.

For ideal dielectrics, the Laplace equation for the electrical potential can be solved using Legendre polynomials, an exercise that can be found in electromagnetism textbooks. Gauss's law gives the boundary condition for the potential at the surface of the sphere ($r = a$)

$$\varepsilon_m \mathbf{E}_m \cdot \hat{\mathbf{n}} - \varepsilon_p \mathbf{E}_p \cdot \hat{\mathbf{n}} = 0 \Rightarrow \varepsilon_m \left.\frac{\partial \phi_m}{\partial r}\right|_{r=a} - \varepsilon_p \left.\frac{\partial \phi_p}{\partial r}\right|_{r=a} = 0 \qquad (62)$$

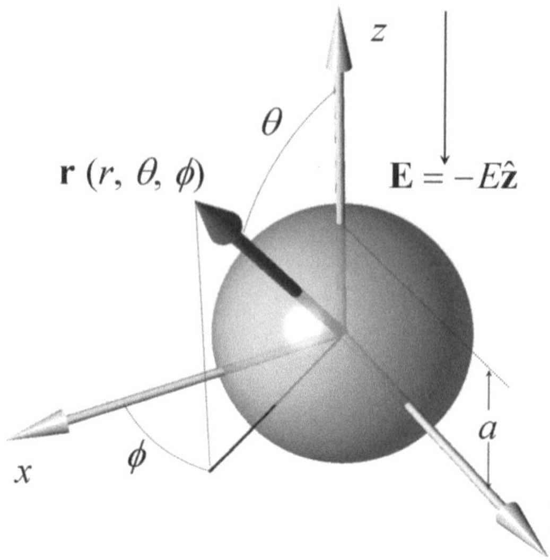

Figure 11. Schematic diagram of a spherical particle showing the relationship between cartesian and spherical polar coordinates. The applied uniform electric field is assumed to lie along the z-axis, making the problem axially symmetric about this axis.

where the subscripts p and m refer to particle and suspending medium. The solutions for the potential inside and outside the dielectric sphere are

$$\phi_p = -\left(\frac{3\tilde{\varepsilon}_m}{\tilde{\varepsilon}_p + 2\tilde{\varepsilon}_m}\right) Er\cos\theta \tag{63}$$

$$\phi_m = \left[\left(\frac{\tilde{\varepsilon}_p - \tilde{\varepsilon}_m}{\tilde{\varepsilon}_p + 2\tilde{\varepsilon}_m}\right)\frac{a^3}{r^3} - 1\right] Er\cos\theta \tag{64}$$

The external potential in the medium can be re-written as

$$\phi_m = Ea^3\left(\frac{\tilde{\varepsilon}_p - \tilde{\varepsilon}_m}{\tilde{\varepsilon}_p + 2\tilde{\varepsilon}_m}\right)\frac{\cos\theta}{r^2} - Er\cos\theta \tag{65}$$

This is the scalar sum of the potential due to the applied field (second term on the right hand side) and the potential of a dipole moment (first term, right hand side). By inspection, the dipole moment is

$$\mathbf{p} = 4\pi\varepsilon_m\left(\frac{\tilde{\varepsilon}_p - \tilde{\varepsilon}_m}{\tilde{\varepsilon}_p + 2\tilde{\varepsilon}_m}\right)a^3\mathbf{E} \tag{66}$$

This is the effective dipole moment of the sphere. It is sometimes rewritten in terms of the volume of the sphere ν and a complex effective polarisability $\tilde{\alpha}$ so that

$$\mathbf{p} = v\tilde{\alpha}\mathbf{E} \tag{67}$$

Using this equation, the magnitude of the dipole for a particle can be calculated. Taking for example a field of 10^6 V/m, the dipole moment of a particle such as a cell is of the order of 10^{-19} Cm. This is equivalent to a charge of 10^{-14} C located at the poles of the cell.

From equations (66) and (67) the *effective polarisability* is therefore

$$\tilde{\alpha} = 3\varepsilon_m \left(\frac{\tilde{\varepsilon}_p - \tilde{\varepsilon}_m}{\tilde{\varepsilon}_p + 2\tilde{\varepsilon}_m} \right) = 3\varepsilon_m \tilde{f}_{CM} \tag{68}$$

The magnitude of the polarisability, and therefore the effective dipole moment of the particle, is frequency dependent. This dependence is described by the factor

$$\tilde{f}_{CM}(\tilde{\varepsilon}_p, \tilde{\varepsilon}_m) = \frac{\tilde{\varepsilon}_p - \tilde{\varepsilon}_m}{\tilde{\varepsilon}_p + 2\tilde{\varepsilon}_m} \tag{69}$$

This is referred to as the *Clausius-Mossotti factor.* It is complex, describing a relaxation in the effective polarisability of the particle with a relaxation time of

$$\tau_{MW} = \frac{\varepsilon_p + 2\varepsilon_m}{\sigma_p + 2\sigma_m} \tag{70}$$

The angular frequency $\omega_{MW} = 2\pi f_{MW} = 1/\tau_{MW}$ is often referred to as the *Maxwell-Wagner relaxation frequency* since the dispersion in the dipole moment is caused by interfacial polarisation. The real part of the Clausius-Mossotti factor reaches a low frequency limiting value of $(\sigma_p - \sigma_m)/(\sigma_p + 2\sigma_m)$, i.e. it depends solely on the conductivity of the particle and suspending medium. Conversely, the high frequency limiting value is $(\varepsilon_p - \varepsilon_m)/(\varepsilon_p + 2\varepsilon_m)$ and the polarisation is dominated by the permittivity of the particle and suspending medium. The imaginary part is zero at high and low frequencies, and at the relaxation frequency f_{MW} has a value

$$\frac{1}{2}\left(\frac{\varepsilon_p - \varepsilon_m}{\varepsilon_p + 2\varepsilon_m} - \frac{\sigma_p - \sigma_m}{\sigma_p + 2\sigma_m} \right) \tag{71}$$

The frequency variation of the real and imaginary parts of the Clausius-Mossotti factor is shown in figure (12). In general, the real part of the Clausius-Mossotti factor is bounded by 1 and -1/2, and the imaginary part bounded by -3/4 and 3/4.

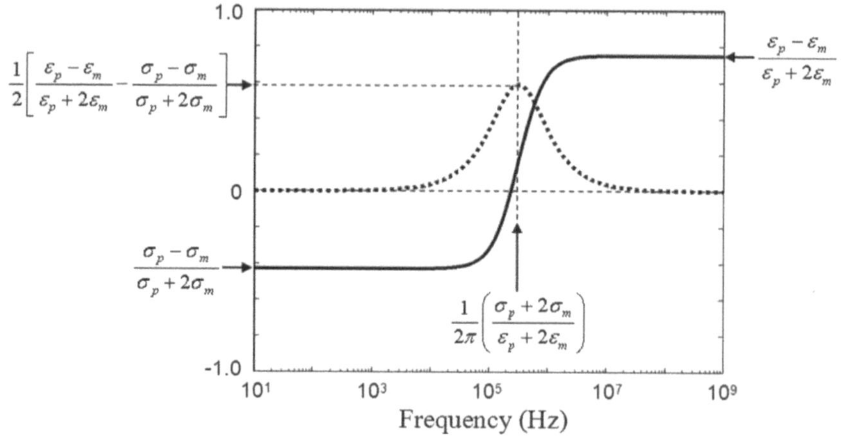

Figure 12. Plot of the real (solid line) and imaginary (dotted line) parts of the Clausius-Mossotti factor as a function of frequency for arbitrary values of permittivity and conductivity. The plot describes a single Debye relaxation.

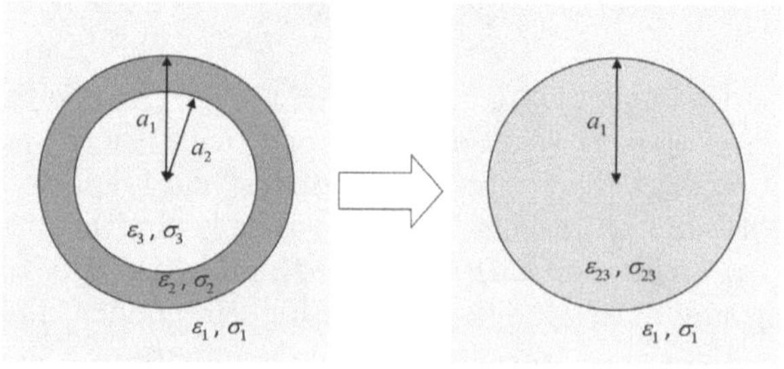

Figure 13. Schematic diagram of cross section of a single shelled particle with two internal permittivities and conductivities. The shell model allows this internal structure to be analysed as a homogeneous sphere with a new value for permittivity and conductivity, both of which are now frequency dependent.

6.4 The shell model for biological particles

Biological particles, such as cells and some types of viruses, have a complicated internal structure. The common approach to theoretically modelling cells is to use a concentric multi-shell model (Irimajiri et al., 1979; Gimsa et al., 1991; Huang et al., 1992; Jones, 1995). The simplest case is that of a single shelled particle, which is shown schematically in figure 13), demonstrating the principle of the shell model in that the internal structure and properties are combined to give an equivalent homogeneous particle which is described by the standard equations.

This system has two intrinsic relaxation frequencies, one for each of the two interfaces. In this case the effective polarisability and the dipole moment are given by

$$\tilde{\alpha} = 3\varepsilon_1 \tilde{f}_{CM,23} = 3\varepsilon_m \left(\frac{\tilde{\varepsilon}_{23} - \tilde{\varepsilon}_1}{\tilde{\varepsilon}_{23} + 2\tilde{\varepsilon}_1} \right) \tag{72}$$

$$\mathbf{p} = 4\pi\varepsilon_1 \tilde{f}_{CM,23} a_1^3 \mathbf{E} \tag{73}$$

and the particle complex permittivity $\tilde{\varepsilon}_{23}$ in the Clausius-Mossotti factor $\tilde{f}_{CM,23}$ is given by

$$\tilde{\varepsilon}_{23} = \tilde{\varepsilon}_2 \left[\gamma_{12}^3 + 2\left(\frac{\tilde{\varepsilon}_3 - \tilde{\varepsilon}_2}{\tilde{\varepsilon}_3 + 2\tilde{\varepsilon}_2} \right) \right] \Big/ \left[\gamma_{12}^3 - \left(\frac{\tilde{\varepsilon}_3 - \tilde{\varepsilon}_2}{\tilde{\varepsilon}_3 + 2\tilde{\varepsilon}_2} \right) \right] \tag{74}$$

where the factor $\gamma_{12} = a_1/a_2$. This representation is equivalent to a homogeneous spherical particle of radius a_1 and permittivity $\tilde{\varepsilon}_{23}$ given by equation (74). The frequency variation of effective polarisability and magnitude of the dipole moment is again described by the Clausius-Mossotti factor, in this case $\tilde{f}_{CM,23}$, the real and imaginary parts of which are plotted in figure (14)

The model can be extended to multiple shells by using equation (74) sequentially for each shell. Figure (15) shows an example with two shells and the variation of the real and imaginary parts of $\tilde{\varepsilon}_{23}$ for this particular case.

6.5 The effective dipole moment of an ellipsoidal particle

The more general case of an ellipsoidal particle is also relevant since many particles are not spherical. The full calculation of the electrical potential around the particle and the corresponding dipole moment can be found in a number of textbooks (Stratton, 1941; Schwarz et al., 1965; Griffin, 1970; Miller and Jones, 1994; Jones, 1995).

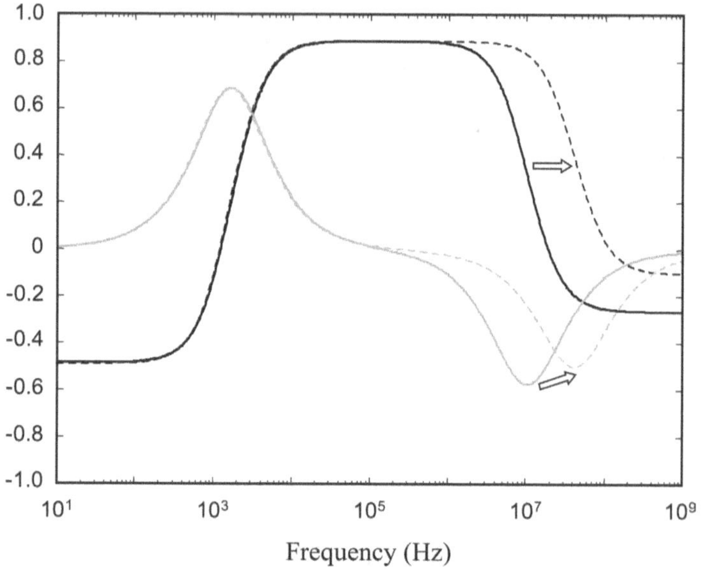

Figure 14. Plot of the real (black lines) and imaginary (grey lines) parts of the Clausius-Mossotti factor for a single shell (solid lines) and a two shell (dashed lines). Typical values for a human biological cell were used.

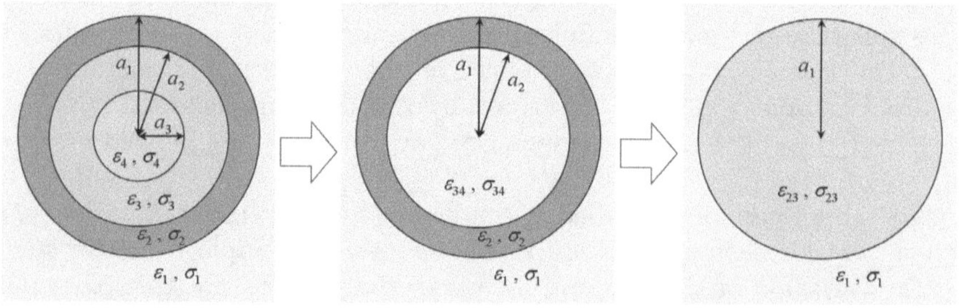

Figure 15. Schematic diagram of cross section of a multi shelled particle with various internal permittivities and conductivities. The shell model allows this internal structure to be analysed by consecutively treating each shell as a homogeneous sphere with a new value for permittivity and conductivity, until the problem is reduced to a homogeneous sphere.

Consider a homogeneous, ellipsoidal, dielectric particle where the half lengths of the major axes are a_1, a_2 and a_3. The effective polarisability, $\tilde{\alpha}_n = 3\varepsilon_m \tilde{K}_n$, is different for each principal axis n (where n = 1, 2, 3). $\tilde{K}_n$ is a frequency dependent factor equivalent to the Clausius-Mossotti factor, given by

$$\tilde{K}_n = \frac{\tilde{\varepsilon}_p - \tilde{\varepsilon}_m}{3(A_n(\tilde{\varepsilon}_p - \tilde{\varepsilon}_m) + \tilde{\varepsilon}_m)} \tag{75}$$

A_n is called the *depolarising factor* for the axis n given by

$$A_n = \frac{1}{2} a_1 a_2 a_3 \int_0^\infty \frac{ds}{(s + a_n^2)B} \tag{76}$$

where $B = \sqrt{(s + a_1^2) + (s + a_2^2) + (s + a_3^2)}$ and s is an arbitrary distance for integration.

The dipole moment of the ellipsoidal particle for each different axis can then be written as

$$\mathbf{p}_n = \frac{4\pi a_1 a_2 a_3}{3} \varepsilon_m \left[\frac{\tilde{\varepsilon}_p - \tilde{\varepsilon}_m}{\tilde{\varepsilon}_p + A_n(\tilde{\varepsilon}_p - \tilde{\varepsilon}_m)} \right] \mathbf{E}_n \tag{77}$$

where $\mathbf{E}_n$ is the component of the electric field in the n direction. Each axis of the particle also has an associated relaxation time, given by

$$\tau_n = \frac{A_n \varepsilon_p + (1 - A_n)\varepsilon_m}{A_n \sigma_p + (1 - A_n)\sigma_m} \tag{78}$$

7 Mixture theory

The theory linking the response of the particle mixture to that of the individual particle and the suspending solution was first derived by Maxwell (1954), and an elegant derivation of the mixture formula is given by Jones (1995). Assuming we have N spherical particles dispersed in a spherical region of the suspending medium at a volume fraction φ as shown in figure (16), the Principle of Superposition and equation (66) gives that the total dipole moment of the spherical region must be given by

$$\begin{aligned} \mathbf{p}_{eff} &= N\mathbf{p} \\ &= 4\varepsilon_m f_{CM} a^3 \mathbf{E} \\ &= 4N\pi\varepsilon_m \left(\frac{\tilde{\varepsilon}_p - \tilde{\varepsilon}_m}{\tilde{\varepsilon}_p + 2\tilde{\varepsilon}_m} \right) a^3 \mathbf{E} \end{aligned} \tag{79}$$

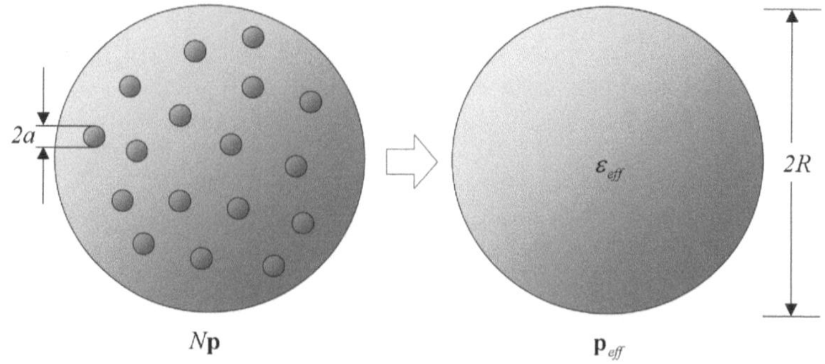

Figure 16. Schematic diagram of how a suspension of N particles can be treated as a homogenous sphere in deriving Maxwell's Mixture Theory.

Similarly equation (66) can be used to give a dipole moment for the whole sphere in terms of the effective dipole as

$$\mathbf{p}_{eff} = 4\pi\varepsilon_m \left(\frac{\tilde{\varepsilon}_{eff} - \tilde{\varepsilon}_m}{\tilde{\varepsilon}_{eff} + 2\tilde{\varepsilon}_m}\right) R^3\mathbf{E} \tag{80}$$

Equating these two expressions gives Maxwell's mixture formula:

$$\begin{aligned} \left(\frac{\tilde{\varepsilon}_{eff} - \tilde{\varepsilon}_m}{\tilde{\varepsilon}_{eff} + 2\tilde{\varepsilon}_m}\right) &= N\frac{a^3}{R^3}\left(\frac{\tilde{\varepsilon}_p - \tilde{\varepsilon}_m}{\tilde{\varepsilon}_p + 2\tilde{\varepsilon}_m}\right) \\ \left(\frac{\tilde{\varepsilon}_{eff} - \tilde{\varepsilon}_m}{\tilde{\varepsilon}_{eff} + 2\tilde{\varepsilon}_m}\right) &= \varphi\left(\frac{\tilde{\varepsilon}_p - \tilde{\varepsilon}_m}{\tilde{\varepsilon}_p + 2\tilde{\varepsilon}_m}\right) \end{aligned} \tag{81}$$

where $\varphi = N\frac{a^3}{R^3}$ is the volume fraction. This is more often written as

$$\tilde{\varepsilon}_{mix} = \tilde{\varepsilon}_m \frac{1 + 2\phi\tilde{f}_{CM}}{1 - \phi\tilde{f}_{CM}} = \tilde{\varepsilon}_m \frac{1 + 2\phi\left(\frac{\tilde{\varepsilon}_p - \tilde{\varepsilon}_m}{\tilde{\varepsilon}_p + 2\tilde{\varepsilon}_m}\right)}{1 - \phi\left(\frac{\tilde{\varepsilon}_p - \tilde{\varepsilon}_m}{\tilde{\varepsilon}_p + 2\tilde{\varepsilon}_m}\right)} \tag{82}$$

This equation gives the dielectric increment due to the particles and this is added to the permittivity of the suspending medium to give the total complex permittivity of the suspension. This is valid under the condition that the volume fraction is small (i.e. $\varphi \ll 1$). Further developments have been proposed to take into account the effect of high particle concentrations (Hanai et al., 1975; Hanai and Koizumi, 1976; Hanai et al., 1979; Zhang et al., 1993). Dielectric spectroscopy measurements of the variation in the complex permittivity of a suspension of dielectric particles with frequency

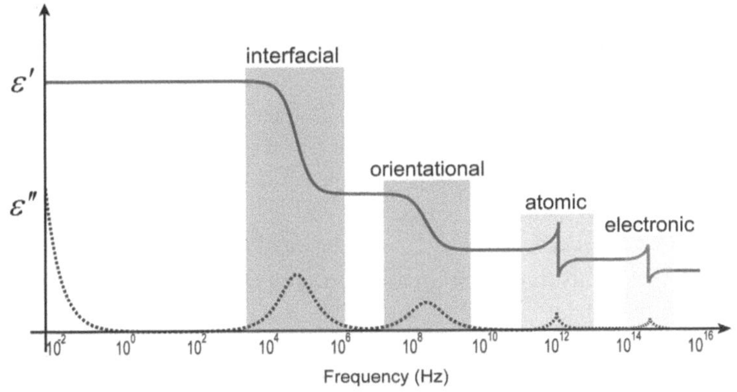

Figure 17. Schematic plot of the many different types of relaxation phenomena which occur in dielectric materials and the approximate frequency ranges in which they are found.

have been the accepted method for determining the effective polarisability and the induced effective dipole moment of individual particles.

8 Conclusion

In this chapter, the basic equations of quasi-electrostatic systems and the behaviour of charges have been discussed. The behaviour of collections of dipoles and the macroscopic properties of dielectrics formed from a collection of molecular dipoles have been discussed. The induced polarisation of particles in AC electric fields have also been discussed and expressions for the dipole moment of particles of different geometries and internal structures have been derived. The successive physical mechanisms have been detailed, with successively complex models being presented. In general, relaxations due to interfacial polarisation occur at frequencies below those for orientational and electronic polarisation. A plot of the real and imaginary parts of the complex permittivity of an arbitrary system, taking into account all the different polarisation mechanisms, would be expected to resemble figure (17)

Bibliography

B. I. Bleaney and B. Bleaney. *Electricity and Magnetism.* Clarendon Press, 1962.

K. S. Cole and R. H. Cole. Dispersion and absorption in dielectrics i. alternating current characteristics. *Journal of Chemical Physics*, 9:341–351, 1941.

J. Gimsa, P.Marszalek, U. Lwe, and T. Y. Tsong. Dielectrophoresis and electrorotation of neurospora slime and murine myeloma cells. *Biophysical Journal*, 60:749–760, 1991.

E. H. Grant, R. J. Sheppard, and G. South. *Dielectric Behaviour of Molecules in Solution.* Clarendon Press, 1978.

J. L. Griffin. Orientation of human and avian erythrocyte in radio-frequency fields. *Experimental Cell Research*, 61:113–120, 1970.

T. Hanai and N. Koizumi. Numerical estimation in a theory of interfacial polarization developed for dipserse systems in higher concentration. *Bull. Inst. Chem. Res. Kyoto Univ.*, 54:248–254, 1976.

T. Hanai, N. Koizumi, and A. Irimajiri. A method for determining the dielectric constant and the conductivity of membrane-bounded particles of biological relevance. *Biophysical Structural Mechanisms*, 1:285–294, 1975.

T. Hanai, K. Asami, and N. Koizumi. Dielectric theory of concentrated suspensions of shell-spheres in particular reference to the analysis of biological cell suspensions. *Bull. Inst. Chem. Res. Kyoto Univ.*, 57:297–305, 1979.

Y. Huang, R. Hlzel, R. Pethig, and X-B. Wang. Differences in the ac electrodynamics of viable and non-viable yeast cells determined through combined dielectrophoresis and electrorotation studies. *Physics in Medicine and Biology*, 37:1499–1517, 1992.

A. Irimajiri, T.Hanai, and A. Inouye. A dielectric theory of "multi-stratified shell" model with its application to a lymphoma cell. *Journal of Theoretical Biology*, 78:251–269, 1979.

T. B. Jones. *Electromechanics of Particles.* Cambridge University Press, 1995.

J. D. Kraus. *Electromagnetics.* McGraw-Hill, 1984.

P. Lorrain, D. R. Corson, and F. Lorrain. *Fundamentals of Electromagnetic Phenomena.* W.H. Freeman, 2000.

J. C. Maxwell. *A treatise on Electricity and Magnetism.* Dover Press, 1954.

R. D. Miller and T. B. Jones. Electro-orientation of ellipsoidal erythrocytes: Theory and experiment. *Biophysical Journal*, 64:1588–1595, 1994.

H. Morgan and N. G. Green. *AC Electrokinetics: Colloids and Nanoparticles.* Research Studies Press, 2003.

R. Pethig. *Dielectric and Electronic Properties of Biological Materials.* John Wiley and Sons, 1979.

G. Schwarz, M. Saito, and H. P. Schwan. On the orientation of non-spherical particles in an alternating electric field. *Journal of Chemical Physics*, 43: 3562–3569, 1965.

J. A. Stratton. *Electromagnetic Theory.* McGraw-Hill, 1941.

H. Z. Zhang, K. Sekine, T. Hanai, and N. Koizumi. Dielectric observations on polystyrene microcapsules and the theoretical analysis with reference to interfacial polarisation. *Colloid and Polymer Science*, 261:381–389, 1993.

Dielectrophoresis and AC Electrokinetics

Nicolas G. Green

School of Electronics and Computer Science, University of Southampton, Highfield, Southampton, SO17 1BJ, United Kingdom

Abstract This chapter discusses the forces produced by the interaction of non-uniform electric fields with the induced moments of the particle, focussing mainly on dielectrophoresis. Other AC electrokinetic effects will be discussed: a field with rotating field vector gives rise to a torque on a particle, referred to as electrorotation; and a field with a spatially varying phase (rather than magnitude) gives rise to travelling wave dielectrophoresis. Expressions will be derived for the different forces and torques and describe the resulting movement of the particles. The frequency dependent behaviour will be discussed and presented as a mechanism of determining the dielectric properties of a particle, as well as manipulating and separating different types of particle.

1 Introduction

Electrokinetics in general refers to the movement of particles suspended in fluids due to the action of electric fields. The interaction of non-uniform electric fields, i.e. one which has a spatial variation in magnitude, with the induced moments of the particle produces a movement termed dielectrophoresis. In general AC electric fields are used since the dielectrophoretic force has a non-zero time average.

In general, any field non-uniformity produces electrokinetic motion: a field with rotating field vector gives rise to a torque on a particle and rotation referred to as electrorotation; and a field with a spatially varying phase (rather than magnitude) gives rise to travelling wave dielectrophoresis. In the following sections, we derive expressions for the different forces and torques and describe the resulting movement of the particles. The frequency dependent behaviour will be discussed and presented as a mechanism of determining the dielectric properties of a particle, as well as manipulating and separating different types of particle.

1.1 A note on Electrophoresis

Electrophoresis is the movement of a particle with a non-zero net charge produced by the Coulomb force. Biological particles and most particles which suspend well in water have a finite surface charge density (usually negative, due to the presence of acid groups on the surface) and observation of the movement of these particles in a uniform electric field is used both to characterise and also to separate particles.

The Coulomb force on a particle is given by the product of the electric field and the charge on the particle

$$\mathbf{F}_{EP} = Q\mathbf{E} = \int_S \sigma_q dS\mathbf{E} \tag{1}$$

where Q is the total charge on the particle which, if the particle has a surface charge density σ_q, is given by the integral of this charge density over the closed surface S of the particle. In an AC electric field, the movement due to this force is oscillatory with zero time-average.

Many particles, including biological particles, when suspended in an electrolyte, are surrounded by an electrical double layer consisting of counter-ionic charges. This screens the particle surface charge, so that the force on the particle is not described exactly by equation (1).

2 Force on an induced dipole: Dielectrophoresis (DEP)

Dielectrophoresis is the motion of a particle produced by the interaction of a non-uniform electric field with the induced effective dipole moment of the particle (Pohl, 1978) as discussed in the previous chapter. If the field is uniform, the force on each of the two poles of the dipole is equal and opposite and there is no movement. If the field is non-uniform, however, the two forces are not equal and the particle moves, which will be demonstrated using a simple point and plane geometry (figure 1) as an illustration.

When the particle is more polarisable (figure 2a) more charge is induced on the inside of the interface and the dipole points in the same direction as the applied field. Both poles experience a force attracting them towards their facing electrode but the pole facing the point electrode is in a higher field strength region and experiences a stronger force. As a result, the particle is pulled in the direction of increasing field strength, an effect called negative dielectrophoresis.

If the particle is less polarisable (figure 2b), more charge is induced on the outside of the interface in the fluid medium and the dipole points in

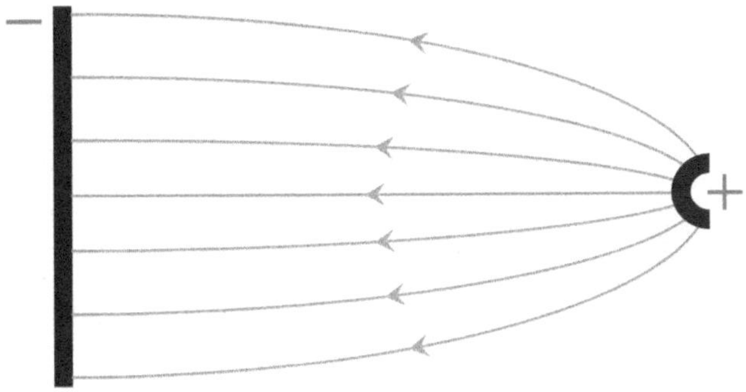

Figure 1. Schematic diagram of the simple non-uniform electric field produced by a point and plane electrode geometry.

the opposite direction to the field. Both poles experience a repulsive force from the facing electrode but again, the pole facing the point electrode experiences a stronger force and the particle is pushed in the direction of decreasing field strength. This is referred to as negative dielectrophoresis.

The figure also shows the arrangement of charges and the dipole for the potential difference applied in one direction and on the bottom, the other direction. This represents the two halves of an AC cycle, demonstrating that while the charges change sign in each half cycle, as does the direction of the dipole, since the electric field also changes direction, the net force on the particle points in the same direction. This means that since the charge is *induced* by the field, the Coulomb force in this case is second order with the field and the time averaged dielectrophoretic force is non-zero. In AC fields therefore, dielectrophoresis has a non-zero effect and can be used across a wide range of frequencies.

2.1 Translational force on a dipole in a non-uniform field

The expression for the force can be derived following Jones (1995). Figure (3) shows a dipole $\mathbf{p} = Q\mathbf{d}$ in a non-uniform electric field $\mathbf{E}$. As shown by the figure, the two charges each experience a different value of electric field so that the dipole will experience a net force given by

$$\mathbf{F} = Q\mathbf{E}(\mathbf{r} + \mathbf{d}) - Q\mathbf{E}(\mathbf{r}) \tag{2}$$

where the vector $\mathbf{r}$ is the position of the negative charge, as shown in the figure. If the length of the vector $\mathbf{d}$ is small compared with a typical dimension of the non-uniformity of the electric field, $\mathbf{E}$ can be expanded around

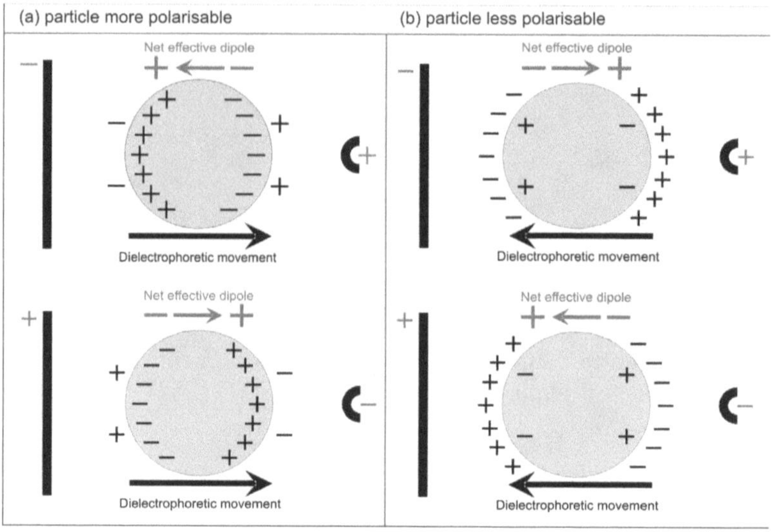

Figure 2. Schematic diagram of the polarisation and the induced dipole produced for the two cases: (a) particle more polarisable than the medium, with the induced dipole parallel to the field and (b) particle less polarisable than the medium, with the induced dipole anti-parallel to the field. When the dipole is parallel to the field, the Coulomb forces on the two poles are towards the nearest electrode but since the field on the point electrode side is stronger the particle is pulled in that direction. When the dipole is anti-parallel, the Coulomb forces on the two poles are repulsive but the field on the point electrode side is still stronger, the particle is repelled from that electrode. The top two diagrams show the arrangement of charges for the potential applied in one direction and the bottom two the arrangement when the potential is applied the other way around.

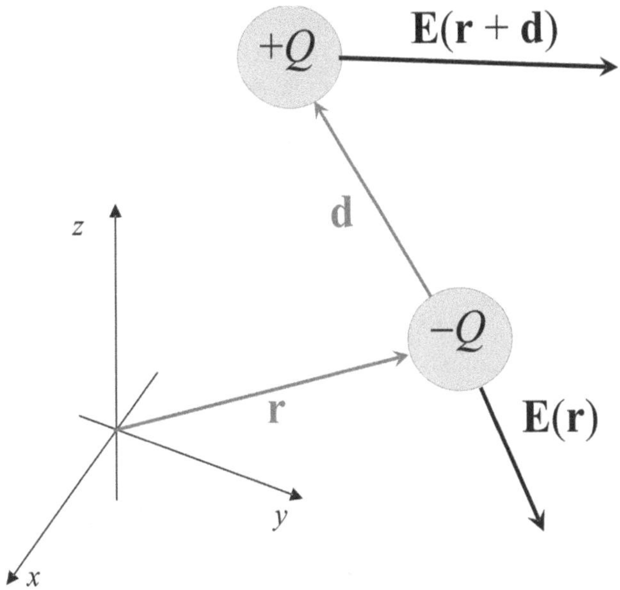

Figure 3. An arbitrary dipole in an arbitrary electric field experiences a difference electric field (magnitude and direction) on each of it's two poles.

$\mathbf{r}$ using the vector Taylor series

$$\mathbf{A}(\mathbf{x} + \mathbf{dx}) = \mathbf{A}(\mathbf{x}) + \mathbf{dx} \cdot \nabla \mathbf{A} + \text{higher order terms} \tag{3}$$

and the force on the dipole becomes

$$\mathbf{F} = Q\mathbf{E}(\mathbf{r}) + Q(\mathbf{d}.\nabla)\mathbf{E} + \text{higher order terms} - Q\mathbf{E}(\mathbf{r}) \tag{4}$$

or

$$\mathbf{F} = Q\mathbf{E}(\mathbf{r}) + Q\left(d_x\frac{\partial}{\partial x} + d_y\frac{\partial}{\partial y} + d_z\frac{\partial}{\partial z}\right)\mathbf{E} + \text{higher order terms} - Q\mathbf{E}(\mathbf{r}) \tag{5}$$

Substituting for the dipole moment and neglecting the higher order terms gives the well-known expression for the force on the dipole

$$\mathbf{F}_{DEP} = (\mathbf{p} \cdot \nabla)\mathbf{E} \tag{6}$$

This equation tells us that the force is zero if the electric field is uniform, i.e. only in a non-uniform field can dielectrophoresis occur.

Equation (6) gives the total force on the particle only if the length of the dipole is smaller than a typical dimension of the field non-uniformity, an assumption referred to as the dipole approximation. In other words, if the magnitude of the electric field does not vary significantly across the dimensions of the dipole, this expression for the force is correct. If this assumption does not hold or the particle is in a field null, then multipole force terms must be considered. A full derivation of the higher order terms is an involved process and has been performed exhaustively by Jones (1985, 1995); Jones and Washizu (1994); Washizu and Jones (1996). An alternative and more exact method is to use the Maxwell stress tensor. This requires an exact solution for the potential in the system and the integration of the stress tensor around the particle to give the force (Jones, 1995; Wang et al., 1994).

2.2 Dielectrophoresis in an AC field

This section will derive the force on the particle considering the case of an electric field with a spatially varying field magnitude and constant phase, which is the common method of deriving the dielectrophoretic force. In the next sections, the more general case of electric field with a spatially varying magnitude and phase will be considered (Morgan and Green (2003)).

Assuming an applied potential of a single frequency ω, then the time dependent values in the system can be represented using phasors. An arbitrary, harmonic potential can be defined as $\phi(\mathbf{x}, t) = \mathrm{Re}[\tilde{\phi}(x)e^{i\omega t}]$ where $i^2 = -1$, $\mathbf{x}$ is the position, the tilde indicates a complex phasor ($\tilde{\phi} = \phi_R + i\phi_I$) and Re[...] indicates the real part of.

The electric field is then given by $\mathbf{E}(\mathbf{x}, t) = \mathrm{Re}[\tilde{\mathbf{E}}(\mathbf{x})e^{i\omega t}]$ where the vector $\tilde{\mathbf{E}} = -\nabla\tilde{\phi} = -(\nabla\phi_R + i\nabla\phi_I)$ is the corresponding phasor. If the phase is constant across the system, the field phasor can be assumed to be real without loss of generality (i.e. $\tilde{\mathbf{E}} = \mathbf{E} = -\nabla\phi_R$).

The dipole moment of the particle is therefore $\tilde{\mathbf{p}} = \upsilon\tilde{\alpha}\mathbf{E}e^{i\omega t}$ and the time-averaged force on the particle is (from equation (6))

$$\langle \mathbf{F}_{DEP} \rangle = \frac{1}{2} \mathrm{Re}[(\tilde{\mathbf{p}} \cdot \nabla)\mathbf{E}^*] \tag{7}$$

where * indicates complex conjugate. In this case, the phasors are real, and the time-averaged force is

$$\langle \mathbf{F}_{DEP} \rangle = \frac{1}{2} \upsilon\, \mathrm{Re}[\tilde{\alpha}](\mathbf{E} \cdot \nabla)\mathbf{E} \tag{8}$$

Using the following vector identity

$$\nabla(\mathbf{A} \cdot \mathbf{B}) = (\mathbf{A} \cdot \nabla)\mathbf{B} + (\mathbf{B} \cdot \nabla)\mathbf{A} + \mathbf{B} \times (\nabla \times \mathbf{A}) + \mathbf{A} \times (\nabla \times \mathbf{B}) \tag{9}$$

and the fact that the field is irrotational ($\nabla \times \mathbf{E} = 0$) gives

$$\langle \mathbf{F}_{DEP} \rangle = \frac{1}{4} \upsilon \operatorname{Re}[\tilde{\alpha}] \nabla (\mathbf{E} \cdot \mathbf{E}) \tag{10}$$

This is the dielectrophoretic force, which is commonly written as

$$\langle \mathbf{F}_{DEP} \rangle = \frac{1}{4} \upsilon \operatorname{Re}[\tilde{\alpha}] \nabla |\mathbf{E}|^2 \tag{11}$$

or using root-mean-square (RMS) values for the field:

$$\langle \mathbf{F}_{DEP} \rangle = \frac{1}{2} \upsilon \operatorname{Re}[\tilde{\alpha}] \nabla |\mathbf{E}_{rms}|^2 \tag{12}$$

which is also the expression for the force in a DC field. Here we will not use RMS designations and equation (11) will be used as the expression for the DEP force, where E is the amplitude of the electric field.

Inspection of this equation shows that the dielectrophoretic force depends on the volume of the particle and the gradient of the field magnitude squared (proportional to the energy density in the field). These two parameters can vary greatly in typical experimental conditions. The force also depends on the real part of the effective polarisability and therefore, on the permittivity and conductivity of both the particle and the suspending medium, as well as the frequency of the applied electric field.

For a spherical particle, the variation in the magnitude of the force with frequency is given by the real part of the Clausius-Mossotti factor (equation (68) of previous chapter). Substituting for the effective polarisability of a sphere (equation (67) of previous chapter), the full expression for the time-averaged DEP force is

$$\langle \mathbf{F}_{DEP} \rangle = \pi \varepsilon_m a^3 \operatorname{Re}\left[\frac{\tilde{\varepsilon}_p - \tilde{\varepsilon}_m}{\tilde{\varepsilon}_p + 2\tilde{\varepsilon}_m} \right] \nabla |\mathbf{E}|^2 \tag{13}$$

where the parameters have been defined previously.

2.3 Dielectrophoretic behaviour

Examination of the DEP force expressions shows that the magnitude of the force depends on several important factors.

First, it depends on the particle volume (a^3), an important factor as the range of particle radii of interest is broad: 100 μm down to 10nm, which is 12 orders of magnitude in volume.

Second it depends on the permittivity of the suspending medium, which is usually relatively high, for example $\sim 80\varepsilon_o$ for water, but can be as low as $\sim 2\varepsilon_o$ in oil.

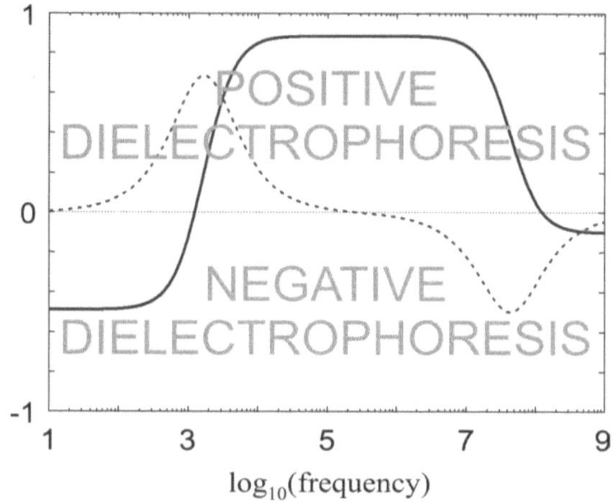

Figure 4. The Clausius-Mossotti factor for a shelled particle similar to a biological cell plotted as a function of frequency separately as real (solid line) and imaginary (dashed line) parts. Where the real part is positive, the particle will experience positive DEP in a non-uniform field and where it is negative, it will experience negative DEP.

Third, it depends on the gradient of the field strength squared, which is proportional to the energy of the particle in the field. This is the main controllable parameter for handling smaller and smaller particles: the gradient in the field can be increased by reducing the dimensions of the electrodes, thereby increasing the curvature of the electric field.

Lastly, the real part of the Clausius-Mossotti factor defines the frequency dependence and direction of the force. Positive DEP occurs if the polarisability of the particle is greater than the suspending medium ($\mathrm{Re}[\tilde{f}_{CM}] > 1$) and the particle moves towards regions of high electric field strength. Negative DEP occurs if the polarisability of the particle is less than the suspending medium ($\mathrm{Re}[\tilde{f}_{CM}] < 1$) and the particles are repelled from regions of high field strength. The force depends through the Clausius-Mossotti factor, on the permittivity and conductivity of both the particle and the suspending medium, and therefore on the frequency of the applied electric field.

Figure (4) shows a typical plot of the Clausius-Mossotti factor as a function of frequency. Where the real part is positive, this particle will experience positive DEP and will be attracted to regions of high field strength. Likewise if the real part is negative, then the particle will experience nega-

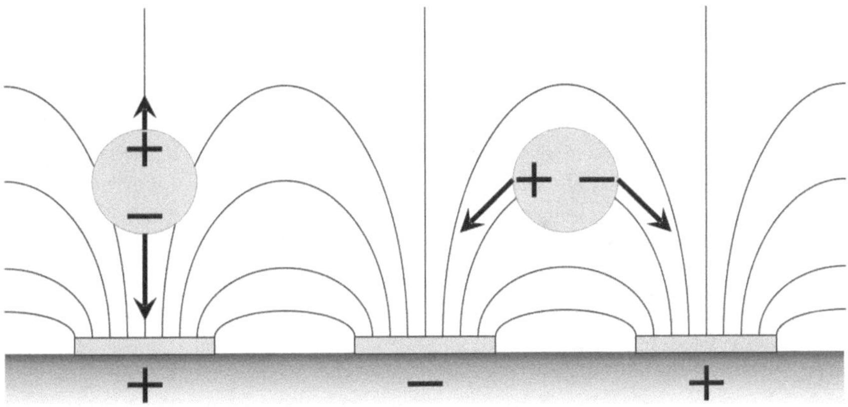

Figure 5. Schematic cross-section diagram of a simple array of interdigitated electrodes with alternate potentials applied generating a non-uniform electric field between each consecutive pair. This is a commonly used electrode design and has two types of planes of symmetry: a plane of even symmetry in the middle of each electrode and a plane of odd symmetry in the centre of each gap. Also shown are the effective dipole moments of two particles placed on these planes of symmetry and the approximate forces on the poles of each dipole.

tive DEP. For the particle shown in the figure, which is typical of a biological cell, it experiences negative DEP at low frequencies, positive DEP in the middle frequency range and negative DEP at high frequencies.

2.4 A note on the direction of the DEP force

It was state previously that the dielectrophoretic force arises from a difference in the strength of the field acting on the two halves of the dipole induced in the particle. This is the picture most commonly used to explain dielectrophoresis because of its simplicity. However, it is also a trifle misleading about the direction of the DEP force in that it implies that the direction is towards the electrode. In fact the force is driven by the curvature of the electric field not the magnitude.

To illustrate, consider the electrode array drawn schematically in figure (5). This is the interdigitated electrode array, which will be discussed in more detail later, consisting of a row of finger electrodes with alternate signals applied along the array. The diagram shows roughly the field lines generated by this array in cross-section demonstrating the repeated pattern and the planes of symmetry in the centres of the electrodes and the gas between the electrodes. Two particles are drawn on these planes of symmetry along with the orientation of the induced dipole for positive DEP. For the particle on the left, the situation is similar to the simple picture outlined previously. The pole closest to the electrode experiences a stronger force than the pole away from the electrode and therefore moves towards the electrode. In this case, the electric field, the dipole moment and the DEP force are all lined up. However, the reason that the field strength is greater close to the electrode and weaker further away in the vertical direction (away from the electrodes entirely) is due to the curvature of the field bending it round to the next electrode.

In addition, for the second particle over the centre of the gap, the field strengths at the two poles of the dipole are equal since the field is symmetrical. The reason that there is a DEP force here is that the field is *curved* across the particle and the field vectors point downwards on either side of the particle. The sum of these two vector forces is a resultant that points straight down towards the substrate between the electrodes, not towards an electrode. The force maps for this electrode design and others will be discussed later.

2.5 Dielectrophoresis in a fluid

A fluid exerts a viscous drag force $\mathbf{F}_\eta$ on a particle as it moves given by:

$$\mathbf{F}_\eta = -f\mathbf{v} \tag{14}$$

where $\mathbf{v}$ is the velocity of the particle and f is the friction factor of the particle. When the particle is subjected to an external force, for example the dielectrophoretic force, it will accelerate until the drag force is equal to the applied force after which it moves at the terminal velocity for the applied force. This movement is described by:

$$\mathbf{v} = \frac{\mathbf{F}_{DEP}}{f}\left(1 - e^{-\frac{t}{\tau_m}}\right) \tag{15}$$

where $\tau_m = m/f$ is referred to as the momentum relaxation time. For times much greater than this time, the particle moves at the terminal velocity:

$$\mathbf{v}_{DEP} = \frac{\mathbf{F}_{DEP}}{f} \tag{16}$$

For biological cells at the largest scale where dielectrophoresis is applied, the momentum relaxation time is of the order of microseconds, signifying that the particles are always observed to move at a velocity proportional to the applied dielectrophoretic force. In general, the dielectrophoretic velocity is

$$\mathbf{v}_{DEP} = \frac{\upsilon \, \mathrm{Re}[\tilde{\alpha}] \nabla |\mathbf{E}|^2}{f} \tag{17}$$

with *dielectrophoretic mobility* μ_{DEP} defined as the terms in front of the electric field term so that

$$\mu_{DEP} = \frac{\upsilon \, \mathrm{Re}[\tilde{\alpha}]}{f} \tag{18}$$

For a spherical particle, the friction factor is $f = 6\pi\eta \, a$ where η is the dynamic viscosity of the fluid so that the dielectrophoretic velocity is

$$\mathbf{v}_{DEP} = \frac{\pi a^3 \varepsilon_m \, \mathrm{Re}[\tilde{f}_{CM}] \nabla |\mathbf{E}|^2}{6\pi\eta a} \tag{19}$$

and the mobility is

$$\mu_{DEP} = \frac{a^2 \varepsilon_m \, \mathrm{Re}[\tilde{f}_{CM}]}{6\eta} \tag{20}$$

Examination of these equations reveals that the observable effect of dielectrophoresis only scales with radius squared not cubed, a fact that allows smaller particles to be successfully manipulated.

3 Dielectrophoresis in a field with a spatially dependent phase

For a general AC field, such as that generated by the application of multiple potentials of different phase, the derivation of the dielectrophoretic force is more involved. The electric field in this case is $\mathbf{E}(\mathbf{x}, t) = \mathrm{Re}[\tilde{\mathbf{E}}(\mathbf{x}) e^{i\omega t}]$, where the vector $\tilde{\mathbf{E}} = -\nabla\tilde{\phi} = -(\nabla\phi_R + i\nabla\phi_I)$ is the corresponding complex phasor. The expression for the time-averaged force on the particle can then be derived from equation (7) noting that the vectors now consist of complex components. The equation for the force is

$$\langle \mathbf{F}_{DEP} \rangle = \frac{1}{2} \mathrm{Re}[(\tilde{\mathbf{p}} \cdot \nabla)\tilde{\mathbf{E}}^*] = \frac{1}{2} \upsilon \, \mathrm{Re}[\tilde{\alpha}(\tilde{\mathbf{E}} \cdot \nabla)\tilde{\mathbf{E}}^*] \tag{21}$$

Using the following vector identity and the fact that the electric field is irrotational ($\nabla \times \tilde{\mathbf{E}} = 0$)

$$\begin{aligned} \nabla(\mathbf{A} \cdot \mathbf{B}) &= (\mathbf{A} \cdot \nabla)\mathbf{B} + (\mathbf{B} \cdot \nabla)\mathbf{A} + \mathbf{B} \times (\nabla \times \mathbf{A}) + \mathbf{A} \times (\nabla \times \mathbf{B}) \\ \nabla(\tilde{\mathbf{E}} \cdot \tilde{\mathbf{E}}^*) &= (\tilde{\mathbf{E}} \cdot \nabla)\tilde{\mathbf{E}}^* + (\tilde{\mathbf{E}}^* \cdot \nabla)\tilde{\mathbf{E}} + \tilde{\mathbf{E}}^* \times (\nabla \times \tilde{\mathbf{E}}) + \tilde{\mathbf{E}} \times (\nabla \times \tilde{\mathbf{E}}^*) \end{aligned} \tag{22}$$

giving

$$\nabla(\tilde{\mathbf{E}} \cdot \tilde{\mathbf{E}}^*) = (\tilde{\mathbf{E}} \cdot \nabla)\tilde{\mathbf{E}}^* + (\tilde{\mathbf{E}}^* \cdot \nabla)\tilde{\mathbf{E}} \tag{23}$$

as well as the following vector identity and the fact that the field has zero divergence (Gauss's law) in a homogeneous dielectric

$$\begin{aligned} \nabla \times (\mathbf{A} \times \mathbf{B}) &= (\mathbf{B} \cdot \nabla)\mathbf{A} - (\mathbf{A} \cdot \nabla)\mathbf{B} + (\nabla \cdot \mathbf{B})\mathbf{A} - (\nabla \cdot \mathbf{A})\mathbf{B} \\ \nabla \times (\tilde{\mathbf{E}} \times \tilde{\mathbf{E}}^*) &= (\tilde{\mathbf{E}}^* \cdot \nabla)\tilde{\mathbf{E}} - (\tilde{\mathbf{E}} \cdot \nabla)\tilde{\mathbf{E}}^* + (\nabla \cdot \tilde{\mathbf{E}}^*)\tilde{\mathbf{E}} - (\nabla \cdot \tilde{\mathbf{E}})\tilde{\mathbf{E}}^* \end{aligned} \tag{24}$$

giving

$$\nabla \times (\tilde{\mathbf{E}} \times \tilde{\mathbf{E}}^*) = (\tilde{\mathbf{E}}^* \cdot \nabla)\tilde{\mathbf{E}} - (\tilde{\mathbf{E}} \cdot \nabla)\tilde{\mathbf{E}}^* \tag{25}$$

then the force expression (equation 21) becomes on combination and substitution of equations (23) and (25)

$$\langle \mathbf{F}_{DEP} \rangle = \frac{1}{4} \upsilon \, \mathrm{Re}[\tilde{\alpha}] \nabla (\tilde{\mathbf{E}} \cdot \tilde{\mathbf{E}}^*) - \frac{1}{2} \upsilon \, \mathrm{Im}[\tilde{\alpha}] (\nabla \times (\tilde{\mathbf{E}} \times \tilde{\mathbf{E}}^*)) \tag{26}$$

where Im [...] is the imaginary part of the function. This can be re-written as

$$\langle \mathbf{F}_{DEP} \rangle = \frac{1}{4} \upsilon \, \mathrm{Re}[\tilde{\alpha}] \nabla \left| \tilde{\mathbf{E}} \right|^2 - \frac{1}{2} \upsilon \, \mathrm{Im}[\tilde{\alpha}] (\nabla \times (\mathrm{Re}[\tilde{\mathbf{E}}] \times \mathrm{Im}[\tilde{\mathbf{E}}])) \tag{27}$$

where $\left|\tilde{\mathbf{E}}\right|^2 = \left|\mathrm{Re}[\tilde{\mathbf{E}}]\right|^2 + \left|\mathrm{Im}[\tilde{\mathbf{E}}]\right|^2$. If there is no spatially varying phase, the phasor of the electric field can be taken to be real (i.e. $\mathrm{Im}[\tilde{\mathbf{E}}] = 0$), the second term on the right hand side of equation (27) is then zero and the expression for the force becomes that of equation (11). However, if there is a spatially varying phase, as in the case of travelling wave dielectrophoresis, the complete force expression must be used. The first term in the force expression depends on the frequency in the same manner as equation (11) and has a similar form. The second term in the force equation depends, however, on the imaginary part of the effective polarisability, or rather the imaginary part of the Clausius-Mossotti factor. This force is zero at high and low frequencies, rising to a maximum value at the Maxwell-Wagner interfacial relaxation frequency.

Note that the form of equation (27) given here is different from that generally given in the literature (e.g. Wang et al. (1994)) but is equivalent and lends itself readily to numerical simulation. Another point to note is that in the presence of electrode polarisation, the potential across the medium is no longer real, it becomes complex and has an imaginary component which is not parallel with the real part. Therefore, equation (11) cannot be used to determine the dielectrophoretic force and equation (27) should be used instead.

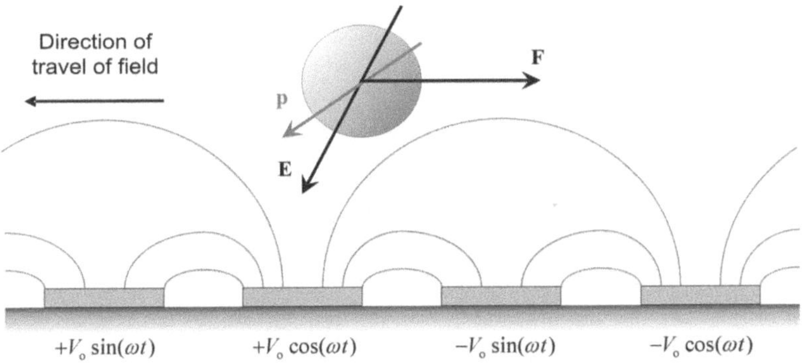

Figure 6. Schematic diagram of a linear travelling wave dielectrophoresis array and the consecutive phase-shifted signals required to generate the travelling electric field. Also shown are the approximate field lines for time $t = 0$, the electric field and the dipole moment induced in the particle together with the force on the particle.

3.1 Travelling wave Dielectrophoresis (twDEP)

As mentioned previously, one example of a phase varying field that is used in the literature is the travelling electric field, as illustrated in figure (6). The sequentially phase-shifted AC voltages generate an electric field with a spatially dependent phase.

From equation (27) the complete expression for the force consists of two components, one of which is the DEP force and another that depends on the phase of the field. In a travelling electric field, both of these components exist, so that the full DEP force is given by

$$\langle \mathbf{F}_{DEP} \rangle = \frac{1}{4} \upsilon \,\mathrm{Re}[\tilde{\alpha}] \nabla \left| \tilde{\mathbf{E}} \right|^2 - \frac{1}{2} \upsilon \,\mathrm{Im}[\tilde{\alpha}] (\nabla \times (\mathrm{Re}[\tilde{\mathbf{E}}] \times \mathrm{Im}[\tilde{\mathbf{E}}])) \tag{28}$$

The two components can be considered separately; a dielectrophoretic force, given by the first part of the r.h.s. and an additional twDEP force which propels the particle in the opposite direction to the moving field vector, given by the second half of the r.h.s. of this equation. Generally, for twDEP to be effective, the frequency and conductivity must be chosen to fulfil two criteria: (a) the particle is levitated above the electrodes by a negative DEP force and (b) the imaginary part of the Clausius-Mossotti factor is non-zero so that the particle moves along the track. These conditions are illustrated in figure (7), which shows a plot of the real and imaginary parts of the Clausius-Mossotti factor calculated for a latex particle suspended in

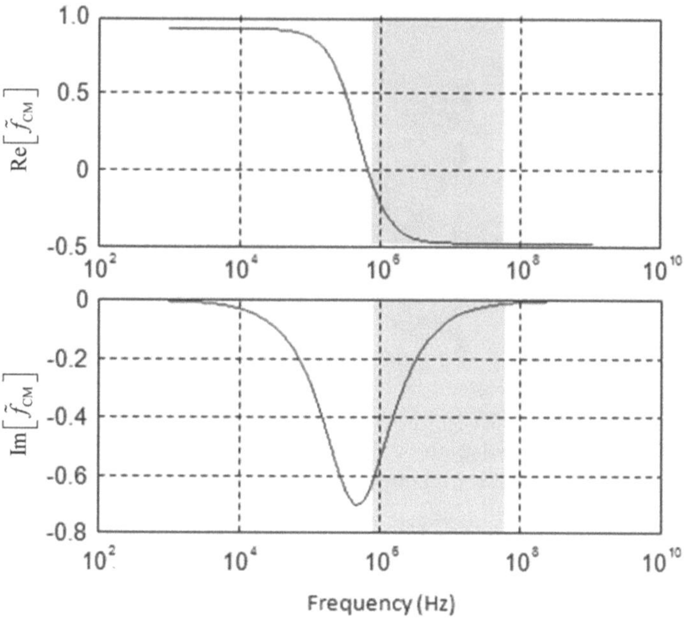

Figure 7. Plot of the real and imaginary parts of the Clausius-Mossotti factor for a solid homogeneous particle plotted as a function of frequency. In the frequency range shown by the grey area, particles in a travelling wave array will experience sufficient dielectrophoretic repulsion to levitate, and a large enough imaginary part to produce a finite translational force along the array.

a very low conductivity medium. Examination of the frequency dependence of both the real and imaginary parts shows three distinct regions, indicated by the shaded parts of the plot. At low frequencies, particles are pulled onto the electrode array by positive DEP forces, so that although the imaginary part is non-zero, twDEP does not occur in practice. At very high frequencies, particles are levitated above the electrodes, but now the imaginary part is zero so that no twDEP occurs. Only in the mid frequency range, where both the real part is substantially high and the imaginary part is non-zero, can twDEP occur.

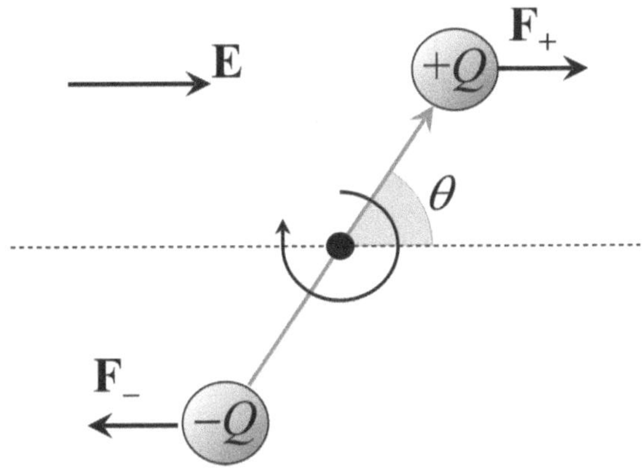

Figure 8. Schematic diagram of the Coulomb force acting on the two poles of a fixed dipole when there is a angle between the dipole moment and the applied field. The two forces act in the direction of the field and therefore act to rotate the dipole into alignment with the field.

4 Torques: electrorotation and electro-orientation

In addition to the force on a dipole, if the dipole moment is at an angle to the applied electric field, a torque will be produced acting to align the two. While in AC electrokinetic systems, the dipoles are induced on particles and are not fixed, two related phenomena result from this type of torque: electrorotation and electro-orientation. The expressions given here are all first order only. Expressions for the electrorotational torque incorporating higher order terms have been covered both by Washizu and Jones (1996).

When a dipole sits in a uniform electric field, each charge on the dipole experiences an equal and opposite force tending to align the dipole parallel to the field, i.e. it experiences a torque. Immediately after applying the electric field, it takes a finite amount of time for the dipole moment to become aligned with the field vector; in other words a time delay exists between the establishment of the field and the formation of the dipole. Therefore, if the field vector changes direction, the induced dipole moment vector must realign itself with the electric field vector, causing particle rotation.

Figure (8) shows the forces on the two poles of a dipole and the resultant

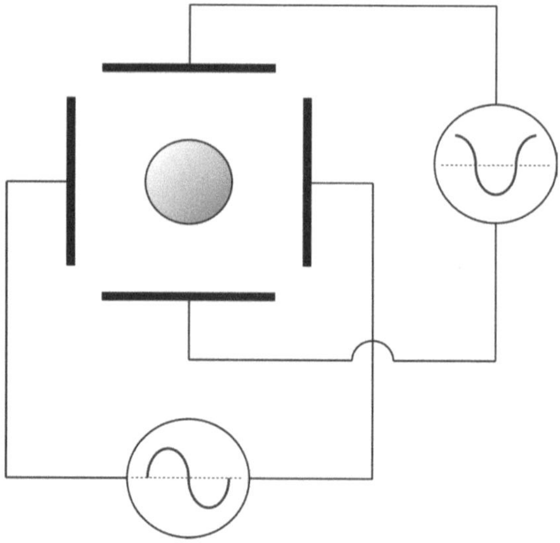

Figure 9. A schematic diagram of an electrorotation setup. Four signals, successively $90^t exto$ out of phase are applied to four electrodes encircling the particle.

rotation. The torque on this dipole is found from:

$$\begin{aligned} \mathbf{\Gamma}_{ROT} &= \frac{\mathbf{d}}{2} \times Q\mathbf{E} + \frac{-\mathbf{d}}{2} \times (-Q\mathbf{E}) \\ &= Q\mathbf{d} \times \mathbf{E} \\ &= \mathbf{p} \times \mathbf{E} \end{aligned} \tag{29}$$

This torque is at the root of both electrorotation, which occurs in a rotating electric field, and electroorientation which occurs when the particle is non-spherical.

4.1 Electrorotation (ROT)

Normally, electrorotation is used to analyse particle by applying a rotating or circularly polarised electric field created by the superposition of two 90° phase-shifted AC signals as shown schematically in figure (9). The electric field induces a dipole which continuously tries to align itself with the rotating electric field vector. As a result, the particle experiences a constant torque and rotates asynchronously around its axis. The torque is zero when the phase angle between the particle's polarisation vector and the applied

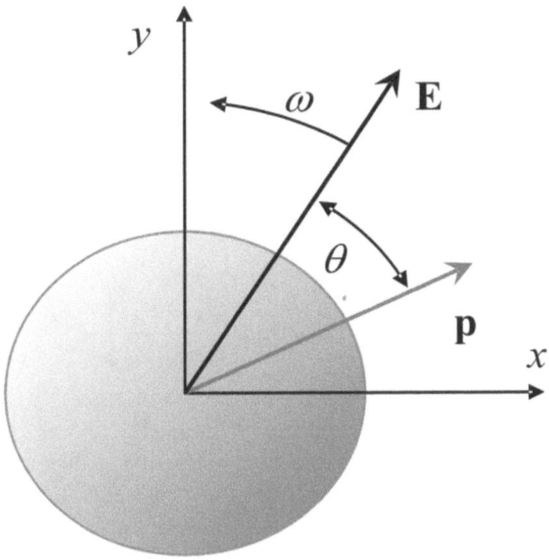

Figure 10. Schematic diagram showing how the induced dipole moment of a particle lags behind a rotating applied electric field.

field is zero and maximum when the phase angle is 90° as shown in figure (10). If the induced dipole moment lags behind the field, then the direction of rotation is with the field and vice versa for a moment that leads the field.

For a harmonic AC electric field, the time-averaged first order torque on the particle is

$$\langle \mathbf{\Gamma}_{ROT} \rangle = \frac{1}{2} \mathrm{Re}[\mathbf{p} \times \mathbf{E}^*] \tag{30}$$

Following the definitions given previously for the electric field and the induced effective dipole moment, the torque can be written as

$$\langle \mathbf{\Gamma}_{ROT} \rangle = \frac{1}{2} \upsilon \, \mathrm{Re}[\tilde{\alpha}(\tilde{\mathbf{E}} \times \tilde{\mathbf{E}}^*)] = -\upsilon \, \mathrm{Im}[\tilde{\alpha}](\mathrm{Re}[\tilde{\mathbf{E}}] \times \mathrm{Im}[\tilde{\mathbf{E}}]) \tag{31}$$

A quick consideration of a locally rotating field vector:

$$\begin{aligned} \tilde{\mathbf{E}} = E(\hat{\mathbf{x}} - i\hat{\mathbf{y}}) &\Rightarrow \begin{cases} \mathrm{Re}[\tilde{\mathbf{E}}] = E\hat{\mathbf{x}} \\ \mathrm{Im}[\tilde{\mathbf{E}}] = -iE\hat{\mathbf{y}} \end{cases} \\ \mathrm{Re}[\tilde{\mathbf{E}}] \times \mathrm{Im}[\tilde{\mathbf{E}}] &= \det \begin{pmatrix} \hat{\mathbf{x}} & \hat{\mathbf{y}} & \hat{\mathbf{z}} \\ E & 0 & 0 \\ 0 & -E & 0 \end{pmatrix} = -E^2\hat{\mathbf{z}} \end{aligned} \tag{32}$$

gives the normal expression for the ROT torque:

$$\langle \mathbf{\Gamma}_{ROT} \rangle = -v\,\mathrm{Im}[\tilde{\alpha}]|\mathbf{E}|^2\hat{\mathbf{z}} \tag{33}$$

and for a spherical particle:

$$\langle \mathbf{\Gamma}_{ROT} \rangle = -4\pi\varepsilon_m a^3\,\mathrm{Im}\left[\frac{\tilde{\varepsilon}_p - \tilde{\varepsilon}_m}{\tilde{\varepsilon}_p + 2\tilde{\varepsilon}_m}\right](\mathrm{Re}[\tilde{\mathbf{E}}] \times \mathrm{Im}[\tilde{\mathbf{E}}]) \tag{34}$$

and

$$\langle \mathbf{\Gamma}_{ROT} \rangle = -4\pi\varepsilon_m a^3\,\mathrm{Im}\left[\frac{\tilde{\varepsilon}_p - \tilde{\varepsilon}_m}{\tilde{\varepsilon}_p + 2\tilde{\varepsilon}_m}\right]|\mathbf{E}|^2\hat{\mathbf{z}} \tag{35}$$

The electrorotational torque for a particle is governed by the permittivity of the suspending medium, the volume of the particle and the applied field magnitude squared. More importantly, the torque is frequency dependent through the effective polarisability. For example, for the spherical shelled particle previously shown in figure (4), the imaginary part of the Clausius-Mossotti factor peaks at the Maxwell-Wagner interfacial frequency so that a single electrorotational peak is found for each interface. In this case, one of the peaks is co-rotational with the field and one is counter-rotational.

Similar to the situation for dielectrophoresis, in a viscous fluid, the rotation rate arising from the electrorotational torque has an acceleration phase and a terminal rate phase given by:

$$\mathbf{R}_{ROT} = \frac{\mathbf{\Gamma}_{ROT}}{f_\theta} \tag{36}$$

where $f_\theta = 8\pi\eta\, a^3$ is the rotational friction factor and for a sphere,

$$\mathbf{R}_{ROT} = -\frac{\varepsilon_m|\mathbf{E}|^2}{2\eta}\,\mathrm{Im}\left[\tilde{f}_{CM}\right]\hat{\mathbf{z}} \tag{37}$$

Note that the steady state rotation rate is now independent of particle size, depending only on the differences in the dielectric properties.

Examination of equation (31) shows a similar field phase dependency to travelling wave dielectrophoresis. In fact, travelling wave dielectrophoresis and electrorotation always occur together, making electrorotational observations across a broad range of frequencies challenging.

4.2 Electroorientation

Electroorientation is the orientation of a non-spherical particle with the electric field. Without going into much detail, using the dipole moment for each axis from the previous chapter (equation (77)), with α, β, γ representing the directions of the three principal axes and the particle volume given by $v = (4/3)\pi a_\alpha a_\beta a_\gamma$:

$$\begin{aligned}\tilde{\mathbf{p}}_\alpha &= 3v\varepsilon_m \tilde{K}_\alpha \tilde{\mathbf{E}}_\alpha \\ \tilde{\mathbf{p}}_\beta &= 3v\varepsilon_m \tilde{K}_\beta \tilde{\mathbf{E}}_\beta \\ \tilde{\mathbf{p}}_\gamma &= 3v\varepsilon_m \tilde{K}_\gamma \tilde{\mathbf{E}}_\gamma\end{aligned} \tag{38}$$

There are then three torque components:

$$\begin{aligned}\langle \mathbf{\Gamma} \rangle_\alpha &= \frac{1}{2}\operatorname{Re}\left[\tilde{\mathbf{p}}_\beta \times \tilde{\mathbf{E}}_\gamma^* - \tilde{\mathbf{p}}_\gamma \times \tilde{\mathbf{E}}_\beta^*\right] \\ \langle \mathbf{\Gamma} \rangle_\beta &= \frac{1}{2}\operatorname{Re}\left[\tilde{\mathbf{p}}_\gamma \times \tilde{\mathbf{E}}_\alpha^* - \tilde{\mathbf{p}}_\alpha \times \tilde{\mathbf{E}}_\gamma^*\right] \\ \langle \mathbf{\Gamma} \rangle_\gamma &= \frac{1}{2}\operatorname{Re}\left[\tilde{\mathbf{p}}_\alpha \times \tilde{\mathbf{E}}_\beta^* - \tilde{\mathbf{p}}_\beta \times \tilde{\mathbf{E}}_\alpha^*\right]\end{aligned} \tag{39}$$

which can be manipulated to give

$$\begin{aligned}\langle \mathbf{\Gamma} \rangle_\alpha &= 2v\varepsilon_m (L_\gamma - L_\beta) E_\beta E_\gamma \operatorname{Re}\left[\tilde{K}_\beta - \tilde{K}_\gamma\right] \\ \langle \mathbf{\Gamma} \rangle_\beta &= 2v\varepsilon_m (L_\alpha - L_\gamma) E_\gamma E_\alpha \operatorname{Re}\left[\tilde{K}_\gamma - \tilde{K}_\alpha\right] \\ \langle \mathbf{\Gamma} \rangle_\gamma &= 2v\varepsilon_m (L_\beta - L_\alpha) E_\alpha E_\beta \operatorname{Re}\left[\tilde{K}_\alpha - \tilde{K}_\beta\right]\end{aligned} \tag{40}$$

The electroorientational torque has a complicated frequency dependent behaviour arising from the sum of the three individual torque components, which can in principle result in a non-spherical particle aligning with any of its major or minor axes parallel to the applied field.

5 Particle-particle interaction

The basic derivations of the AC electrokinetic forces so far have assumed that the particles are isolated entities, or at least are sufficiently separated from each other that they do not interact. However, in many situations this is not true and particle-particle interactions must be considered.

From basic electrostatics, when two or more particles with the same sign of charge are suspended in an insulating medium, Coulomb's law states that they will always repel each other. However, the situation in an electrolyte is a little more complicated, since there is free charge which moves in the applied electric field and screens the particle's charge. The consequence is that the field produced by the particle charge rapidly decays with distance into the suspending medium, so that any long-range electrostatic interaction

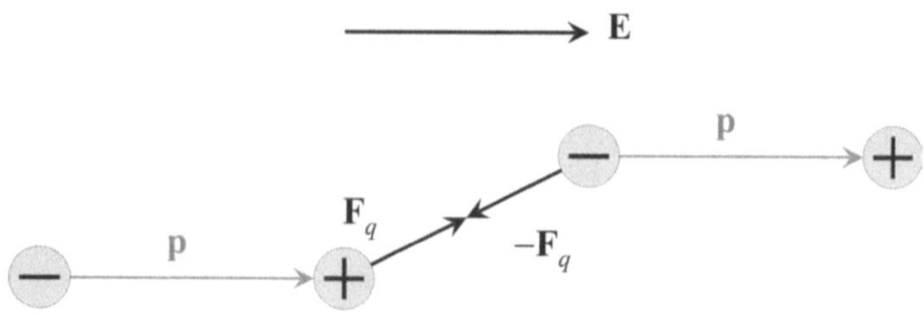

Figure 11. Schematic diagram of the attractive force between two fixed dipoles aligned by an applied uniform electric field.

with other particles does not occur. The only interaction that can occur is when the particles are very close; in this case both electrostatic interactions and Van der Waals forces contribute to the total interaction force between the particles. These effects are important in determining the stability of colloids.

5.1 Two fixed dipoles

As a qualitative example, consider two fixed dipoles suspended in a uniform electric field. Both dipoles align with the field in the same direction, as shown schematically in figure (11). If the dipoles are close enough, each of the charges in one dipole experiences a different force from the two charges in the other dipole. The positive charge of one dipole "feels" an attractive force from the negative charge on the second dipole and vice versa. The action of this attractive force is to pull the dipoles together to form a chain. The rapid decay of the dipole field with distance ensures that the force imbalance rapidly goes to zero with increasing separation of the dipoles.

5.2 Induced dipoles

In the case of polarisable particles and induced dipoles rather than fixed dipoles, then similar effects occur. A quantitative approach to the issue of particle-particle interaction can be found, for example, in (Jones, 1995) and (Giner et al., 1999). This section presents only a qualitative summary using field simulations for illustration.

The behaviour of a particle in an electric field can be phenomenologically predicted by examining the electric field distribution. This is equally true for examining the behaviour of two particles interacting as the formation of the induced dipole produces a non-uniform field around a particle. Particle-particle interactions can therefore be thought of as one particle experiencing

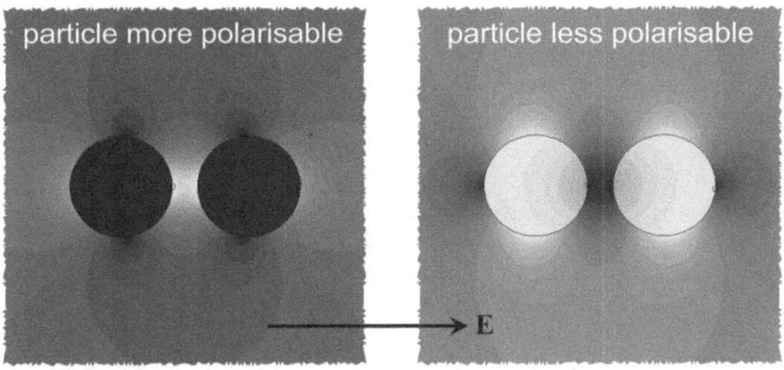

Figure 12. The magnitude of the electric field around two particles aligned along a horizontal applied electric field. In (a) both particles are much more polarisable than the surrounding medium i.e. they experience positive DEP. In (b) both particles are much less polarisable than the medium and experience negative DEP.

dielectrophoresis in the non-uniform field produced by the other particle and vice-versa. Two dimensional simulations can be used to illustrate some interesting phenomena of particle-particle interaction.

For two particles aligned along the direction of a uniform electric field, the distribution in the field magnitude around the particles is shown in figure (12). There are two interesting cases to consider. The first is shown in figure (12a), where both particles are much more polarisable than the surrounding medium (e.g. conducting particles in a relatively low conductivity electrolyte at low frequencies). The field strength inside the particle is low and there are low field strength regions on either side of the particles. However, there are high field strength regions at the poles of the particles, along the axis of the applied field. This is particularly obvious in the region between the two particles. Since both particles experience positive dielectrophoresis, they will always move towards high field regions and, as this figure shows, they will therefore move towards each other. The second case is when both particles are much less polarisable than the suspending medium, as shown in figure (12b). The strong field regions are at the sides of the particles and the field strength outside the particles at the poles is low. Since both particles experience negative dielectrophoresis, they should move away from high field regions and therefore move towards each other. In both cases the particles will line up along the direction of the electric field and, if there is a large number of particles, chains will form.

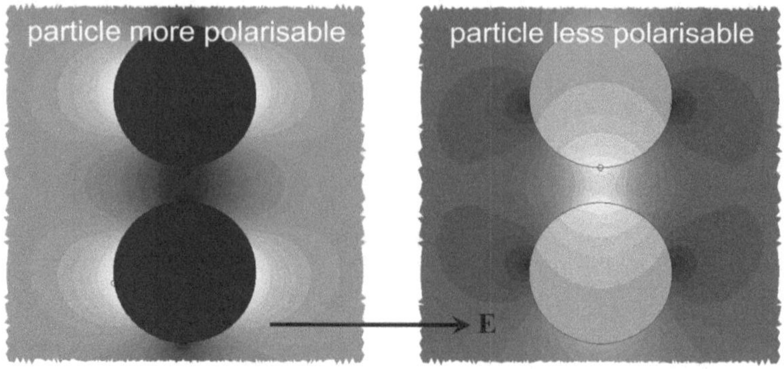

Figure 13. The magnitude of the electric field around two particles aligned perpendicular to a horizontal applied electric field. In (a), both particles are much more polarisable than the surrounding medium and experience. positive DEP. In (b) both particles are much less polarisable than the suspending medium and experience negative DEP.

For the same two particles aligned perpendicular to the applied field, figure (13a) shows the electric field magnitude around the more polarisable particles. The regions of strong electric field are again at the poles of the particles and along the direction of the field. This time there is a significant region of low field strength between the two particles. Since the particles experience positive dielectrophoresis, they will be pushed away from the low field region; in other words they repel each other. This arrangement is therefore unstable; the particles move into the arrangement shown in figure (12a). The same situation occurs for the less polarisable particles shown in figure (13b). These particles experience negative dielectrophoresis and are attracted to the low field regions, which in this case are at the poles along the field axis. They are also pushed away from the high field region, which is between the particles. The stable configuration is therefore that shown in figure (12b).

There is a third case that can be imagined: two different particles, one more polarisable and the second less polarisable than the electrolyte. In this case, one particle experiences positive dielectrophoresis and the other negative dielectrophoresis. The electric field magnitude is shown in figure (14), for two particles aligned along the field or perpendicular to the applied field. When the particles are aligned along the field, figure (14a), the more polarisable particle is attracted towards the high field points around the other particle, which are located at the sides of the particle along the axis

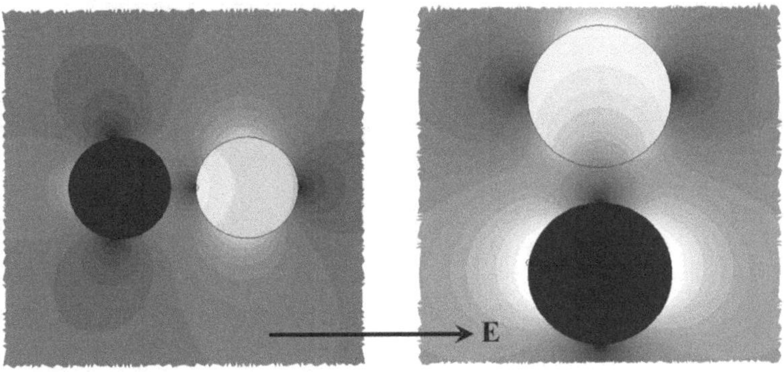

Figure 14. The magnitude of the electric field around two different particles, one more polarisable and the other less polarisable than the medium, in a horizontal applied electric field. Two arrangements are shown, (a) the particles aligned parallel to the field and (b) the two particles aligned perpendicular to the field.

perpendicular to the applied field. It is also repelled from the low field region between the two particles. The less polarisable particle on the right is likewise attracted to the sides of the particle along the perpendicular axis, since the low field regions around the more polarisable particle are there. The dielectrophoretic movement suggests that alignment along the field is unstable in this situation and that alignment across the field is preferred. Examining (14b), which shows the two particles aligned perpendicular to the field, it can be seen that the more polarisable (lower) particle would be attracted towards the less polarisable (upper) particle as there are high field regions between the two. The less polarisable particle is also attracted to the more polarisable particle because of the low field region between the particles. The stable arrangement is, therefore, alignment along the axis perpendicular to the applied field.

To summarise, for two or more particles of similar nature, the arrangement is the formation of long chains aligned along the applied field lines. If, however, there is a mixture of particle types, some of which experience positive and some negative dielectrophoresis, chains of like particles would form along the field lines. In addition, alternate unlike particles would align perpendicular to the field lines. This behaviour is illustrated and shown experimentally in figure (15), showing the different chains comprising different particle types, and showing a still video image of latex particles and yeast cells interacting as has been outlined. This has also been observed in

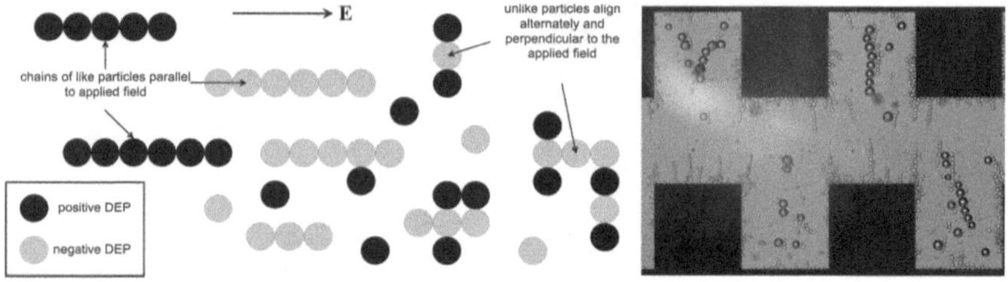

Figure 15. Schematic diagram and still video image of how a mixture of particles experiencing both positive and negative dielectrophoresis align in an applied electric field **E**. Like particles form pearl chains along the field lines and unlike particles align perpendicular to the applied field and alternately according to the type of DEP the particle is experiencing. The still video image shows 15microm particles experiencing negative DEP and yeast cells moving under positive DEP.

the literature: (Giner et al., 1999).

Bibliography

V. Giner, M. Sancho, R. S. Lee, G. Martinez, and R. Pethig. Transverse dipolar chaining in binary suspensions induced by rf fields. *Journal of Physics D: Applied Physics*, 32:1182–1186, 1999.

T. B. Jones. Multipole corrections to dielectrophoretic force. *IEEE Transactions on Industrial Applications*, IA-21:930–934, 1985.

T. B. Jones. *Electromechanics of Particles.* Cambridge University Press, 1995.

T. B. Jones and M. Washizu. Multipolar dielectrophoretic force calculation. *Journal of Electrostatics*, 33:187–198, 1994.

H. Morgan and N. G. Green. *AC Electrokinetics: Colloids and Nanoparticles.* Research Studies Press, 2003.

H. A. Pohl. *Dielectrophoresis.* Cambridge University Press, 1978.

X-B. Wang, Y. Huang, F. F. Becker, and P. R. C. Gascoyne. A unified theory of dielectrophoresis and travelling wave dielectrophoresis. *Journal of Physics D: Applied Physics*, 27:1571–1574, 1994.

M. Washizu and T. B. Jones. Generalized multipolar dielectrophoretic force and electrorotational torque calculation. *Journal of Electrostatics*, 38: 199–211, 1996.

Fundamentals of Electrowetting and Applications in Microsystems

Pablo García-Sánchez † and Frieder Mugele ‡

† Dpto. Electrónica y Electromagnetismo. Facultad de Física. Universidad de Sevilla. Sevilla (Spain).

‡ Physics of Complex Fluids, MESA+ and IMPACT Institutes. Faculty of Science and Technology, University of Twente. Enschede (The Netherlands)

Abstract Electrowetting has become widely used to control the wettability of solid surfaces in microsystems. In this chapter, we briefly introduce basic concepts of wetting and we discuss in detail the fundamental physics behind the electrowetting phenomenon. We compare the different theoretical approaches to the electrowetting equation, i.e. the thermodynamic derivation and the electromechanical interpretation. The effects of using AC signals are discussed and the limits of validity of the electrowetting equation for increasing voltage are presented (contact angle saturation and contact line instabilities). In the second part of this chapter, we review applications where electroweting has shown itself as a powerful tool, like electrowetting-based displays and lenses. Special attention is dedicated to the use of electrowetting in microfluidic devices.

1 Introduction

As dimensions of physical systems are smaller, the surface-to-volume ratio increases rapidly and surface forces become, relatively, more important. Thus, the surface tension (i.e. the tendency of a given interface to minimize its area) has an increasingly importance as we reduce the volume of liquid we are dealing with. For example, the shape of droplets resting on solid substrates depends on the wettability and topography of the latter. The droplet morphology adjusts itself to find the configuration of minimum energy. As shown in Gau et al. (1999), several metastable states can exist for complex topographies and the liquid can abruptly change for one morphology to another when some parameteres are varied (see Figure 1a)). Two inmiscible fluids coflowing in a microchannel also illustrate the importance of wettability in microsystems. Figure 1b) shows a phase diagram of the flow patterns observed by Dreyfus et al. (2003). They studied two distinct situations. When the outer fluid completely wets the walls, they found well structured flow patterns. However, if the outer fluid partially wets the walls, no stable flow patterns are identified.

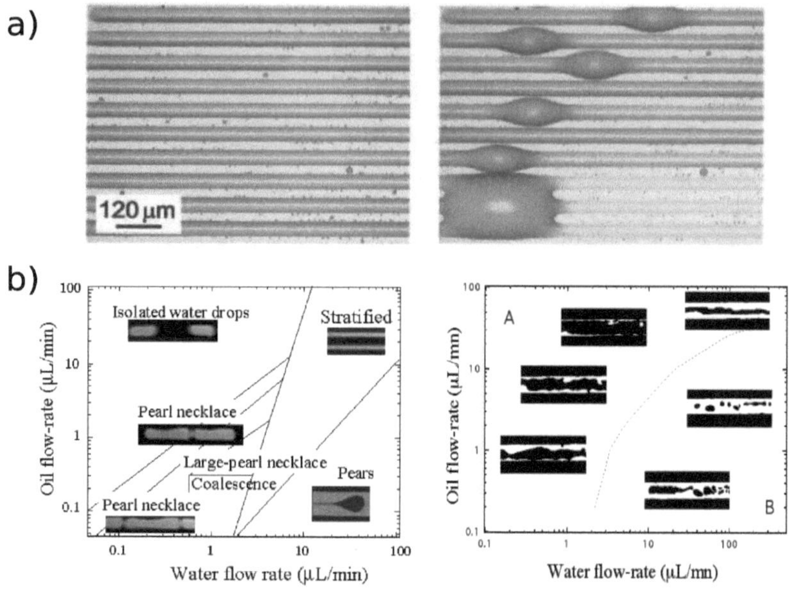

Figure 1. Two examples on the importance of wall wettability in microfluidic systems. a) An array of hydrophilic stripes on a hydrophobic substrate. For a low amount of water the stripes are uniformly covered by the liquid, forming channels of constant cross section (Left). For higher volumes, the channels develope a single bulge as soon as the contact angle exceeds a certain characteristic value (Right), (reproduced with permission from Gau et al. (1999), © 1999 AAAS). b) Flow patterns observed for two inmiscible coflowing fluids in a microchannel. If the outer fluid completely wets the walls, well structured flow patterns are found (Left). However, if the outer fluid only partially wets the walls, no clear flow structures are identified (Right), (reproduced with permission from Dreyfus et al. (2003), © 2003 American Physical Society).

From these examples, it is clear that the control of surface tension is a powerful tool for manipulating fluids in the micrometer range. Surface tension can be modified in several ways (chemically, gradients in temperature, electrically...). In this chapter we show how the wetting of solid substrates by conducting liquids can be tuned by adequately placing electrodes, a phenomenon that has been named *electrowetting* (Mugele and Baret, 2005; Quilliet and Berge, 2001). Electrical control of the effective surface tension presents several advantages against other methods, *electrowetting* provides reversible changes of the contact angle in excess of 90° with actuation speeds on the scale of milliseconds. This capability has made of electrowetting the most popular tool for "digital" microfluidic systems, which are based on the manipulation of discrete drops in a microfluidic chip (Fair, 2007). Manipulation of drops includes generating, moving, merging, splitting and mixing. In the second part of this chapter we will show how electrowetting has become a versatile tool for all these operations. In the first part of the chapter, we will briefly introduce the basic concepts of wetting and describe in detail the fundamentals of electrowetting.

2 Physics of Electrowetting

Electrowetting generally refers to the reduction of the contact angle upon application of an electric field. In this section we aim to give a global view of the current knowledge on the fundamental physics behind this phenomenon. We do not only describe electrowetting by introducing the "classical" thermodynamic derivation of Lippmann equation. Important consequences arise when the phenomenon is studied more in detail and, therefore, we also introduce the recent electromechanical interpretation as well as other aspects that make electrowetting a much richer and intriguing problem. We start the section by introducing premilinary concepts as surface tension and contact angle. We have also included a note about the work by Lippmann, where the relation between surface tension and electric field was described for the first time.

2.1 Basics of Wetting

Surface Tension. Young-Laplace Law. As already stated in the introduction to the chapter, the surface tension γ [1]can be intuitively understood as the tendency of a given interface to minimize its area and, therefore, is a crucial quantity when working with small amount of liquids (drops). Physically, the surface tension is a property of a given interface between

[1]In most texts, surface tension is denoted either by σ or γ. In other chapters of this text σ refers to the liquid conductivity. Therefore, we choose γ for surface tension.

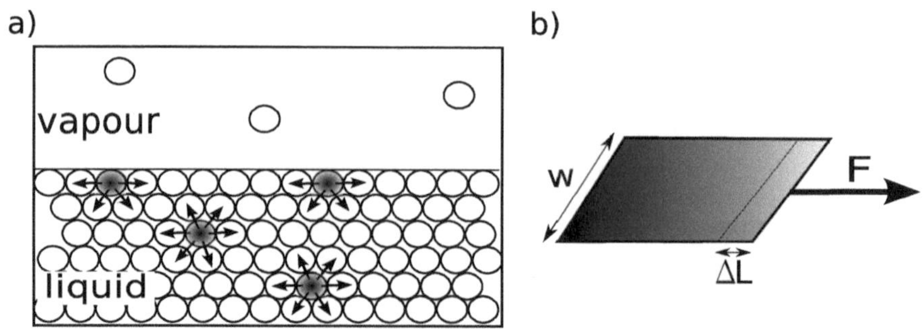

Figure 2. a)Microscopic interpretation of surface tension. Molecules at the surface cannot form as many bonds with neighbours as the molecules in the bulk can form. b)Surface tension as a force per unit length. The force required to strecth the liquid film is $F = \gamma/2w$.

two materials and it is defined as the Gibbs free energy G per unit area of that interface A, at given pressure p and temperature T:

$$\gamma \equiv \left(\frac{\partial G}{\partial A}\right)_{p,T} \tag{1}$$

SI units for surface tension are $[\gamma] = \mathrm{J \cdot m^{-2}} = \mathrm{N \cdot m^{-1}}$ and, as a typical value, the surface tension for air-water at $20\,°\mathrm{C}$ and $1\,\mathrm{atm}$ is $72\,\mathrm{mN \cdot m^{-1}}$.

Fig. 2a) shows the microscopic interpretation of the origin of surface tension between a liquid and a gas. Molecules in the bulk of the liquid experience intermolecular forces, forming chemical bonds (i.e. hydrogen bridges, etc) between neighbours. Molecules at the surface cannot form as many bonds as those in the bulk and, therefore, bringing one molecule from the bulk to the interface implies the rupture of some bonds that are not later created. A certain amount of energy is then required and that is the origin of surface tension: it costs energy to form a surface.

It is also convenient to interpret surface tension as a force per unit length. In Fig. 2b), the force F acts on a liquid film over a distance ΔL, changing its total surface area. The work exerted on the film corresponds to $W = F\Delta L$ and it must be equal to the increase of surface energy of the stretched film $G_{surf} = 2\gamma w \Delta L$, where the factor 2 is because the film has a surface both on the top and on the bottom. Surface tension can be then written as $\gamma = F/2w$, with the meaning of a force per unit length trying to minimize the liquid surface.

As a consequence of surface tension, a discontinuity in hydrostatic pressure P arises at curved interfaces between inmiscible fluids. This jump in pressure is given by the Young-Laplace law, which reads:

$$\Delta P = \gamma \left(\frac{1}{R_1} + \frac{1}{R_2} \right) \tag{2}$$

where R_1 and R_2 are the radii of curvature of the interface.

Contact Angle. Young-Dupré equation. In electrowetting, we usually deal with droplets on solid substrates. In most cases, the droplets are aqueous salt solutions with a typical size of the order of 1 mm or less [2]. The enviroment fluid can be either air or another immiscible liquid, frequently an oil, that we designate vapour phase. The *contact angle*, θ, is a property of the contact line between the three inmiscible phases, the three-phase contact-line (TCL) (see Figure 3a)), and it is defined as the angle between the solid/liquid and the liquid/gas interface. For a droplet at rest[3], and in the absence of an external electric field, the value of θ is designated as θ_Y, the *Young angle.* As stated by the Young-Dupré equation, θ_Y is determined by surface tensions of the three interfaces: solid-liquid γ_{sl}, liquid-vapour γ_{lv} and solid-vapour γ_{sv}.

$$cos(\theta_Y) = \frac{\gamma_{sv} - \gamma_{sl}}{\gamma_{lv}} \tag{3}$$

Expression 3 can be found from figure 3(b) imposing that, at equilibrium, the energy has to be stationary with respect to any infinitesimal diplacement of the TCL. Following de Gennes (1985), a displacement dx involves a change in free energy per unit length, w, given by:

$$\frac{1}{w} dG = \gamma_{sl} \mathrm{dx} - \gamma_{sv} \mathrm{dx} + \gamma_{lv} cos(\theta) \mathrm{dx} \tag{4}$$

and the equilibrium condition implies $\gamma_{sl} - \gamma_{sv} + \gamma_{lv} cos(\theta) = 0$.

Since surface tension can be understood as a force per unit length, a mechanistic interpretation of the equilibrium contact angle can be obtained if we impose equilibrium of these forces at the TCL: Vertical components are equilibrated, while the balance of horizontal components reads

[2] for this droplet size, we can neglect gravity versus surface tension

[3] for example, the advancing and receding angle in a moving drop can be different, see section on contact angle hysteresis

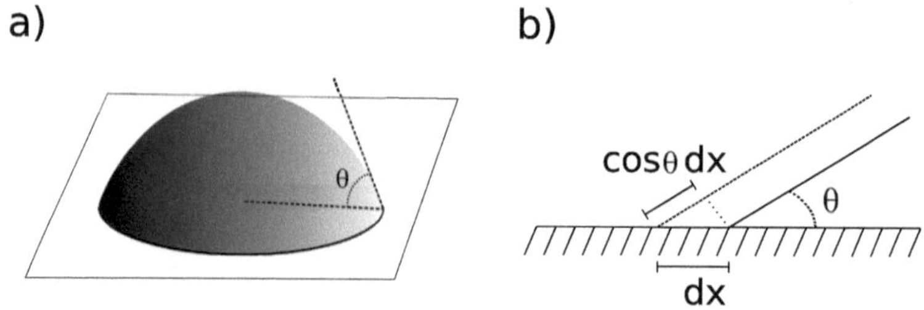

Figure 3. a)The contact angle θ is defined as the angle between the solid/liquid and the liquid/gas interfaces. b) Small displacement dx of the contact line from the equilibrium position at a contact angle θ.

$\gamma_{sl} - \gamma_{sv} + \gamma_{lv} cos(\theta) = 0$, which corresponds with expression 3.

Systems with contact angles $\theta < 90^o$ are often called hydrophilic, while those with $\theta > 90^o$ are called hydrophobic. Complete wetting refers to the case $\theta = 0$ and superhydrophobicity is usually employed when $\theta > 150^o$.

2.2 The Electrowetting Equation

Electrowetting is generally used to denote the reduction of the contact angle upon application of an electric field. The most common electrowetting setup is that in Figure 4a), where a drop of conducting liquid (salty water) is placed on top of one electrode coated with an insulating material. The dielectric layer was first used by Berge (1993) and the advantage is twofold: (1) it avoids Faradaic reactions that would otherwise occur at bare electrodes, allowing to increase significantly the applied voltage, and (2) highly hydrophobic dielectric layers are usually employed because Young's angle is high and, therefore, the electrowetting effect can be used to decrease it within a wide range. The wire on top of the droplet is used to establish the voltage difference, for instance, by means of a battery. Figure 4b) shows the dependence of the contact angle on the magnitude of the voltage. For a wide range of values, the reduction of the contact angle is correctly described by the so-called *Electrowetting Equation*, which reads:

$$\cos(\theta) = \cos(\theta_Y) + \frac{1}{2}\frac{\varepsilon_d \varepsilon_0 U^2}{d\gamma_{lv}} \tag{5}$$

where θ_Y is the contact angle when no electrical signal is applied (Young angle), d and ε_d are, respectively, the thickness and dielectric constant of the insulating layer and U the electrical potential applied to the electrodes. For voltages higher than $\sim 250\,\mathrm{V}$ in Figure 4, the contact angle does not follow this trend and, instead, tends to a constant value. This is known as contact angle saturation and it will be discussed in section 2.6. The remaining of this section is dedicated to the interpretation of equation 5.

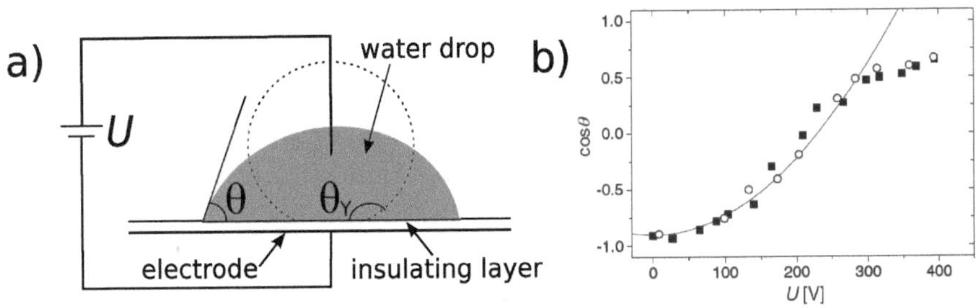

Figure 4. a) Generic electrowetting setup: a wire is immersed in a sessile drop that rests on top of an electrode covered by an insulator. At zero voltage the contact angle θ_Y is high (dashed line). The contact angle is reduced θ upon application of a voltage U. b) Typical electrowetting curve. The cosine of contact angle shows a parabolic increase with voltage up to a certain value above which saturation is observed (reproduced with permission from Mugele and Baret (2005), © 2005 Institute of Physics).

2.3 Electrochemical derivation

The electrowetting equation (5) is usually denoted as Young-Lippmann equation because it can be derived by combining Young equation (3) and Lippmann equation for the reduction of solid-liquid interfacial tension. Lippmann investigated the capillary depression of mercury in electrolyte-filled glass capillaries at variable voltages U, the electrocapillary effect (see section 2.7), and interpreted his results as an effective reduction of the mercury-electrolyte surface tension. Upon application of an electric field, electrical charges accumulate at the mercury-electrolyte interface, building up an electrical double layer (EDL) and accumulating electrostatic energy as in a capacitor. This redistribution of charge is a spontaneous proccess leading to a reduction of free energy of the interface and, consequently, to a reduction of surface tension. We can therefore write $\rho_{sl} = -\partial\gamma_{sl}^{eff}/\partial U$, where ρ_{sl} is the surface charge. Considering a constant capacitance per unit area C of the electrical double layer, $\rho_{sl} = CU$ and $\gamma_{sl}^{eff}(U) = \gamma_{sl} - (1/2)CU^2$.

Inserting this relation into Young's formula for the contact angle, we find:

$$\cos(\theta) = \cos(\theta_Y) + \frac{1}{2}\frac{CU^2}{\gamma_{lv}} \tag{6}$$

which coincides with equation 5 if the capacitance of the EDL is written as $C = \varepsilon_w \varepsilon_0 / d_H$, with ε_w the dielectric constant of the electrolyte and d_H the typical thickness of the EDL.

It is interesting to note that we have found the Lippmann relation for a metal-electrolyte interface, while equation 5 was written for electrowetting on a dielectric (EWOD). Young-Lippmann equation for EWOD was first derived by Berge (1993) and it is also demonstrated by minimizing the free energy. In this case the electrical double layer builds up at the insulator-droplet interface and the total capacitance corresponds to the association in series of this double layer and the insulator, with capacitance per unit area given by $\varepsilon_d \varepsilon_0 / d$, where ε_d is the permittivity of the insulator and d its thickness. The capacitance of the insulator is usually much smaller than the capacitance of the EDL, which can then be neglected. Therefore, equation 6 is also valid for EWOD if the specific capacitance of the electrical double layer C is replaced by that of the insulating layer.

This thermodynamic interpretation provides a correct description of electrowetting at macroscopic scales, i.e. the contact angle changes accordingly to Young-Lippman equation. In a previous section, the Young angle was interpreted as the necessary angle for mechanical equilibrium of the TCL. This view is still valid if we assume that the effective solid-liquid interfacial tension decreases upon application of the electric field and a smaller contact angle is required to find a new equilibrium.

2.4 Electromechanical interpretation and fine structure of the three-phase contact line

The thermodynamic approach in the previous section represents, probably, the simplest way of finding the dependence of the contact angle on the applied voltage. However, it does not provide a mechanical picture of how electrical forces modify the shape of the droplet. Furthermore, it does not clarify what is the origin of the effective or equivalent solid-liquid interfacial tension. Recent theoretical and numerical works have addressed this issue (Buehrle et al., 2003; Mugele and Buehrle, 2007) and it has been shown that the reduction of the contact angle can be understood from the balance

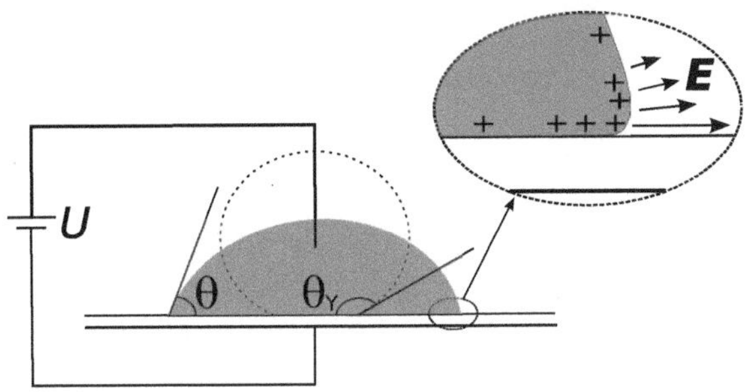

Figure 5. Drop of conductive liquid in a generic electrowetting setup. The zoomed region shows the air-liquid interface near the three-phase contact line. The electric field (arrows) acts on induced charges at the interface and, as a result, the drop surface is bent.

of electrostatic and Laplace pressures. In effect, let us consider Figure 5 where the electrical force is shown to deform the air-liquid interface. The stresses of electrical origin on the liquid can be found from the Maxwell Stress tensor (Stratton (1941)):

$$T_{ij} = \varepsilon(E_i E_j) - \frac{1}{2}\varepsilon\delta_{ij}|\mathbf{E}|^2 \tag{7}$$

where δ_{ij} is the Kronecker delta and $i, j = x, y, z$. Since the liquid is considered a perfect conductor, tangential stresses at the liquid surface are zero. Being n_j the components of the normal vector to the liquid surface, the normal component of the stress (electrical pressure) is $T_{ij}n_j = P_{el} = (\varepsilon/2)|\mathbf{E}(\mathbf{r})|^2$ and, in equilibrium, must be balanced by the Laplace pressure (eq. 2):

$$P_{el} = (\varepsilon/2)|\mathbf{E}(\mathbf{r})|^2 = \Delta P_L = 2\gamma\kappa(\mathbf{r}) \tag{8}$$

where $\kappa(\mathbf{r})$ is the local mean curvature of the drop surface at $\mathbf{r}$, $\kappa(\mathbf{r}) \equiv (1/2)(1/R_1 + 1/R_2)$.

Buehrle et al. (2003) solved eq. 8 and found the drop profile for an air-liquid interface on a dielectric layer of thickness d and assuming that the contact angle remains equal to Young angle θ_Y, i.e. the electric field does not change the local contact angle, see Fig. 6a). The drop shape asymptotically approaches the apparent contact angle θ_A, which is the angle we observe in the macroscopic scale. Fig. 6b) shows the numerical results for a given apparent contact angle and increasing electrowetting number $\eta = \varepsilon_d\varepsilon_0 U^2/(2\gamma_{lv}d)$. In this way, the theoretical profile for a system with a given Young angle θ_Y is revealed by finding the curve for that θ_Y and the corresponding electrowetting number.

Interestingly, these computations predict that the curvature of the drop surface changes within typical distances of the thickness of the insulator. This was experimentally tested by Mugele and Buehrle (2007) with electrodes coated with insulating layers of different thicknesses. Figures 7a),b) and c) show experimental drop shapes close to the contact line for electrowetting number $\eta = 1$ and three different thicknesses: 10, 50, 150 μm. In each case, a constant curvature is approached for distances larger than the thickness of the insulator (represented by the white bar). Figure 7d), e) and f) correspond to a constant thickness of 150 μm and increasing electrowetting number. Young's angle remains unchanged. These results reconcile the thermodynamic and electromechanical interpretations of electrowetting: *The local contact angle does not depend on the applied voltage, only the apparent contact angle for distances larger than the typical length for the electric field.* The "true" solid-liquid interfacial tension remains constant. This has been recently confirmed down to the nanoscale with an experimental setup based on a surface force apparatus (Gupta et al., 2010).

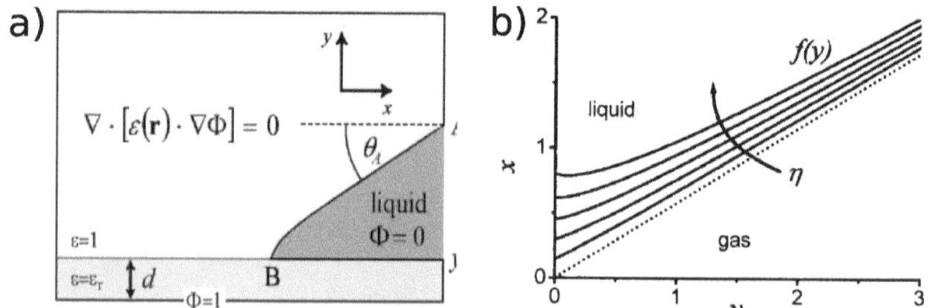

Figure 6. a)Schematic representation for the problem of finding the equilibrium drop shape. The apparent contact angle θ_A is fixed. b)Equilibrium surface profile for $\theta_A = 60^o$ and increasing electrowetting number $\eta = \varepsilon_d \varepsilon_0 U^2/(2d\gamma_{lv})$. Reproduced with permission from Buehrle et al. (2003), © 2003 American Physical Society.

Another important result of the electromechanical interpretation is that the total force per unit length of the contact line does not depend on the drop shape. Following Jones (2002), we calculate the total force on the liquid by integrating the Maxwell stress tensor (eq. 7) on the surface Σ in

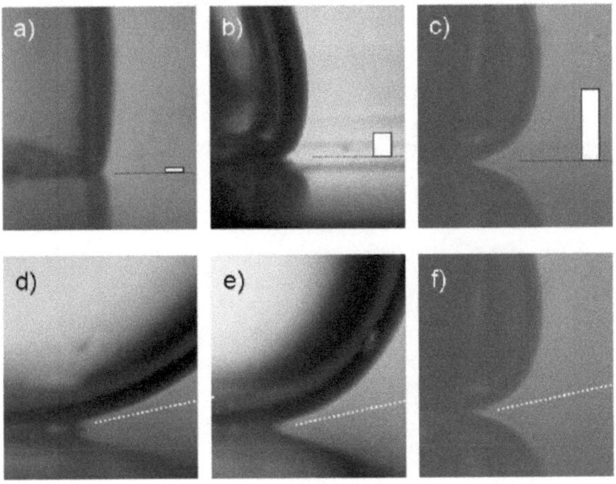

Figure 7. a),b) and c) are experimental drop shapes close to the contact line for $\eta = 1$ and $d = 10, 50, 150\,\mu$m, respectively. A constant curvature is approached for distances larger than the thickness of the insulator, represented by the white bar. d), e) and f) correspond to a constant thickness of $150\,\mu$m and increasing electrowetting number, $\eta = 0, 0.5, 1$. Young angle remains unchanged. Reproduced with permission from Mugele and Buehrle (2007), © 2007 IOP Publishing Ltd.

fig. 8.

$$\mathbf{f} = \int_{\Sigma} T_{ij} n_j dS \tag{9}$$

The surface can be chosen large enough so that the electric field on it is zero everywhere but in the line A to B in the insulator (the liquid is assumed to be a perfect conductor). This contribution yields a net force on the drop in the x direction. The total force per unit length of contact line w results:

$$f/w = (1/2)\varepsilon_d \varepsilon_0 (U^2/d) \tag{10}$$

Note that the total force per unit length is not dependent on the drop shape, as mentioned above. Expression 10 can be found either by integrating the Maxwell stress tensor or by using an electric-circuit model with lumped elements (Jones, 2002, 2005). The same methods can be also applied to find the net electrical force on dielectric liquids. In any case, no precise location for the electrical force is specified and the net force is not related to a change in contact angle. Thus, as pointed out by Jones (2002), the term electrowetting should be restricted to the change in contact angle, while the net force on the center of mass of the drop is an independent phenomenon, which should more conveniently be referred to as liquid dielectrophoresis. Also, a solid dielectric body partially inserted between two electrodes experiences a force pulling the body into the gap.

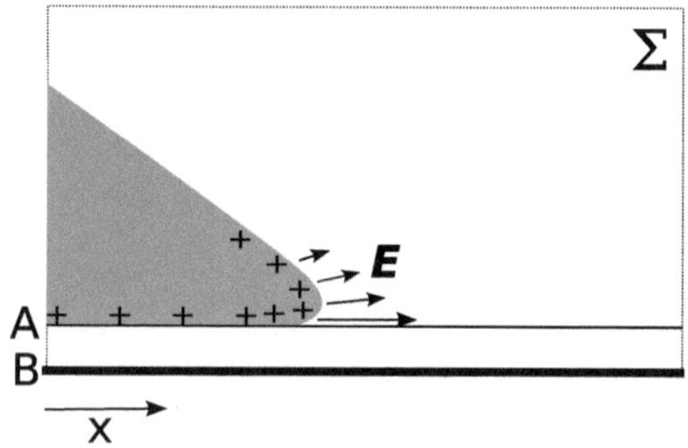

Figure 8. The total force per unit length of contact line can be computed by integrating the Maxwell stress tensor on the surface Σ.

2.5 Effect of AC signals and internal flows in AC electrowetting experiments

The theoretical analysis in the previous section corresponds to static conditions. The electrical signal was assumed to be constant (DC voltage) and the equilibrium droplet shape was found. Electrowetting also occurs with AC signals, the liquid response depends only on the time average of the applied voltage for signals with periods shorter than the hydrodynamic response time of the droplet[4]. The energy minimization argument is also applicable for AC voltages, the contact angle is given by eq. 5 but the amplitude of the voltage U has to be replaced by the root-mean-square value U_{rms}:

$$\cos(\theta) = \cos(\theta_Y) + \frac{1}{2}\frac{CU_{rms}^2}{\gamma_{lv}} \tag{11}$$

However, at frequencies around the typical hydrodynamic response of the droplet, surface standing waves are generated on the droplets, see Figure 9. In Oh et al. (2008), these oscillations patterns were theoretically reproduced by assuming an electrical force at the TCL with frequency double of

[4]for millimeter-sized droplets, AC signals at frequencies exceeding a few hundred hertz

the electric field.

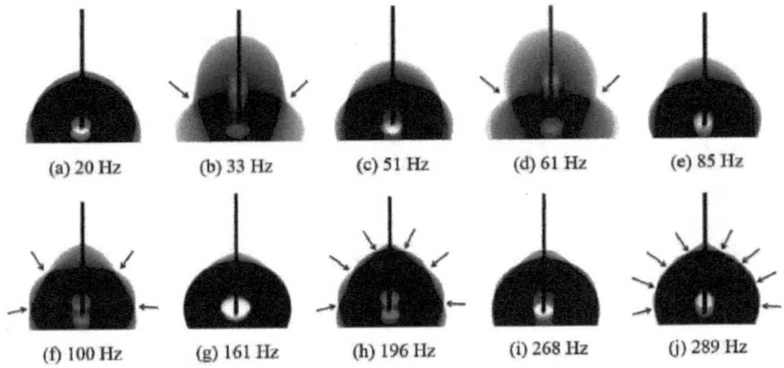

Figure 9. Surface drop oscillations patterns in AC electrowetting. Images are obtained after superimposing 50 frames (reproduced with permission from Oh et al. (2008), © 2008 American Chemical Society).

For even higher frequencies, typically above kHz, the liquid may not longer behave as a perfect electrical conductor. Jones et al. (2004) experimentally demonstrated this effect in a modification of Pellat's experiment that includes dielectric coatings on the vertical electrodes, see Figure 10. They measured the height of rise of different liquids between the two coated electrodes upon application of AC voltages and they showed that this height h decreases for increasing frequency above a critical value, ω_c, which depends on the liquid conductivity . Figure 10a) also includes the RC-circuit model used by Jones et al. (2004) to describe the dependence of h with the applied frequency. The liquid is modelled as resistor in parallel with a capacitor and the dielectric layers as capacitors. The total electrical force on the liquid can be derived from this circuit model and the height of rise is found by balancing this force with gravity. Figure 10b) shows the reduced height of rise, $K(\omega) = h(\omega)/U_{rms}^2$ (where U_{rms} is the rms-amplitude of the applied AC signal) for liquids of different conductivities. The circuit model correctly predicts the frequency dependence. Interestingly, they also observed a saturation of h for high voltages which they related to the saturation of contact angle (section 2.6).

Electrowetting experiments with AC signals in the common needle-substrate configuration are also affected by the finite value of the liquid conductivity. For example, Mugele et al. (2005) observed that the contact angle depends both on the frequency of the signal and on the position of the

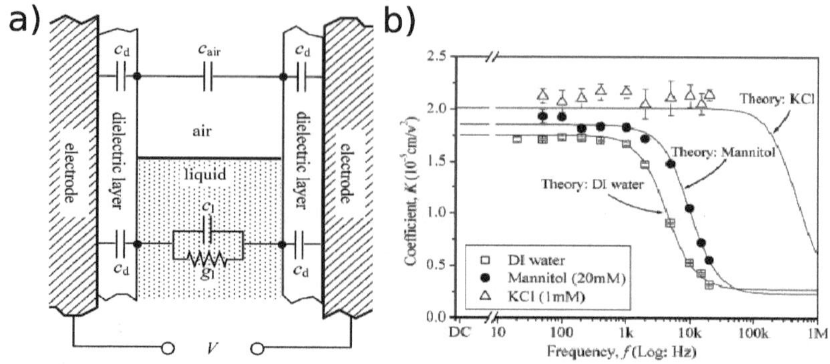

Figure 10. a)Variation of Pellat experiment where the electrodes are coated with a dielectric layer. The circuit model is used to compute the net electrical force on the liquid. b)Frequency dependence of reduced height of rise $K(\omega) = h(\omega)/U_{rms}^2$ for three different liquids, (reproduced with permission from Jones et al. (2004), © 2004 American Chemical Society).

needle. Kumar et al. (2006) studied a system where the complications of the needle-substrate geometry were avoided, see Figure 11a). The drop was sandwiched between two planar electrodes separated by a distance l. One of the electrodes was coated with a dielectric layer of thickness d and the contact angle was measured as a function of amplitude and frequency of the AC voltage. Figure 11b), c) and d) show measurements for three different conductivities: 5, 70 and 300 μS/m, respectively. No effect of the frequency up to 100 kHz was observed for the highest conductivity (Figure 11d)), however, for the other two conductivities, the contact angle was strongly dependent on frequency. As in the previous example of Figure 10, a circuit model can be used to analyze the electrical problem. As shown in Figure 11e), the liquid is modelled as a resistor in parallel with a capacitance, while the dielectric layer is modelled by a capacitor. According to Young-Lippmann equation, the contact angle must be a function of the rms voltage drop in the dielectric layer, which we can write as a fraction of the rms-amplitude of the applied signal, $f(\omega) \cdot U_{rms}$. The contact angle then becomes a function of the frequency of the applied signal ω. The circuit model in Figure 11e) correctly describes this frequency behaviour.

Another effect of using AC signals with potential applications is the appearance of internal flows. For frequencies of the order of the drop hydrodynamic resonance (as in Figure 9), the drops periodically oscillate between

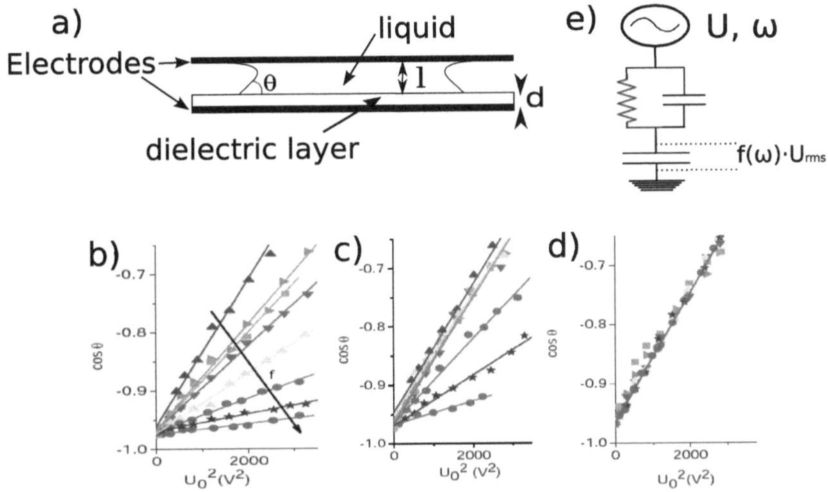

Figure 11. a)Scheme of experimental setup used by Kumar et al. (2006) for studying the effect of AC voltages. b), c) and d) are measurements of the contact angle versus the voltage amplitude for three different conductivities: 5, 70 and 300 μS/m, respectively. In each case, the frequency of the signal was gradually increased from DC (blue triangles) to 100 kHz (red circles). e) Circuit model to describe the frequency dependence in AC electrowetting

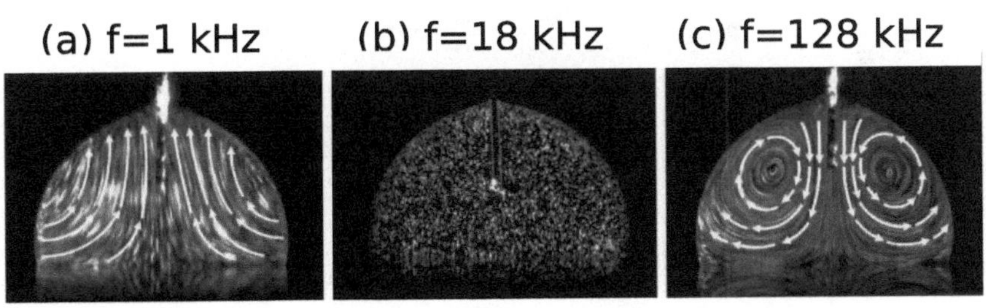

Figure 12. Internal flows in AC electrowetting. a)For frequencies around the typical hydrodynamic resonance of the drop, capillary waves occur at the drop surface which, in turn, generate fluid flow. b)The motion disappears when the frequency is increased. c)At higher frequencies, depending on liquid conductivity, the fluid flow is generated by electrothermal effect (images reproduced with permission from Ko et al. (2008), © 2008 American Chemical Society).

states of high and low contact angle. Despite the periodicity of this motion, there is a symmetry breaking in the drop shape between the spreading and the receding phase, which causes a time-averaged net flow inside the drop. This motion can be observed in Figure 12a), where a superposition of several frames shows the trajectory of fluorescent beads suspended in the liquid. These beads act as flow tracers, indicating the direction and magnitude of the generated flows. Mugele et al. (2006) demonstrated efficient mixing within droplets by this mechanism. Figure 13 shows different stages of the mixing process: (a) the fluorescent dye is injected on top of the drop and, (d) rapid mixing is achieved upon application of AC signals. The time interval between (a) and (d) is around 2 seconds. This flow can be described using a model based on capillary wave-driven Stokes drift (Staicu and Mugele, unpublished).

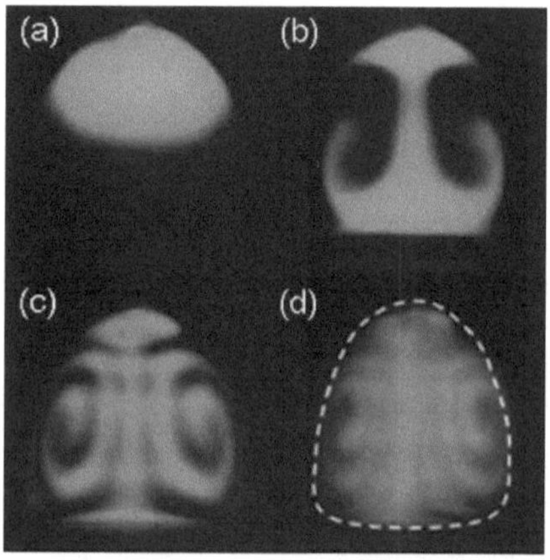

Figure 13. Fluorescent dye is used to visualize different stages of the mixing achieved with AC electrowetting. The frequency of the signal was around 80 Hz. (Reproduced with permission from Mugele et al. (2006), © 2006 American Institute of Physics).

Figure 12b) shows that the fluid flow ceases if the frequency is increased. However, for AC signals of somewhat higher frequencies (depending on liquid conductivity) a different type of fluid flow arises, Figure 12c). The origin of the flow in this case is due to the so-called electrothermal effect (García-Sánchez et al. (2010); Lee et al. (2009)). The liquid is no longer a perfect conductor at these frequencies, as mentioned above. The electric field then penetrates and Ohmic currents generate changes in temperature by Joule effect which in turn give rise to gradients in liquid conductivity. The action of the electric field on these conductivity gradients induces free charge in the liquid bulk. Thus, a net electrical body force appears, driving the flow. Nichols and Gardeniers (2007) used the high-frequency flows for demonstrating mixing of drops in a microfluidic device.

2.6 Limiting effects: contact angle saturation and contact line instabilities

The Young-Lippmann equation predicts a parabolic dependence of the contact angle on the applied voltage. However, experiments show that the contact angle does not follow this trend for sufficiently high voltages. As we noted in previous sections, θ saturates to a certain value and no further decrease with increasing voltage is observed, this phenomenon is known as *Contact Angle Saturation.* For example, Figure 4b) shows Lippmann behaviour (parabolic) up to voltages around 250 volts. The contact angle slightly decreases if higher voltages are applied and it remains around 60°. This saturation angle depends on specific properties of the materials and is usually around 30° and 80° (Berge, 1993; Vallet et al., 1999; Welters and Fokkink, 1998; Peykov et al., 1998).

Another interesting observation at high voltages is the luminescence of the contact line when using salty solutions (Vallet et al., 1999), see Figure 14. During these experiments, the authors monitored the electrical current and found a series of short spikes simultaneous to the emission of light pulses, indicating discrete discharge processes. They also showed that the emission spectrum changes for different ambient gases, as expected if the light comes from the dielectric breakdown of the gas. Contact angle saturation and light emission appear around the same voltage and it seems clear that both phenomena are caused by the strong divergence of the electric field near the three-phase contact line.

Dielectric breakdown of the insulating layer is one consequence of increasing electric fields. The relation between contact angle saturation and dielectric breakdown has been studied by Papathanasiou and co-workers (Papathanasiou and Boudouvis, 2005; Papathanasiou et al., 2008; Drygiannakis et al., 2009). Using a standard needle-substrate configuration, these authors found leakage currents through the insulating layer for voltages around and above those of contact angle saturation. These currents indicate the breakdown of the insulator which, consequently, does not longer behave

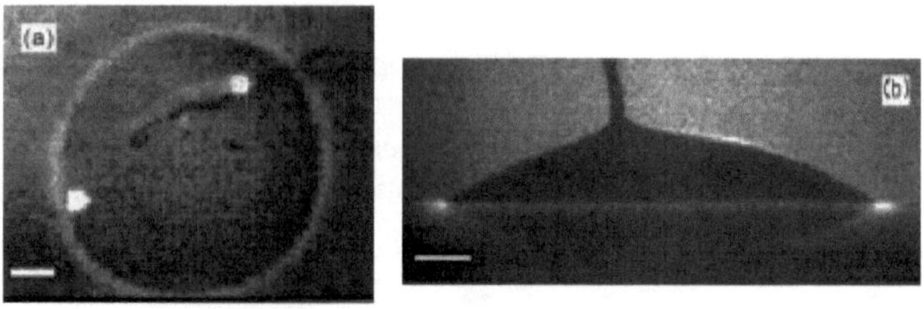

Figure 14. Light emission from the contact line at high voltages with salty solutions. The insulating layer is made of a film of PTFE 50 μm thick and the voltage amplitude is 950 V. a) top view and b) side view. The scale bars are 1 mm (reproduced with permission from Vallet et al. (1999), © 1999 Springer).

as such. In another work (Drygiannakis et al., 2009), the authors presented a model where elements of the dielectric are considered conductive when the local electric field exceeds the dielectric strength of the material. Results are shown in Figure 15a), finger-like structures of conductive material appear, increasing their size for increasing voltages. The resulting electric field at the contact line is lower than in the case of a perfect insulator and, consequently, the change of contact angle is diminished. They used this model to fit experimental measurements, showing a very good agreement in a wide range of voltages, Figure 15b).

A different approach to the phenomenon of contact angle saturation was proposed by Peykov et al. (1998). The authors stated that the reduction of the effective solid-liquid surface tension, as predicted by the thermodynamic interpretation of electrowetting, should be limited because a system with negative surface tension is unstable. In other words, Lippmann equation should fail for voltages at which the solid-liquid surface tension changes sign. Despite the model is in good agreement with several experimental observations (Quinn et al., 2005; Berthier et al., 2007; Berry et al., 2006), the validity of this criterion is not well founded for severeal reasons (Mugele, 2009), mainly because the true solid-liquid interfacial tension remains unchanged, as shown in section 2.4.

Another limiting effect at high voltages is the emission of small drops from the three-phase contact line when working with low conductivity liquids, like pure water or ethanol-water mixtures. This phenomenon was first observed by Vallet et al. (1999) and reproduced in Mugele and Herminghaus (2002), (see Figure 16). In the same article, Vallet et al. (1999) proposed

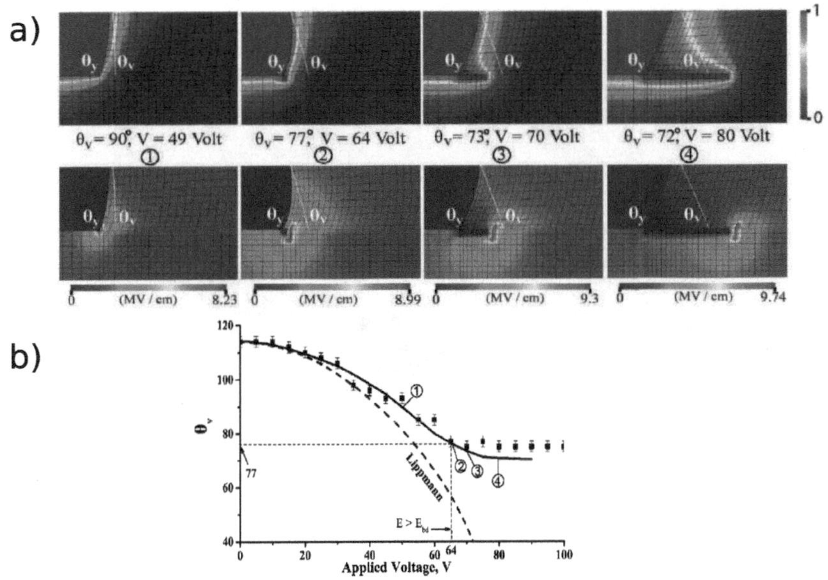

Figure 15. a)Electric potential distribution (up) and electric field (down) near the contact line for four different values of the applied voltage. b) Contact angle versus applied voltage for a substrate of SiO_2 with thickness 1 μm and coated with a teflon layer of 20 nm. The model fits experimental values above saturation: points 1,2,3 and 4 correspond with those indicated in (a). (Reproduced with permission from Drygiannakis et al. (2009), © 2009 American Chemical Society).

that the accumulation of charge at the contact line can generate transverse modulations on its shape. For sufficiently high voltages, these modulations can become unstable and small drops are ejected. Interestingly, a full numerical analysis of the 3D system by Fontelos and Kindelan (2008) shows instabilities that break the axial symmetry of the drop, however, the structure of the three-phase contact line is not correctly described in their model.

As a summary of this section, several effects limit the validity of the Young-Lippmann equation when the voltage increases. A unified description of the different phenomena is difficult, since specific material properties determine the behaviour of the different systems. In any case, electrowetting at high voltages is an open field for fundamental research.

Figure 16. Contact line instability at high voltage amplitudes. Small drops are emitted from the contact line. The inset shows a zoomed view of the contact line area. (Reproduced with permission from Mugele and Baret (2005), © 2005 IOP Publishing Ltd).

2.7 Note: Electrocapillary effect and Lippmann experiment

First experiments relating capillary effects and electric fields are due to Gabriel Lippmann, who published his results in 1875 with the name "*Relations entre les Phénomenes Électriques et Capillaires*" (Lippmann, 1875). Lippmann not only described the fundamental research that he performed, he also envisioned applications as, for example, an electrometer that was later used in the first electrocardiogram. We refer to Mugele and Baret (2005) for a complete translation to English of Lippmann's original work. In this note, we concentrate on describing the basic findings that led him to the foundations of the electrocapillary phenomenon, i.e. the basis of modern electrowetting.

Figure 17a) shows a scheme of the experimental setup used by Lippmam. The mercury in reservoir (A) penetrates into the glass capillary (GG'), where it undergoes a depression. Diluted sulfuric acid fills the upper part of this capillary tube, which is connected to the beaker (B) filled with the same acidic solution. Sulfuric acid is a better electrical conductor than pure water and it also wets glass better. Jurin's law establishes a relation between the interfacial tension and the capillary depression h (see fig. 17b)) in a capillary tube of radius R ($\rho gh = 2\gamma/R$), assuming a constant contact angle of mercury on glass of 180°.

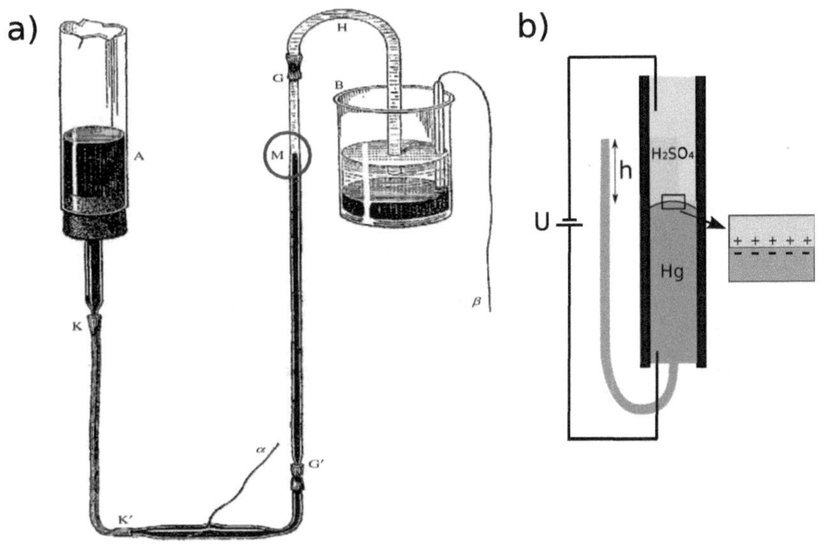

Figure 17. a)Experimental setup used by Lippmann. A voltage difference is applied to wires α and β and, as a result, the position of meniscus (M) changes (reproduced from Lippmann (1875)). b)Schematic representation of the phenomenon. Electrical charges accumulate at the mercury/acidic-solution interface, changing the surface energy and, consequently, the surface tension.

Platinum wires (α and β) make electrical connections. At the bottom of the beaker B lies a layer of mercury where the wire β is immersed. In this way, Lippmann could vary the voltage drop at the mercury-water interface (M) by applying a voltage difference between the platinum wires. The result of varying this voltage is that the position of the meniscus M changes and, as a consequence of Jurin's law, a change of interfacial tension is inferred. Lippmann readily interpreted this observation as a coupling between capillarity and electrical phenomena, as originally established in his paper as the first law:
the capillary constant at the mercury/diluted sulfuric acid interface is a function of the electrical difference at the surface.

Lippmann used both the terms capillary constant and interfacial tension indistinctily. He recognized that the redistribution of charge at the water-mercury interface was responsible for the change in surface energy and then in interfacial tension, Fig. 17b). As in section 2.2, we can write the surface charge as $\rho_{sl} = -\partial\gamma^{eff}/\partial U$, and considering a constant capacitance per unit area C of the electrical double layer, $\rho_{sl} = CU$ and $\gamma^{eff}(U) = \gamma - (1/2)CU^2$. Interestingly, Grahame (1947) used a similar setup for determining the differential capacitance of the electrical double layer at the interface between mercury and different electrolytes.

3 Applications of Electrowetting in microsystems

From electronic paper to liquid lenses and Lab-on-a-Chip devices, electrowetting is showing itself as a succesfull strategy for performing all kinds of operations with liquids in the field of microtechnology. As shown in the preceding section, electrowetting is a flexible method for controlling surface tension and, as a consequence, applications are numerous and increasing everyday. In this section, we make an overview of the most relevant applications, providing basic references to the reader and putting more emphasis on applications where fundamental physics appears.

3.1 Electrowetting for microfluidic operations: dispensing, splitting, merging, mixing and transporting of drops

Microfluidics, defined as the control and manipulation of minute amounts of liquids within microchannels, can be broadly divided in two groups: continuous microfluidics and digital microfluidics. While the former is based on continuous streams of liquids, digital microfluidics[5] refers to the use of discrete amounts of one liquid (droplets) dispersed in another liquid or air. By using discrete unit-volume droplets, a microfluidic function can be reduced to a set of repeated basic operations (Cho et al., 2003). Figure 18 shows a scheme of a generic "digital-microfluidic circuit" and the operations required for performing Lab-on-a-Chip functions: Drops have to be generated from liquid reservoirs (dispensing). We might want to split a given drop or, perhaps, to merge two of them. Mixing of the two drops is likely to be the operation required after merging. Also, moving a drop from one place to another (transporting) is, obviously, another important operation.

[5] the name digital microfluidics is given in analogy to digital electronics, where electrical signals are represented with discrete voltage levels or "bits".

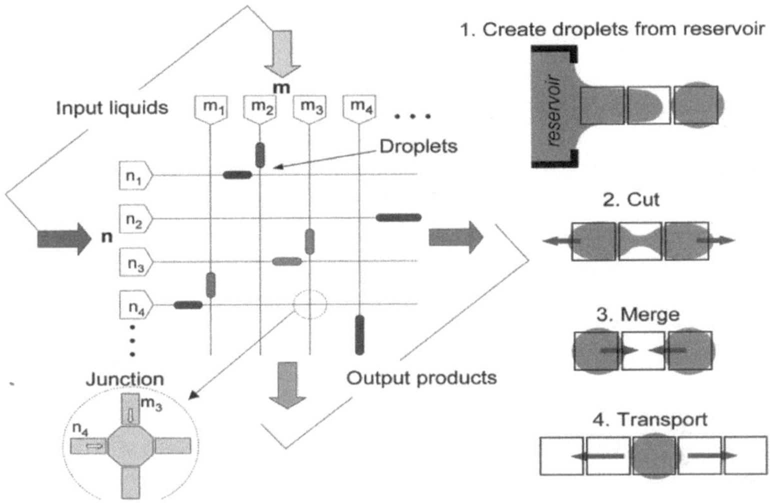

Figure 18. Generic digital-microfluidic circuit. Drops are generated from the reservoirs ($m \times n$), later transported over the circuit and combined to perform Lab-on-a Chip functions. (Right) Basic operations with drops are also shown (reproduced with permission from Cho et al. (2003), © 2003 IEEE).

Among the different ways of actuating droplets, electrowetting is the most common platform for performing digital microfluidics. Arrays of individually addressable electrodes allow the realization of all basic operations in configurable systems and with fast actuation times. Fair (2007) and Miller and Wheeler (2009) review the recent achievements in application of electrowetting to droplet-based microfluidics. We introduce here the basic electrode structures that have been proposed and demonstrated in the literature. For example, Figure 19 shows a schematic diagram of the standard setup for translating drops (Pollack et al., 2000). The drop is sandwiched between two plates coated with insulating layers. At the top there is an electrode (usually a transparent ITO electrode to view through) coated with a hydrophobic insulator and connected to ground. The bottom plate consists of a series of electrodes coated by a hydrophobic layer that can be individually activated. In this way, as shown in Figure 19, the contact angle of the drop on top of a specific electrode can be modified and, consequently, a net force is applied on the centre of mass of the drop [6]. By sequentially activating the electrodes, Pollack et al. (2000) demonstrated drop velocities up to 30 mm/s for signals of 40-50 V. In Pollack et al. (2002), the authors demonstrate formation, mixing, and splitting of drops with volume ranging from nanoliters to microliters by using the same kind of actuator of Figure 19. Scaling the drop velocity with the different quantities is a current topic of research. In principle, the force on the drop has to overcome the contact angle hysteresis (section 3.2) and the net velocity will result from a balance of the electrostatic force with the contact line friction and bulk viscous forces. However, the agreement between experiments and theoretical predictions is poor.

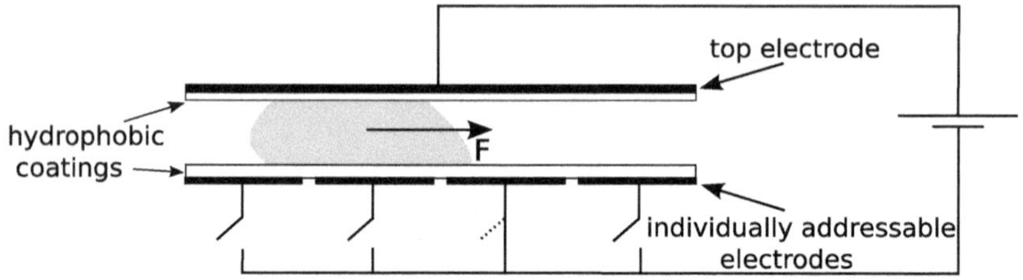

Figure 19. Schematic diagram of an array of electrodes for moving drops. The top plate is the reference electrode and electrodes at the bottom can be sequentially activated.

Mixing can be achieved by translating the drops along electrode arrays (Paik et al., 2003). Viscous drag with the walls generates flows within the drops. Efficient mixing has also been shown by exploiting EW-induced internal flows (see Figure 13 in section 2.5 and description therein). These flows generate efficient mixing on top of a single electrode, with the advantage that no lateral translations are required. However, a wire is required to be immersed in the drops. Malk et al. (2010) have recently shown EW-induced internal flow with coplanar electrodes. Yi and Kim (2006) also demonstrated all basic operations with coplanar electrode structures, i.e. electrodes only on the bottom plate. Electrowetting microfluidics on a sin-

[6] in accordance to Jones (2002), this net force on the centre of mass should be referred as liquid dielectrophoresis, while electrowetting should be restricted to the change of contact angle. However, the nomenclature is confusing in literature and most authors working in drop manipulation use the term electrowetting to name the net motion of the drop.

gle planar substrate is also posible, see for instance Cooney et al. (2006).

3.2 Electrowetting-induced reduction of contact angle hysteresis

It was mentioned in the preceding section that we must overcome the *contact angle hysteresis* before translating drops on solid substrates. In effect, the triple contact line gets pinned because of random forces caused by surface heterogenities and, as a consequence, the TCL is immobile not only when the CA is Young's angle but within a given range (de Gennes, 1985):

$$\theta_{\text{rec}} < \theta < \theta_{\text{adv}} \tag{12}$$

θ_{adv} is called the *advancing contact angle* and it is measured when the liquid-solid contact area increases. θ_{rec} is the *receding contact angle* and it is measured when this area shrinks, see Figure 20a). *Contact angle hysteresis* is defined as the difference between the two, $\theta_{\text{adv}} - \theta_{\text{rec}}$, and it is a measure of the amplitude range of the pinning forces, as depicted in Figure 20b).

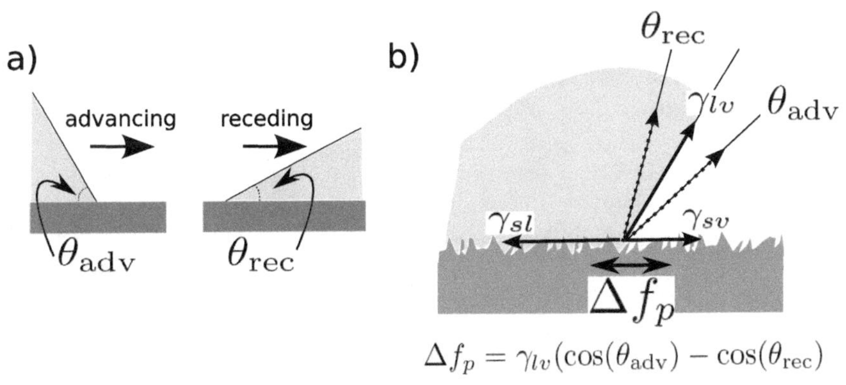

Figure 20. a) Advancing contact angle is the minimum CA at which the solid-liquid increases. Oppositely, the receding contact angle corresponds to the maximum CA at which the solid-liquid shrinks. b) Solid surface with roughnes and the role of random pinning forces: its variability range gives rise to a range of CA's within the contact line is immobile.

Droplet motion can only occur when both the CA on the leading edge of the droplet exceeds the advancing contact angle and, also, when the CA on the trailing edge is smaller than the receding one. Evidently, diminishing the CA hysteresis is desirable when moving drops. For example, it has been shown that immersing the liquid droplet into an oil environment greatly reduces the hysteresis. Contact angle hysteresis can be also diminished if, as suggested by Johnson and Dettre (1964), vibrational energy is added to the drop and, thus, the energy barriers introduced by the surface heterogeneity can be overcome. For example Volpe et al. (2002) used acoustic excitation for shaking the drops.

When moving drops with electrowetting, a threshold voltage is generally required to overcome the pinning forces on the solid substrate, see e.g. Pollack et al. (2000). However, Li and Mugele (2008) showed that the contact angle hysteresis almost disappears with increasing AC voltage, while it remains constant for DC. Their experimental setup consisted in a drop of NaCl water solution on top of a teflon-coated ITO electrode. The contact line was forced to advance/retract by infusing/withdrawing liquid with a syringe at a sufficiently low rate to avoid dynamics effects. The drop was electrically grounded by means of the syringe needle. Figure 21 shows measurements of the advancing and receding CA for DC (open symbols) and AC voltages (filled symbols) as a function of the electrowetting number ($\eta \equiv (\varepsilon_0 \varepsilon_d U^2)/(2d\gamma_l v)$). For DC voltage, it can be seen that the difference between θ_{adv} (open diamonds) and θ_{rec} (open triangles) remains constant. However, for AC signals the difference diminishes for increasing η, i.e. amplitude of the voltage.

The reduction in CA hysteresis with AC voltage can be explained if we consider that the electrical force on the triple contact line can be written as $f_{el}(t) = \eta(1 - \cos(2\omega t))$ (Li and Mugele, 2008). This force contains a DC component equal to η and oscillating part of the same amplitude and frequency 2ω. As mentioned in section 2.5, millimeter-size drops cannot follow the electric field oscillations for frequencies higher than a few tens of Hertz and the apparent contact angle is given by U_{rms}. However, contact line pinning occurs at a much smaller scale at which the liquid can respond much faster, e.g. for frequencies in the kHz range (as in Li and Mugele (2008)). The electrical force f_e pulls from the contact line in the advancing direction and therefore, it helps to advance but not to recede.

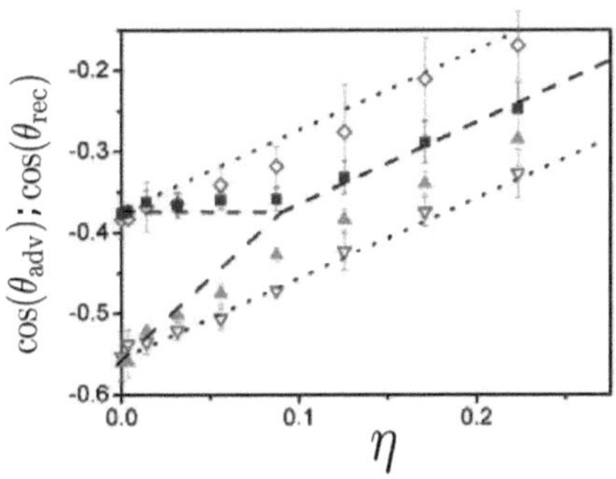

Figure 21. Cosine of advancing and receding CA's versus electrowetting number. Open symbols are experimental results for DC voltages and dotted lines correponds to the theoretical model. Filled symbols are measurements with AC signals and dashed line is the expectation of the theoretical model. Reproduced with permission from Li and Mugele (2008), © 2008 American Institute of Physics.

3.3 Electrowetting on complex surfaces

The wettability and topography of solid substrates determine the possible morphologies of drops resting on solid substrates (Herminghaus et al. (2008) and references therein). The liquid adjusts its shape to a configuration of minimum energy and, as shown in the introduction (Figure 1a)), abrupt morphological transitions can occur when some parameter (e.g. contact angle) is varied in the system. The contact angle can be modified in several ways (chemically, gradients in temperature,...) and, of course, by using electrowetting. [7]

The ability of Electrowetting for modifying the shape of a drop was demonstrated by Klingner and Mugele (2004) using a hydrophobic substrate with a stripe electrode, see Figure 22a). This system was theoretically studied in Brinkmann and Lipowsky (2002) and they found a first-order phase transition between two shapes: a bulge-like and a cylindrical shape. Figure 22b) shows the phase diagram as a function of the reduced volume $\overline{V}$ ($\overline{V} = V/L^3$, where L is the stripe width) and the contact angle on the stripe

[7] In applications, we usually deal with drops much larger than the thickness of the insulator on the electrode. Therefore, for computing the global shape of the drop, we can ignore the fine structure of the triple contact line and only consider the apparent change in wall wettability caused by the applied potential.

θ_s. The region below the continuous line corresponds to states with cylinder shape and the region to the right of the dashed line to bulge-like shapes. The critical point of the transition is $\overline{V} = 2.85$ and $\theta_s = 39.2$. Klingner and Mugele (2004) tested this theory by controlling the wettability of the stripe (θ_s) with electrowetting. When increasing the voltage for a drop of given volume, they observed the transition from the cylinder shape, filled symbols in Figure 22b), to bulges, open symbols. Figure 22c) are snapshots of the side and top views of the experiment.

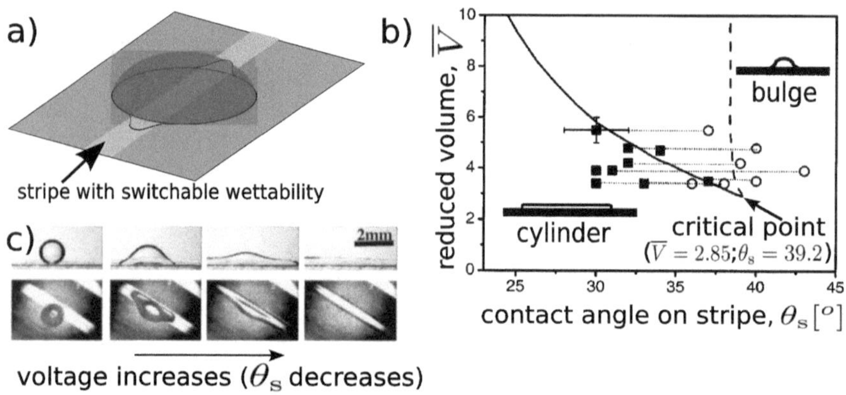

Figure 22. a) Schematic representation of a water droplet resting on a hydrophobic substrate with a stripe electrode. b) Phase diagram: solid and dashed lines represent instability lines for the bulge and cylinder morphologies, respectively. c) Side and top view video snapshots at voltages of 0, 250, 300, and 310 V, respectively. Reproduced with permission from Klingner and Mugele (2004), © 2004 American Institute of Physics.

Krupenkin et al. (2004) demonstrated electrowetting-induced transitions from superhydrophobic (Cassie-Baxter) to hydrophilic (Wenzel) states on textured surfaces, see Figure 23.

Baret et al. (2005) studied the filling of grooves from feeding droplets by using electrowetting. Figure 24 shows images of a droplet on top of the grooves for several voltages. For low voltages, the liquid hardly penetrates the grooves. The curvature of the finger tips decreases with voltage and, above a certain threshold, the pressure equals the pressure in the droplet and filling starts. It might be expected a discontinuous increase of the finger length (from zero to infinity) at the threshold voltage, However, due to

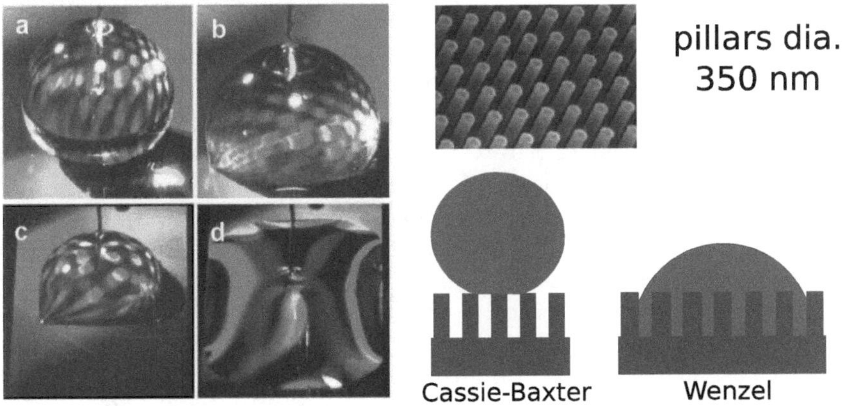

Figure 23. Pictures a-d show the wetting transition when the voltage applied to the drop increases. SEM Image: Array of nanopillars 350 nm diameter. Reproduced with permission from Krupenkin et al. (2004), © 2004 American Chemical Society.

finite-conductivity effects, the voltage drop in the liquid bulk increases with this length, limiting the effect of electrowetting.

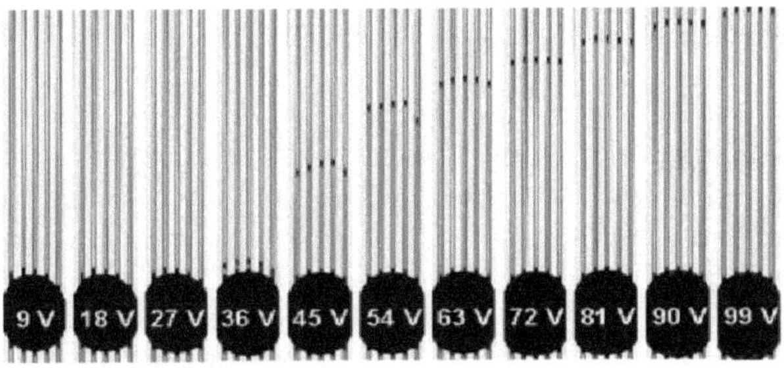

Figure 24. Filling of grooves from feeding droplets by using electrowetting. Above a certain threshold, the pressure equals the pressure in the droplet and filling starts. Reproduced with permission from Baret et al. (2005), © 2005 American Chemical Society.

3.4 Electrowetting-assisted drop generation in microchannels

Drop generation in microchannels has become a major issue in microfluidics research (Stone et al., 2004). Digital microfluidics (section 3.1) relies on our ability to generate drops in a controlled manner. Micrometer-size drops are also used in disparate applications like synthesis of novel materials or performance of logic operations in microfluidic devices (Cheow et al., 2007). Well established microfluidic techniques, such as T-junctions (Thorsen et al., 2001) and flow focusing devices (FFD's) (Anna et al., 2003), are commonly used for generating drops of one liquid (dispersed phase) into the continuous flow of another (continuous phase). These methods allow for high drop-throughput, however, they are not useful for controlling the moment at which the drops are produced. Moreover, the sizes of the drops and their production rate in these devices are fixed for given flow rates of the two liquids.

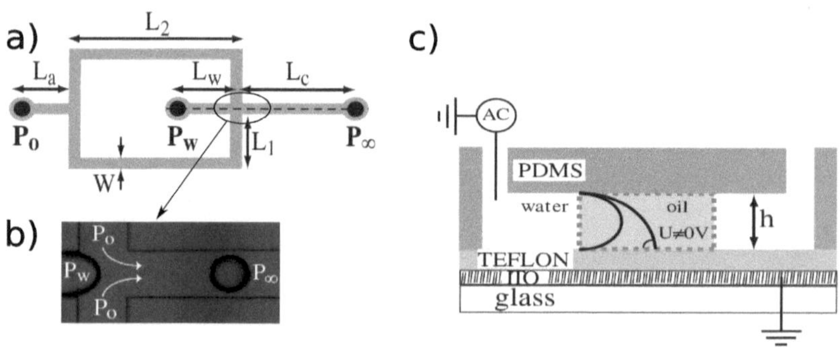

Figure 25. a) Scheme of a microfluidic network for flow focusing. Oil and water are injected at controlled pressures P_O and P_W, respectively. b) Image of the generation of a water drop. c) Side view of the junction area in a FFD including an ITO electrode for EW functionality. The electrode is grounded and the application of AC voltages to water allows for controlling the shape of the water-oil interface. Reproduced with permission from Malloggi et al. (2007), © 2007 IOP Publishing Ltd.

Electrowetting has been proved as a versatile tool for "drop on-demand" generation (Malloggi et al., 2007) and for controlling both the size and production rate (Malloggi et al., 2008; Gu et al., 2008). Figure 25 shows a scheme of a flow focusing device: oil (continuous phase) is injected at pressure P_O at one inlet and water (dispersed phase) is injected at another inlet at pressure P_W. The two fluids meet at the channel junction and water drops are generated in a competition between viscous and capillary stresses, figure 25b). Malloggi et al. (2007) modified this device by embedding teflon-coated ITO electrode at the bottom of the junction, allowing for local control of the wall wettability by means of voltage pulses, Figure 25c). Walls are hydrophobic and, when electrical signals are applied, the shape of the oil-water interface is modified since contact angle of water is reduced.

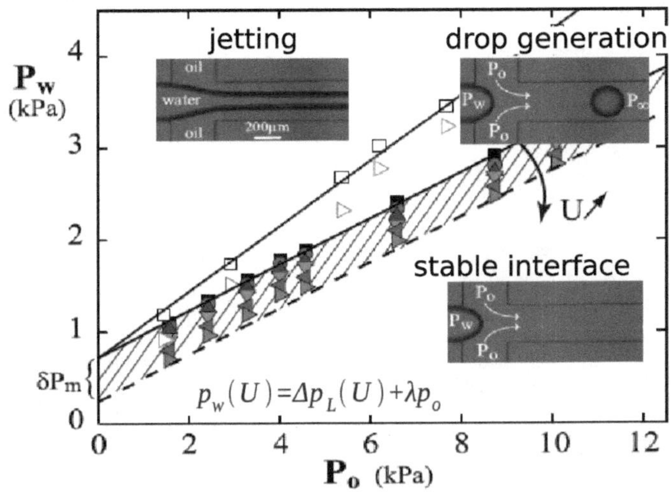

Figure 26. Depending on the values of control parameteres, three different operating regimes are found for a FDD: jetting, drop formation and a stable water/oil interface. The inclusion of electrowetting allows for controlling the onset of drop generation, i.e. the minimum water pressure for a given oil pressure is a function of the voltage amplitude U, hatched area in figure. Reproduced with permission from Malloggi et al. (2007), © 2007 IOP Publishing Ltd.

Figure 26 describes the three basic operation modes as a function of the three control parameteres: P_W, P_O and the voltage amplitude U. A stable water/oil interface is found if the water pressure P_W is too low for a given oil pressure P_O. Oppositely, a jet of water is produced if P_W is much higher than P_O. For intermediate pressures, water drops are formed. Electrowetting can be used to tune this pressure range, as shown in Figure 26 where the hatched area increases for increasing voltage amplitude. In effect, the water pressure to generate a drop has to be sufficient to overcome the oil pressure plus the Laplace pressure due to the interface curvature, $P_W \geq P_O + \Delta P_L$. We reduce the interface curvature by means of electrowetting (Figure 25c)) and, therefore, the minimum water pressure to generate drops is smaller for greater voltage amplitudes U. From an applied perspective, this means that we can choose values of P_W and P_O for having a water/oil stable interface but, when a voltage pulse is applied, drops are generated, i.e. drop-on demand generation. Finally, Figure 27 shows pictures of an EW-enhanced flow focusing device with a small orifice of $\approx 50\mu$m (Gu et al., 2008). The flow rate of oil is fixed and a great variety of drop generation modes are observed depending on the values of P_W and U.

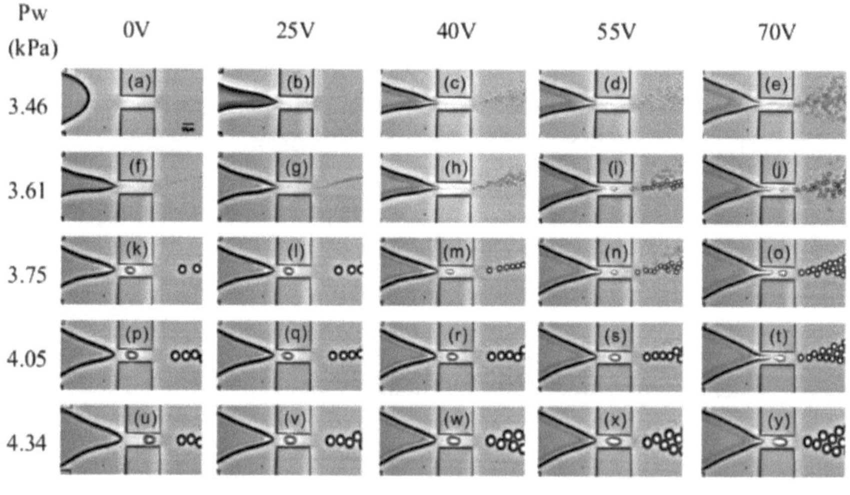

Figure 27. Images of the drop generation in a flow focusing device with electrowetting functionality. The orifice in the device is 50μm and the oil pressure is fixed. The size and production rate of the drops can be tuned by changing the water pressure P_W and voltage amplitude U. Reproduced with permission from Gu et al. (2008), © 2008 American Institute of Physics.

3.5 Electrowetting and optical applications

Applications related to optics is the field where electrowetting has grown most rapidly and may have a major impact in commerciable products. The focal length of an optical lens depends on its curvature. One advantage of a liquid lens, in contrast with solid ones, is that the shape of the liquid is non-permanent and controlling its curvature allows for controlling the focal length. Berge and Peseux (2000) first proposed electrowetting as the way of controlling the curvature of liquid lenses. Figure 28a) shows a scheme of a liquid lens with electrowetting-controlled focal length. Figure 28b) shows the dioptries as a function of the applied voltage. Recently, liquid lenses with variable focal lens have become commercially available and they are being integrated in mobile phones and cameras [8]

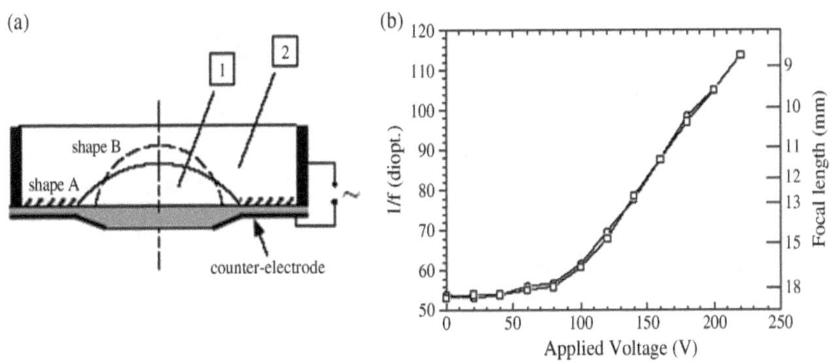

Figure 28. a) Schematic representation of a liquid lens: liquid 1 is an oil and liquid 2 is water. The curvature of the interface can be tuned by applying a voltage (shapes A and B). b) Dioptries of the lens as a function of the voltage amplitude. Reproduced with permission from Berge and Peseux (2000), © 2000 Springer.

Displays based on electrowetting have experienced a major advance in the very few last years. Hayes and Feenstra (2003) introduced the principle of an electroweting display by constructing a single pixel, Figure 29. When the voltage is off, a coloured oil film is present but when the voltage is on, the layer retracts and the pixel changes state. By using three of these films independently actuated, it is possible to switch between different colours

[8] visit *http://www.varioptic.com* for information on EW-based lenses

[9]. The company liquavista currently develops an e-reader technology with video speed and colour capability. Since these displays are based on reflection of ambient light, they can be operated in sunlight without any back lighting (in contrast to LCD displays). Other groups in this field include Heikenfeld (Heikenfeld et al., 2009) and Steckl (You and Steckl, 2010).

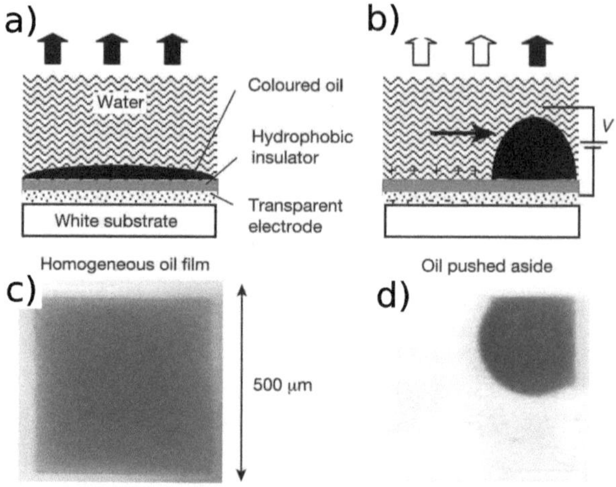

Figure 29. Working principle of an electrowetting display. a) If no voltage is applied, a coloured oil film occupies the whole area on top of the hydrophobic coating. b) When the voltage is applied, the oil retracts and the pixel changes its state. Pictures in c) and d) are photographs corresponding to cases a) and b), respectively. Reproduced with permission from Hayes and Feenstra (2003), © 2003 Macmillan Publishers Ltd: Nature.

3.6 Other applications

Many other applications of electrowetting are being explored and here we can just mentioned a few of particular interest. Figure 30 shows a scheme of a "liquid-state" Field-Effect Transistor demonstrated by Kim and Steckl (2007). The device consists of three electrodes, the one in the middle is for electrowetting and plays the role of a gate electrode in a MOSFET-like transistor. The two on the sides are the source and the drain, respectively. When a certain voltage is applied to the gate electrode, the electrolyte removes the insulating oil and occupies the region between source and drain,

[9]visit *http://www.liquavista.com* for information on EW-based displays

forming a channel for the conduction of electricity.

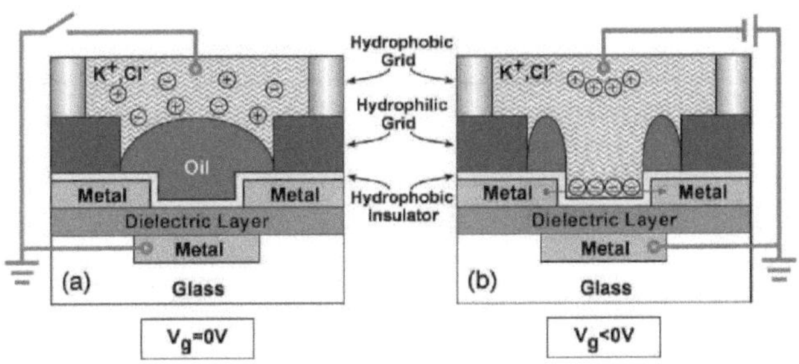

Figure 30. Liquid-state Field Effect Transistor. (a) corresponds to the "Off" state: The insulating oil occupies the region between electrodes. (b) "On" state: when a voltage is applied to the gate electrode, the water occupies the region between electrodes and a conducting channel is formed. Reproduced with permission from Kim and Steckl (2007), © 2007 American Institute of Physics.

Pumping of liquids in microsystems by means of electrowetting was first described by Beni et al. (1982). They reported an electrowetting-based effect that they named Continuous Electrowetting (CEW). Figure 31 shows a scheme of the setup consisting in a slug of mercury inserted in a capillary filled with (H_2SO_4). Note the similarity with the experiment by Lippmann (section 2.7). By applying a few volts, a charge density is induced at the mercury-electrolyte interface. The charge density is not homegeneous and, thus, a gradient in surface tension occurs resulting in a net force in the slug that drives the motion. Yun et al. (2002); Lee and Kim (2000) reported pumping of liquids in microsystems by means of CEW.

As mentioned in the introduction, wall wettability is a fundamental parameter in 2-phase flows in microchannels. Huh et al. (2003) reported an example of how electrowetting can be used to manipulate an air-water two-phase flow. Figure 32 shows a scheme of a flow focusing device where air is used to focus a stream of water. Electrowetting is applied by means of electrodes occupying half of the channel and the water stream is displaced towards the side on top of electrodes.

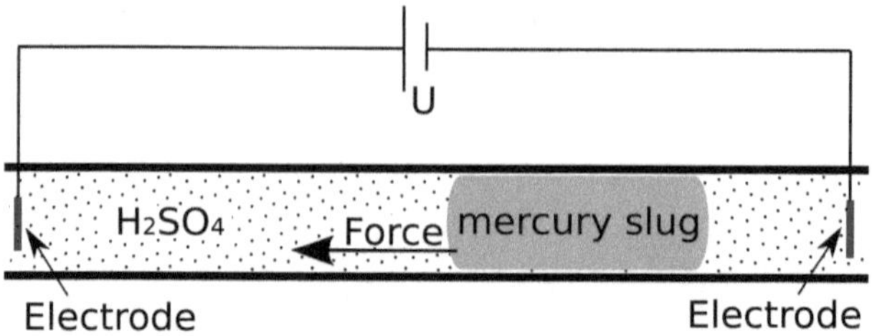

Figure 31. Scheme of the setup used by Beni et al. (1982) for demonstrating the Continuous Electrowetting Effect. A slug of mercury is inserted in a capillary tube and a voltage drop is applied. The surface tension at the electrolyte(H_2SO_4)-slug is not homogeneous and a net force appears on the slug.

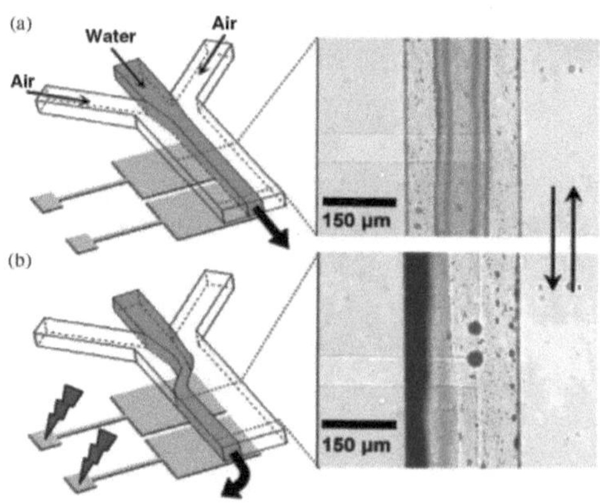

Figure 32. A stream of water is focused by air flowing on the sides (a). Upon application of a voltage to the electrodes, the stream of water is diverted (b). Reproduced with permission from Huh et al. (2003), 2003 American Chemical Society.

4 Conclusions

Major advances in electrowetting on dielectric have occurred in the past two decades, both in theory and in applications. From a theoretical perspective, the development of the electromechanical interpretation has been a major leap forward. It is by now clear that the drop shape is determined by the balance between capillary and electrical stresses and that this balance results in a macroscopic apparent contact angle in agreement with the "classical" thermodynamic approach (section 2.4). The number of applications is growing fast and some devices, like EW-based displays and lenses, are already commercially available products. Electrowetting is commonly used in many Lab-on-a-Chip devices and has become the most versatile tool for digital microfluidic operations.

Bibliography

S. L. Anna, N. Bontoux, and H. A. Stone. Formation of dispersions using flow-focusing in microchannels. *Applied Physics Letters*, 82:364–366, 2003.

J. C. Baret, M. Decre, S. Herminghaus, and R. Seemann. Electroactuation of fluid using topographical wetting transitions. *Langmuir*, 21:12218–12221, 2005.

G. Beni, S. Hackwood, and J. L. Jackel. Continuous electrowetting effect. *Applied Physcis Letters*, 40:912–914, 1982.

B. Berge. Electrocapillarite et mouillage de films isolants par l'eau. *C. R. Acad. Sci. II*, 317:157, 1993.

B. Berge and J. Peseux. Variable focal lens controlled by an external voltage: an application of electrowetting. *European Physical Journal E*, 3:159, 2000.

S. Berry, J. Kedzierski, and B. Abedian. Low voltage electrowetting using thin fluoroploymer films. *Journal of Colloid and Interface Science*, 303: 517–524, 2006.

J. Berthier, P. Dubois, P. Clementz, P. Claustre, C. Peponnet, and Y. Fouillet. Actuation potentials and capillary forces in electrowetting based microsystems. *Sensors and Actuators A*, 134:471–479, 2007.

M. Brinkmann and R. Lipowsky. Wetting morphologies on substrates with striped surface domains. *Journal of Applied Physics*, 92:4296–4306, 2002.

J. Buehrle, S. Herminghaus, and F. Mugele. Interface profiles near three-phase contact lines in electric fields. *Physical Review Letters*, 91:086101, 2003.

L. F. Cheow, L. Yobas, and D. L. Kwong. Digital microfluidics: Droplet based logic gates. *Applied Physics Letters*, 90:054107, 2007.

S. K. Cho, H. Moon, and C. J. Kim. Creating, transporting, cutting, and merging liquid droplets by electrowetting-based actuation for digital microfluidic circuits. *Journal of Microelectromechanical Systems*, 12:70–80, 2003.

C. G. Cooney, C.Y. Chen, M. R. Emerling, A. Nadim, and J. D. Sterling. Electrowetting droplet microfluidics on a single planar surface. *Microfluidics Nanofluidics*, 2:435–446, 2006.

P. G. de Gennes. Wetting: statics and dynamics. *Reviews of Modern Physics*, 57:827–863, 1985.

R. Dreyfus, P. Tabeling, and H. Willaime. Ordered and disordered patterns in two-phase flows in microchannels. *Physical Review Letters*, 90:144505, 2003.

A. I. Drygiannakis, A. G. Papathanasiou, and A. G. Boudouvis. On the connection between dielectric breakdown strength, trapping of charge, and contact angle saturation in electrowetting. *Langmuir*, 25:147–152, 2009.

R. B. Fair. Digital microfluidics: is a true lab-on-a-chip posible? *Microfluidics-Nanofluidics*, 3:245–281, 2007.

M. A. Fontelos and U. Kindelan. The shape of charged drops over a solid surface and symmetry-breaking instabilities. *SIAM, Journal of Applied Mathematics*, 69:126–148, 2008.

P. García-Sánchez, A. Ramos, and F. Mugele. Electrothermally driven flows in ac electrowetting. *Physical Review E*, 81:1(R), 2010.

H. Gau, S. Herminghaus, P. Lenz, and R. Lipowsky. Liquid morphologies on structured surfaces: From microchannels to microchips. *Science*, 283: 73–49, 1999.

D. C. Grahame. The electrical double layer and the theory of electrocapillarity. *Chemical Review*, 41(3):441–501, 1947.

H. Gu, F. Malloggi, S. Vanapalli, and F. Mugele. Electrowetting-enhanced microfluidic device for drop generation. *Applied Physics Letters*, 93: 183507, 2008.

R. Gupta, G. K. Olivier, and J. Frechette. Invariance of the solid-liquid interfacial energy in electrowetting probed via capillary condensation. *Langmuir*, 26:11946–11950, 2010.

R. A. Hayes and B. J. Feenstra. Video-speed electronic paper based on electrowetting. *Nature*, 425:383–385, 2003.

J. Heikenfeld, K. Zhou, E. Kreit, B. Raj, S. Yang, B. Sun, A. Milarcik, L. Clapp, and R. Schwartz. Electrofluidic displays using Young-Laplace transposition of brilliant pigments dispersions. *Nature Photonics*, 3:292–296, 2009.

S. Herminghaus, M. Brinkmann, and R. Seemann. Wetting and dewetting of complex surface geometries. *Annual Reviews of Materials Research*, 38:101–121, 2008.

D. Huh, A. H. Tkaczyk, J. H. Bahng, Y. Chang, H. H. Wei, J. B. Grotberg, C. J. Kim, K. Kurabayashi, and S. Takayama. Reversible switching of high-speed air-liquid two-phase flows using electrowetting assisted flow-pattern change. *Journal American Chemical Society*, 125:14678, 2003.

R. E. Johnson and R. H. Dettre. Contact Angle Hysteresis III. Study of an Idealized Heterogeneous Surface. *The Journal of Physical Chemistry*, 68:1744–1750, 1964.

T. B. Jones. On the Relationship of Dielectrophoresis and Electrowetting. *Langmuir*, 18:4437–4434, 2002.

T. B. Jones. An electromechanical interpretation of Electrowetting. *J. Micromech. Microeng.*, 15:1184–1187, 2005.

T. B. Jones, K. L. Wang, and D. J. Yao. Frequency-dependent electromechanics of aqueous liquids: Electrowetting and dielectrophoresis. *Langmuir*, 20:2813–2818, 2004.

D. Y. Kim and A. J. Steckl. Liquid-state field-effect transistors using electrowetting. *Applied Physics Letters*, 90:043507, 2007.

A. Klingner and F. Mugele. Electrowetting-induced morphological transitions of fluid microstructures. *Journal of Applied Physics*, 95:2918–2920, 2004.

S. H. Ko, H. Lee, and K. H. Kang. Hydrodynamic flows in electrowetting. *Langmuir*, 24:1094–1101, 2008.

T. N. Krupenkin, J. A. Taylor, T. M. Schneider, and S. Yang. From rolling ball to complete wetting: The dynamic tuning of liquids on nanostructured surfaces. *Langmuir*, 20:3824–3827, 2004.

A. Kumar, M. Pluntke, B. Cross, J. C. Baret, and F. Mugele. Finite conductivity effects and apparent contact angle saturation in ac electrowetting. *MRS symposia proceedings*, 899:69–76, 2006.

H. Lee, S. Yun, S. H. Ko, and K. H. Kang. An electrohydrodynamic flow in ac electrowetting. *Biomicrofluidics*, 3:044113, 2009.

J. Lee and C. J. Kim. Surface-tension driven microactuators based on continuous electrowetting. *Journal of Microelectromechanical Systems*, 9: 171, 2000.

F. Li and F. Mugele. How to make sticky surfaces slippery: Contact angle hysteresis in electrowetting with alternating voltage. *Applied Physics Letters*, 92:244108, 2008.

G. Lippmann. Relations entre les phénomenes éléctriques et capillaires. *Annales de Chimie et de Physique*, 5:494, 1875.

R. Malk, Y. Fouillet, and L. Davoust. Rotating flow within a droplet actuated with AC EWOD. *Sensors and Actuators B: Chemical*, 2010.

F. Malloggi, S. A. Vanapalli, H. Gu, D. van den Ende, and F. Mugele. Electrowetting-controlled droplet generation in a microfluidic flow-focusing device. *Journal of Physics: Condensed Matter*, 19:462101, 2007.

F. Malloggi, A. G. Banpurkar, H. Gu, S. Vanapalli, and F. Mugele. Electrowetting: A versatile tool for controlling microdrop generation. *The European Physical Journal E*, 26:91–96, 2008.

E. M. Miller and A. R. Wheeler. Digital bioanalysis. *Anal Bioanal Chem*, 393(2):419–426, 2009.

F. Mugele. Fundamental challenges in electrowetting: from equilibrium shapes to contact angle saturation and drop dynamics. *Soft Matter*, 5: 3377–3384, 2009.

F. Mugele and J. C. Baret. Electrowetting: from basics to applications. *J. Phys.: Condens. Matter*, 17:R705–R774, 2005.

F. Mugele and J. Buehrle. Equilibrium drop surface profiles in electric fields. *Journal of Physics: Condensed Matter*, 19:375112, 2007.

F. Mugele and S. Herminghaus. Electrostatic stabilization of fluid microstructures. *Applied Physics Letters*, 81:2303–2305, 2002.

F. Mugele, A. Klingner, J. Buehrle, D. Steinhauser, and S. Herminghaus. Electrowetting: a convenient way to switchable wettability patterns. *J. Phys.: Condens. Matter*, 17:S559–S576, 2005.

F. Mugele, J. C. Baret, and D. Steinhauser. Microfluidic mixing through electrowetting-induced droplet oscillations. *Applied Physics Letters*, 88: 204106, 2006.

K. P. Nichols and H. J. G. E. Gardeniers. A Digital Microfluidic System for the Investigation of Pre-Steady-State Enzyme Kinetics Using Rapid Quenching with MALDI-TOF Mass Spectrometry. *Analytical Chemistry*, 79:8699–8704, 2007.

J. M. Oh, S. H. Ko, and K. H. Kang. Shape oscillation of a drop in ac electrowetting. *Langmuir*, 24:8379–8386, 2008.

P. Paik, V. K. Pamula, and R. B. Fair. Rapid droplet mixers for digital microfluidic systems. *Lab on a Chip*, 3:28–33, 2003.

A. G. Papathanasiou and A. G. Boudouvis. Manifestation of the connection between dielectric breakdown strength and contact angle saturation in electrowetting. *Applied Physics Letters*, 86:164102, 2005.

A. G. Papathanasiou, A. T. Papaioannou, and A. G. Boudouvis. Illuminating the connection between contact angle saturation and dielectric breakdown in electrowetting through leakage current measurements. *Journal of Applied Physics*, 103:034901, 2008.

V. Peykov, A. Quinn, and J. Ralston. Electrowetting: a model for contact-angle saturation. *Colloid Polym Sci*, 278:789–793, 1998.

M. G. Pollack, R. B. Fair, and A. D. Shenderov. Electrowetting-based actuation of liquid droplets for microfluidic applications. *Applied Physics Letters*, 77(11):1725–1726, 2000.

M. G. Pollack, A. D. Shenderov, and R. B. Fair. Electrowetting-based actuation of droplets for integrated microfluidics. *Lab on a Chip*, 2: 96–101, 2002.

C. Quilliet and B. Berge. Electrowetting: A recent outbreak. *Current Opinion in Colloidal and Interface Science*, 6:34–39, 2001.

A. Quinn, R. Sedev, and J. Ralston. Contact angle saturation in electrowetting. *Journal of Physical Chemistry B*, 109:6268–6275, 2005.

A. Staicu and F. Mugele. *unpublished.*

H. A. Stone, A. D. Stroock, and A. Ajdari. Engineering flows in small devices: Microfluidics toward a lab-on-a-chip. *Annu. Rev. Fluid Mech.*, 36:381411, 2004.

J. A. Stratton. *Electromagnetic Theory.* McGraw-Hill, 1941.

T. Thorsen, R. W. Roberts, F. H. Arnold, and S. R. Quake. Dynamic pattern formation in a vesicle-generating microfluidic device. *Physical Review Letters*, 86:4163–4166, 2001.

M. Vallet, M. Vallade, and B. Berge. Limiting phenomena for the spreading of water on polymer films by electrowetting. *The European Physical Journal B*, 11:583–591, 1999.

C. Della Volpe, D. Maniglio, M. Morra, and S. Siboni. The determination of a stable-equilibrium contact angle on heterogeneous and rough surfaces. *Colloids and Surfaces A: Physicochemical and Engineering Aspects*, 206: 47–67, 2002.

W. J. J. Welters and L. G. J. Fokkink. Fast electrically switchable capillary effects. *Langmuir*, 14:1535–1538, 1998.

U. C. Yi and C. J. Kim. Characterization of electrowetting actuation on addressable single-side coplanar electrodes. *Journal Micromechanical Microengineering*, 16:2053–2059, 2006.

H. You and A.J. Steckl. Three-color electrowetting display device for electronic paper. *Applied Physics Letters*, 97:023514, 2010.

K. S. Yun, I. J. Cho, J. U. Bu, C. J. Kim, and E. Yoon. A surface-tension driven micropump for low-voltage and low-power operations. *Journal of Microelectromechanical Systems*, 11:454–461, 2002.

Electrohydrodynamic Pumping in Microsystems

Antonio Ramos *

Departamento de Electrónica y Electromagnetismo, Universidad de Sevilla, 41012-Sevilla, Spain

Abstract The electrical manipulation of fluids in microsystems has many existing and potential applications. This chapter reviews five different ways of electrohydrodynamic actuation in microdevices. First, we describe the basic equations of Electrohydrodynamics in the microscale, providing some basic concepts of electrical conduction in liquids. We also deal with some basic fluid-mechanical aspects that are common for micropumps. Then, five different electrohydrodynamic micropumps are studied and compared: from those that employ forces in the liquid bulk to those that employ forces in the electrical double layer.

1 Introduction

Microfluidics deals with the pumping, control and manipulation of microlitre to pico-litre volumes of fluids. The typical length of the system ranges from 100 nm to 1 mm. In our case, we are going to deal with liquids in microsystems: sub-continuum effects that are important for gases in microfluidics (or, in general, in nanofluidics) are not important here. The continuum assumption of fluid mechanics is a good approximation in our case since the Knudsen number (molecular mean free path length divided by typical length) is much smaller than one.

Liquid manipulation can be subdivided loosely into: digital microfluidics where the liquid is divided into droplets (e.g. electrowetting) and continuous microfluidics where the liquid is transported inside conduits. In this chapter, we deal with continuous microfluidics.

*This work has been supported by the Spanish Government Ministry MEC (contract FIS2006-03645) and the Regional Government Junta de Andalucía (contract P09-FQM-4584).

Possible applications of microfluidics are (Stone et al., 2004; Laser and Santiago, 2004): dispensing therapeutic agents into the body; cooling of microelectronic devices; handling of liquids in space exploration, where miniaturization is required; micropropulsion for small satellites, as another potential application of microfluidics in space; handling of liquids for the Lab-on-a-chip technology, i.e. micro-devices for chemical or biological analysis of chip-format.

Microfluidic transport can sometimes be achieved by passive mechanisms such as surface tension or gravity. Other applications employ macroscale pumps: syringe pumps, pressure/vacuum chambers and valves. However, many applications would benefit from an on-chip active micropump. According to Laser and Santiago (2004), there are two categories of micropumps: (a) displacement pumps, which exert pressure forces on the working fluid by one or more moving boundaries and (b) dynamic micropumps, which exert forces directly on the liquid, without moving parts. The piezoelectric or the electro-wetting micropumps are examples of displacement pumps. Most of electrohydrodynamic (EHD) micropumps are dynamic pumps: they use electric forces directly on the liquid. The piezo-actuated micropump is an integrated pump design that has been studied intensively over the years. In this, a membrane is displaced to create a pulsating flow that is rectified using valves. However, the moving parts make the fabrication and operation delicate. In this context, EHD micropumps can work as on-chip active pumps with no moving parts. In addition, the local EHD pumping of microflows in otherwise pressurised microsystems can have many applications, such as stirring (Sigurdson et al., 2005) or sorting (Dittrich and Schwille, 2003).

In all EHD dynamic micropumps, electric current flows through the working liquid. Different EHD actuation techniques appear depending on liquid electrical properties (mainly, the electrical conductivity). Here we will analyse five different ways of EHD actuation in microsystems: in three of them electric forces are applied in the liquid bulk (the *injection, conduction* and *induction* pumps (Seyed-Yagoobi, 2005)) and in the other two electric forces are applied in the double layer (the *electroosmotic* and *ac induced-charge electroosmotic* pumps). One of the objectives of the present work is to study different possibilities of EHD pumping, analysing the EHD actuation for different liquid electrical conductivities: from very insulating liquids to strong electrolytes.

This chapter is organised as follows. First, the main equations of Electrohydrodynamics in microsystems are provided, together with some important concepts of electrical conduction in liquids. Fluid-mechanical aspects that are important for micropumps are also given. Then, these concepts are applied to analyse five different strategies of EHD pumping of fluids in microsystems. Finally, the EHD micropumps that have been analysed are compared.

1.1 EHD equations in the microscale

Electrical equations. We consider systems where magnetic effects are negligible in front of electric effects. For these systems, the magnetic energy density $B_0^2/2\mu$ is much smaller than the electric energy density $\varepsilon E_0^2/2$, where B_0 and E_0 are typical magnetic and electric field amplitudes, respectively, and μ and ε are, respectively, the liquid magnetic permeability and electrical permittivity. $B_0^2/2\mu \ll \varepsilon E_0^2/2$ is equivalent to stating that in our system $cB_0 \ll E_0$, with $c = 1/\sqrt{\varepsilon\mu}$ the speed of electromagnetic waves in the liquid.

Let us see when the magnetic field intensity B_0 generated by the currents in our system is much smaller than E_0/c so that we can neglect magnetic effects. From Ampère-Maxwell law

$$\nabla \times \mathbf{B} = \mu \mathbf{j} + \mu\varepsilon \frac{\partial \mathbf{E}}{\partial t} \tag{1}$$

we can estimate $B_0 \sim \max(\mu j l, \mu\varepsilon\omega E l)$, where ω and l are typical frequency and length, respectively. Using Ohm's law, $j = \sigma E$, the ratio of cB_0/E_0 is

$$\frac{cB_0}{E_0} \sim \max(\mu c\sigma l, \omega l/c). \tag{2}$$

For liquid conductivity $\sigma \ll 10$ S/m, typical system length $l \sim 100\,\mu$m, and nonmagnetic fluids, we have $\mu c\sigma l \ll 0.377/\sqrt{\varepsilon_r} < 1$, where ε_r is the relative permittivity of the liquid. In microsystems, $\omega l/c$ is much smaller than one up to frequencies of the order of 10 GHz. In this situation of $cB_0 \ll E_0$, the electric field that drives the current can be considered to be irrotational. In effect, Faraday's law $\nabla \times \mathbf{E} = -\partial \mathbf{B}/\partial t$ can be approximated by $\nabla \times \mathbf{E} = 0$ if

$$\frac{|\partial \mathbf{B}/\partial t|}{|\nabla \times \mathbf{E}|} \sim \frac{\omega B_0}{E_0/l} < \frac{\omega l}{c} \ll 1. \tag{3}$$

Therefore, working with fluids with conductivities smaller than 10 S/m, with applied ac signals of frequencies up to 10 GHz, the electric fields in our microsystem hold the quasi-electrostatic limit of Maxwell equations (Castellanos, 1998; Haus and Melcher, 1989):

$$\nabla \cdot (\varepsilon \mathbf{E}) = \rho \tag{4}$$

$$\nabla \times \mathbf{E} = 0 \tag{5}$$

$$\nabla \cdot \mathbf{j} + \frac{\partial \rho}{\partial t} = 0 \tag{6}$$

with ρ the volume charge density.

Mechanical equations. For incompressible fluids, the mechanical equations reduce to the Navier-Stokes equations

$$\nabla \cdot \mathbf{u} = 0 \tag{7}$$

$$\rho_m \frac{\partial \mathbf{u}}{\partial t} + \rho_m (\mathbf{u} \cdot \nabla)\mathbf{u} = -\nabla p + \eta \nabla^2 \mathbf{u} + \mathbf{f} \tag{8}$$

where ρ_m is the mass density, η the dynamic viscosity, $\mathbf{u}$ the fluid velocity, p the hydrostatic pressure and $\mathbf{f}$ represents any body force density applied to the liquid. In the microscale, the convection term $\rho_m(\mathbf{u} \cdot \nabla)\mathbf{u}$ is usually negligible in front of the diffusion term $\eta\nabla^2\mathbf{u}$. The ratio between convection and diffusion is measured by the Reynolds number $\mathrm{Re} = \rho_m ul/\eta$ and this is much less than one in many situations in microfluidics (Stone et al., 2004). For water and taking typical values in microsystems, $u \sim 10^{-3}$ m/s, $l \sim 10^{-4}$ m, we obtain $\mathrm{Re} \sim 0.1$. Under steady-state conditions, the momentum equation is then

$$0 = -\nabla p + \eta \nabla^2 \mathbf{u} + \mathbf{f}. \tag{9}$$

We are interested in cases where the main body force density is of electrical origin. When an electric field is present in a liquid, electric forces are exerted on it. The electric volume force density is (Stratton, 1941)

$$\mathbf{f}_E = \rho \mathbf{E} - \frac{1}{2} E^2 \nabla \varepsilon + \frac{1}{2} \nabla \left[E^2 \rho_m \left(\frac{\partial \varepsilon}{\partial \rho_m} \right)_T \right] \tag{10}$$

where the first term is the Coulomb force, the second term is the dielectric force, and the third is the electrostriction. The latter is the gradient of a scalar and can be incorporated into a redefined pressure for an incompressible fluid. The Coulomb term is the body force exerted on a medium that contains free charge. The dielectric term is the force exerted on an nonhomogeneous dielectric liquid by an electric field. The Coulomb force is the term primarily responsible for generating pumping in Electrohydrodynamics, although we will see that the dielectric term can sometimes be important.

Temperature equation. Together with the electrical and mechanical equations, the energy equation is required, which can be related to the temperature distribution T through (Castellanos, 1998)

$$\rho_m c_p \left(\frac{\partial T}{\partial t} + \mathbf{u} \cdot \nabla T \right) = k \nabla^2 T + \mathbf{E} \cdot \mathbf{j}'. \tag{11}$$

Here c_p and k are, respectively, the specific heat at constant pressure and the thermal conductivity and $\mathbf{j}'$ is the current density measured in the rest frame of the fluid element. In the previous equation we have neglected the heat generation due to viscous dissipation and considered moderate heat dissipation, where fluid properties such as ρ_m, c_p, k are assumed to be constant.

In many situations in microsystems, heat convection is small compared to heat diffusion, as demonstrated by the small value of the Peclet number. This is, taking typical values $u < 100\,\mu$m/s, $l < 100\,\mu$m,

$$\frac{|\rho_m c_p \mathbf{u} \cdot \nabla T|}{|k \nabla^2 T|} \sim \mathrm{Pe} = \frac{\rho_m c_p u l}{k} < 0.07 \tag{12}$$

for aqueous solutions. Therefore, the temperature equation in the steady-state reduces to

$$0 = k \nabla^2 T + \mathbf{E} \cdot \mathbf{j}'. \tag{13}$$

1.2 Some concepts of electrical conduction in liquids

Electrical current density. Charge carriers in liquids are electrons and ions (Schmidt, 1994). Except for liquid metals or extremely pure dielectric liquids where electrons can survive without being trapped by molecules, charge in liquids is carried by ions in dissolution (Castellanos, 1998).

Free charge density ρ and ion concentration n_i are related as $\rho = \sum_i q_i n_i$, where q_i is the ion charge. Under the action of an electric field, ions acquire a velocity given by $\mathbf{v} = \mu \mathbf{E}$, where μ is the mobility of the ion. A good picture is to consider the ions as spheres affected by viscous friction. This provides an expression for the mobility as $\mu = q/6\pi a \eta$ (Stokes-Einstein relation), where a is the hydrodynamic radius of the solvated ion and η the dynamic viscosity of the surrounding liquid. The ion charge is $q = ze$, where e is the charge of a proton and z is the valence. The mobility of ions in liquids is usually between 10^{-7} to 10^{-9} m^2/Vs. Ion motion is also affected by molecular diffusion and convection, so the electric current in the liquid bulk can be written as

$$\mathbf{j} = \sum_i q_i (n_i \mu_i \mathbf{E} - D_i \nabla n_i + n_i \mathbf{u}) \tag{14}$$

where D_i is the diffusion coefficient of an ion, and $\mathbf{u}$ the velocity of the fluid. Diffusion and mobility coefficients are linked by Einstein's relation,

$D_i/\mu_i = k_B T/q_i$, where k_B is the Boltzmann's constant and T is the absolute temperature.

In many situations common for dielectric liquids in microsystems, the diffusion current is negligible in front of the electro-migration current. If we compare diffusion and migration, $|D\nabla n|/|\mu n\mathbf{E}| \sim D/\mu El$, with l a typical distance. The ratio $D/\mu = k_B T/q \sim 0.025$ V at room temperature and $E \cdot l$ (typical increment of voltage in the bulk) is usually much greater than 0.025 V. However, for electrolytes, the diffusion current can be important close to the electrodes and walls, where double layers are formed.

For electrolytes the convective current is usually much smaller than the conduction current. In effect, the ratio between convective current and electro-migration current is of the order of

$$\frac{\rho u}{\sum_i q_i n_i \mu_i E} \sim \frac{\nabla \cdot (\varepsilon \mathbf{E}) u}{\sum_i q_i n_i \mu_i E} \sim \frac{\varepsilon u}{\sum q_i n_i \mu_i l} \sim \frac{\varepsilon u}{\sigma l} \tag{15}$$

where σ is the equilibrium liquid conductivity. For electrolytes, this ratio is much smaller than one. For insulating dielectric liquids, the ratio between convection and ion-migration velocities, $u/\mu E$, can be greater or smaller than one.

Conservation of species. Species conservation equations for ions and molecules must be included to complete the description of our system (Saville, 1997):

$$\frac{\partial n_i}{\partial t} + \nabla \cdot (n_i \mu_i \mathbf{E} - D_i \nabla n_i + n_i \mathbf{u}) = r_i \tag{16}$$

where r_i is the rate of production due to chemical reactions since the neutral species act as a source for ions in the bulk ($\mu_i = 0$ for neutral species). Some examples are:

- In *unipolar injection*, there is only a single ionic species, and there is no production of charges in the bulk $r_1 = 0$. The ions are produced (or consumed) at the electrodes.
- For strong electrolytes, the majority of ions come from the dissociation of neutral salt and this is fully dissociated. There are no reactions in the bulk ($r_i = 0$ per each ionic species). Electrochemical reactions take place only at the electrodes.
- For a weak binary z-z electrolyte, there are three species: the positive and negative ions, and the neutral molecule that dissociates into them. Ions are produced (or consumed) by reactions at the electrodes and by homogeneous chemical reactions in the bulk,

$$-r_0 = r_+ = r_- = k_d n_0 - k_r n_+ n_- \tag{17}$$

where k_d and k_r are the dissociation and recombination rate constants and n_0, n_+ and n_- are the concentrations of neutral, positive and negative species, respectively.

Liquid conductivity. Ions in dielectric liquids are originated from dissociation of impurities (or electrolytes) in the liquid bulk, or of the liquid itself (Castellanos, 1998). True electrolytes, also called ionophores, are ionic crystals, while potential electrolytes, also called ionogens, react with the solvent to form ions (Castellanos, 1998):

$$(A^+B^-) \leftrightarrows A^+ + B^- \qquad \text{(true electrolyte)},$$

$$C + D \leftrightarrows (A^+B^-) \leftrightarrows A^+ + B^- \qquad \text{(potential electrolyte)}.$$

In both cases, in order to understand some features of conduction, a good model is a liquid with three species: a neutral species that dissociates into ions so that in equilibrium $k_d n_0 = k_r n_+ n_-$.

The conductivity of a dielectric liquid depends strongly on its dielectric constant. Ions recombine if the energy of Coulomb attraction is greater than the thermal energy $k_B T$ at the point of closest approach of the ions. The equilibrium constant $K \equiv k_d/k_r = n_+ n_-/n_0$ is proportional to

$$K \propto \exp\{-\frac{e^2}{4\pi\varepsilon(a_+ + a_-)k_B T}\} \tag{18}$$

where a_+ and a_- are the radius of positive and negative solvated ions, respectively (Saville, 1997). The distance at which Coulomb attraction energy equals $k_B T$ is called the Bjerrun distance, $l_B = e^2/4\pi\varepsilon k_B T$ (Castellanos, 1998). Because the liquid dielectric constant ε_r affects strongly the probability to occur dissociation through the exponential factor $\exp\{-l_B/(a_+ + a_-)\}$, polar liquids (where $\varepsilon_r \gg 2$) tend to have all impurities dissociated. In contrast, in non-polar liquids (where $\varepsilon_r \sim 2$) the equilibrium is displaced towards the neutral species.

Continuity of electrical current. In order to establish a dc current in a liquid, electrons should be transferred from the metal electrodes to the molecules in the liquid and viceversa. This can happen by electron emission from surface irregularities, from electrochemical red-ox reactions at the surface or by other means. From the mathematical point of view, the different mechanisms can be described by similar parameters (Zhakin, 1998). A simple pictures is to consider that neutral molecules can be charged at the electrodes and that ions can be discharged (see figure 1). This together

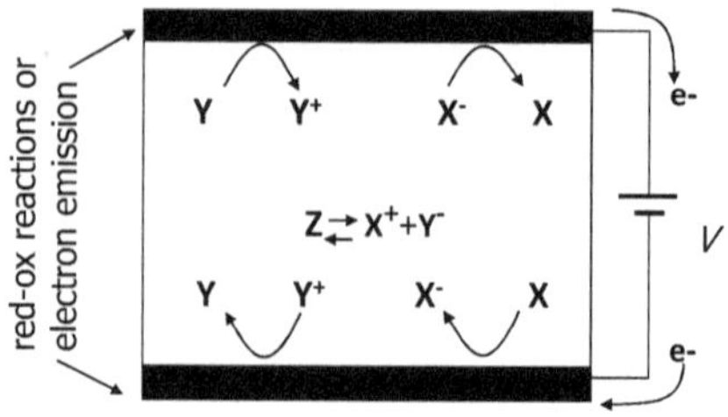

Figure 1. Simple picture of conduction in liquids.

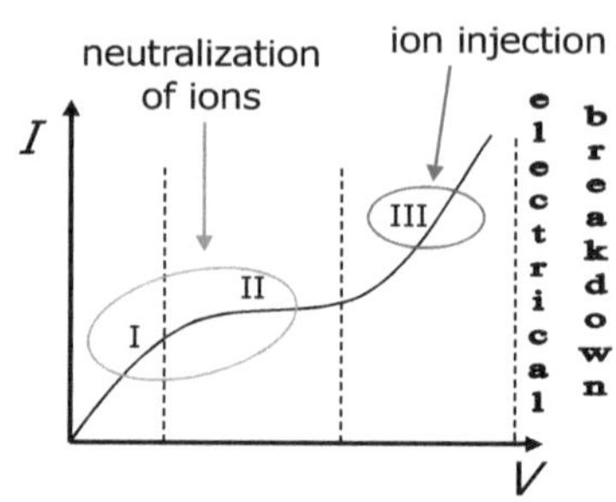

Figure 2. Current-Voltage characteristic for a semi-insulating dielectric liquid.

with the homogeneous reactions of ion production in the bulk form a simple dissociation-injection picture of conduction in liquids (Zhakin, 1998).

The current-versus-voltage characteristic for a semi-insulating dielectric liquid is typically of the form shown in figure 2 (Schmidt, 1994; Zhakin, 1998). It can be roughly divided into three regions. At low voltages (region I), the current is linear with voltage, ions are produced by dissociation of impurities (or electrolytes) in the liquid and they are discharged at the electrodes (at the cathode $X^+ + e^- \rightarrow X$ and at the anode $X^- - e^- \rightarrow X$). As the voltage increases (electric fields around 10^5 V/m), ions generated from the dissociation of impurities can not be created as fast as they are neutralized at the electrodes and the current density reaches saturation (region II). This region II is always present in gases but is ill-defined or may not be present in dielectric liquids. At higher voltages, a steep increase in the current is observed (region III). The beginning of region III can be due to enhancement of dissociation by the electric field strength. Nevertheless, at further increase of the electric field strength, ion injection from the electrodes becomes the dominant process. The ion injected at the cathode $X + e^- \to X^-$ (or at the anode $X - e^- \to X^+$) is neutralized at the opposite electrode. The current usually follows a quadratic law $I \propto V^2$, which is typical of space-charge-limited emission. Finally, at electric fields of sufficient high strength (of the order of 100 MV/m), breakdown takes place. The breakdown electric field is very dependent on the liquid, impurities and geometry.

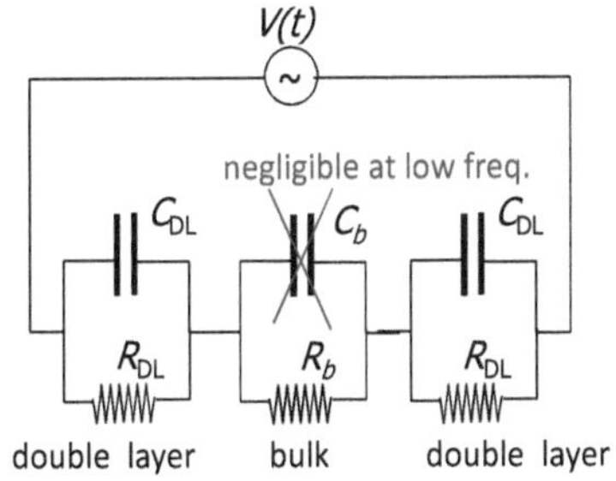

Figure 3. Circuit model for ac currents. The double layer behaves as a capacitor with a resistor in parallel.

In the case of ac currents, the charging of the double layer at the electrodes should be considered (Bard and Faulkner, 2001). In order to understand the electrical behaviour of the double layer at the electrode/liquid interface, this can be modelled as a leaky capacitor (a capacitor and a resistor in parallel) connected to a resistive liquid bulk (see fig. 3). The charging of the capacitor C_{DL} models the charging of the double layer, while the current through the resistor R_{DL} represents the Faradaic current. If the frequency of the applied ac voltage is high enough, the voltage drop across the double layer can be below the threshold voltage for appearance of the electrochemical reactions at the electrodes. Typically, this will happen for angular frequencies much greater than the reciprocal charging time of the double layer, i.e. $\omega \gg 1/R_{\rm bulk}C_{DL}$ (with $R_{\rm bulk}$ and C_{DL} typical values for the bulk resistance and double layer capacitance, respectively). In this situation, the charging of the double layer guarantees the continuity of the electric current with no need of Faradaic currents.

1.3 Some concepts of pumps

At low Reynolds numbers, typical of microsystems, the equations relating fluid velocities to forces and pressure gradients are linear. This linearity allows for the classical representation of microfluidic networks in terms of equivalent electrical circuits: resistors and generators. Flow-rate Q and pressure drop Δp are equivalent to electrical current intensity I and potential drop ΔV, respectively. The force that moves the liquid is commonly concentrated in a small region of the microfluidic system, this constitutes the micropump. This is equivalent to the generator in an electrical circuit. The microchannels are equivalent to resistors. For a microchannel outside the micropump, the volume flow-rate Q is linearly related to the pressure difference Δp between the inlet and outlet of the channel. For a straight channel of circular cross-section we have

$$Q = \Delta p \frac{\pi D^4}{128 \eta L} \tag{19}$$

where D is the diameter and L the length of the channel.

Two important parameters of a pump are: the maximum pressure Δp_{max} and the maximum flow-rate Q_{max} that the pump can generate. The maximum pressure is the pressure difference required to stop the flow that the pump generates. The maximum flow-rate is obtained when the pump actuates without an external load, that is, when the pressure difference between the inlet and outlet of the pump is zero. In many situations, there is a linear relation between flow-rate and back pressure under constant operating conditions

$$Q = Q_{\mathrm{max}} \left(1 - \frac{\Delta p}{\Delta p_{\mathrm{max}}} \right). \tag{20}$$

Power consumption and energy efficiency are important operational parameters of pumps (Laser and Santiago, 2004). When comparing different pumps, we would prefer the one that consumes less power for obtaining a given flow rate or a given pressure. The *energy efficiency* is defined as the maximum delivered hydraulic power divided by the consumed (input) power

$$\mathrm{eff} = \frac{(\Delta p \, Q)_{\mathrm{max}}}{P_{\mathrm{in}}} \tag{21}$$

where P_{in} is the input power. When, under constant operating conditions, there is a linear relationship between Δp and Q, the maximum energy efficiency can be calculated as

$$\mathrm{eff} = \frac{1}{4} \frac{Q_{\mathrm{max}} \Delta p_{\mathrm{max}}}{P_{\mathrm{in}}}. \tag{22}$$

Let us consider as an example a homogeneous volume force f applied to a liquid in a pipe of diameter D and length L. The generated pressure is fL, which is also the one required to stop the flow $\Delta p_{\max} = fL$. The maximum flow-rate is $Q_{\max} = fL\pi D^4/128\eta L$. The flow-rate if there is a back pressure Δp is

$$Q = (fL - \Delta p)\frac{\pi D^4}{128\eta L} = Q_{\max}\left(1 - \frac{\Delta p}{\Delta p_{\max}}\right). \qquad (23)$$

The minimum input power should be the mechanical power that the force is providing $P_{\text{in}} = \int fuL\,dS = fLQ$, where u is the liquid velocity and S is the cross-sectional surface area. Therefore, the maximum energy efficiency is

$$\text{eff} = \frac{1}{4}\frac{Q_{\max}\Delta p_{\max}}{P_{\text{in}}} = \frac{1}{4}\frac{\Delta p_{\max}\,Q_{\max}}{fLQ_{\max}/2} = \frac{1}{2} \qquad (24)$$

where the set $Q = Q_{\max}/2$ and $\Delta p = \Delta p_{\max}/2$ gives the maximum hydraulic power.

2 Electrohydrodynamic Micropumps

2.1 Injection micropump

The injection micropump is also known as ion-drag pump. Ions emitted from one electrode (the emitter) are collected in the other electrode (the collector). Since ions move at the terminal velocity $\mathbf{v} = \mu\mathbf{E}$ (ion inertia is negligible) the electric force acting on them is transmitted directly to the liquid by viscous friction. The typical working fluid for this pump is a very insulating liquid. Stuetzer (1959, 1960) and Pickard (1963a,b) studied theoretically and experimentally the performance of this kind of pumps. In microsystems, Richter, Sandmier and co-workers (Richter and Sandmaier, 1990; Richter et al., 1991) fabricated ion-drag micropumps made of two opposite metallic grids through which the pumped fluid moves. Ion-drag micropumps made of co-planar microelectrodes have been studied by Ahn and Kim (1998) and Darabi et al. (2002).

Theoretical analysis. Let us consider two opposite metallic grids subjected to a potential difference V and placed inside a tube of circular cross-section of diameter D(see fig. 4). Inside the tube there is an insulating liquid that is pumped by the ion-drag mechanism. We consider that only one type of ions exists and these are injected at $x = 0$ and collected at $x = L$. We, therefore, restrict ourselves to study the case of unipolar injection. Ions are injected from the emitter, rather than from the collector, either because of

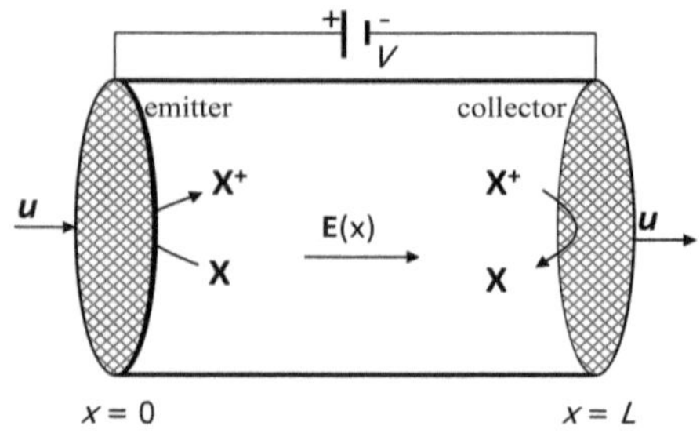

Figure 4. Scheme of injection pump.

the difference between molecule-ion energy barriers at emitter and collector or because the local electric field at the emitter is much greater than the one at the collector due to the presence of sharp edges or points. We further simplify the problem by considering a one-dimensional model and assuming that the diffusion of ions is negligible. The latter is a good approximation for dielectric liquids since the ratio of diffusion to electromigration current is of the order of $k_B T/eV \ll 1$ (see section 1.2). In the one-dimensional approximation, the electric field, the liquid velocity, the current, etc. are directed along the axis of the pipe.

For this geometry, the total electric force per unit area exerted on the liquid (the generated pressure) is

$$\Delta p_g = \int_0^L \left(\rho E - \frac{E^2}{2}\frac{\partial \varepsilon}{\partial x}\right) dx = \int_0^L \frac{1}{2}\frac{\partial \varepsilon E^2}{\partial x}\, dx = \left.\frac{\varepsilon E^2}{2}\right|_{x=L} - \left.\frac{\varepsilon E^2}{2}\right|_{x=0} \tag{25}$$

where we have used that $\rho = \partial(\varepsilon E)/\partial x$. The generated pressure is also the pressure difference required to stop the flow that the pump generates, i.e. the maximum pressure $\Delta p_{\max}$. The result says that the generated pressure is equal to the difference between electrical energy density at the outlet and inlet of the pump. This is a straightforward result using the Maxwell stress tensor and it is not restricted to the injection pump case. The objective now is to find the electric field in the system.

In the steady-state, the equations for the electric field are

$$0 = \frac{\partial j}{\partial x} = \frac{\partial \rho(\mu E + u)}{\partial x} \tag{26}$$

$$\frac{\partial E}{\partial x} = \frac{\rho}{\varepsilon}. \tag{27}$$

In this one-dimensional model, u is the average velocity in a cross-section of the pipe and is a constant of x due to mass conservation. The value of u is obtained *a posteriori* after finding the generated pressure and imposing a given back pressure. The current density is then

$$j = \rho(\mu E + u) = (\mu E + u)\frac{\partial \varepsilon E}{\partial x}. \tag{28}$$

From here we can obtain the electric field strength as

$$E(x) = \left[\frac{2jx}{\varepsilon\mu} + \left(E_0 + \frac{u}{\mu}\right)^2\right]^{1/2} - \frac{u}{\mu} \tag{29}$$

and the charge density as

$$\rho(x) = \frac{j/\mu}{\left[(E_0 + u/\mu)^2 + 2jx/\varepsilon\mu\right]^{1/2}} \tag{30}$$

where $E_0 = E(0)$ is the electric field at the emitter. In the previous expressions $u > 0$, which is the case when the liquid is moved by the electric field. The current j for given u and E_0 can be determined from $V = \int_0^L E(x)\,dx$. A certain charge injection law should be given in order to eliminate E_0.

The expressions say that the charge density decreases from the emitter to the collector, while the field strength E increases. Injection of large amounts of charge has a limit, since the charge already present in the system repels the incoming charge. When the field at the emitter is too low, charge can no longer be removed from the emitter, and the pump reaches its space charge limit (Crowley et al., 1990). For space-charge-limited emission (SCLE), the electric field at the emitter is much smaller than the average field, $E_0 \ll V/L$, and this is the boundary condition that we are going to consider to close our problem ($E_0 \approx 0$). The result is an implicit equation for j

$$V = \frac{\varepsilon\mu}{3j}\left\{\left[\frac{2jL}{\varepsilon\mu} + \left(\frac{u}{\mu}\right)^2\right]^{3/2} - \left(\frac{u}{\mu}\right)^3\right\} - \frac{uL}{\mu}. \tag{31}$$

Once we know the electric current we can calculate the generated pressure as a function of V and u. The generated or maximum pressure for space-charge-limited emission (SCLE) becomes

$$\Delta p_{\max} = \frac{1}{2}\varepsilon E_L^2 = \frac{\varepsilon}{2}\left\{\left[\frac{2jL}{\varepsilon\mu} + \left(\frac{u}{\mu}\right)^2\right]^{1/2} - \frac{u}{\mu}\right\}^2 \tag{32}$$

where $E_L = E(L)$ is the field at the collector. We can now obtain the liquid velocity u as a function of the external back pressure $\Delta p_{\rm out}$ at different applied potentials V, taking into account that $\Delta p_{\rm max}$ is balanced by the pump internal hydrodynamic resistance plus the external back pressure: $\Delta p_{\rm max} = R_{\rm in} Q + \Delta p_{\rm out}$, where $R_{\rm in} = 128\eta L/\pi D^4$ is the hydrodynamic resistance and $Q = u\pi D^2/4$ is the flow-rate.

Let us analyse two limits: a) the liquid velocity is much smaller than the ion migration velocity ($u \ll \mu V/L$); and b) the liquid velocity is much greater ($u \gg \mu V/L$). Here V/L is an estimate of the electric field in the pump. When $u \ll \mu V/L$, the current density is $j \approx \rho\mu E$ and the final expressions for the current density, electric field, generated pressure and maximum flow-rate are

$$j = \frac{9}{8}\frac{\mu\varepsilon V^2}{L^3}, \tag{33}$$

$$E(x) = \frac{3V}{2L}\sqrt{\frac{x}{L}}, \tag{34}$$

$$\Delta p_{\rm max} = \frac{9}{8}\frac{\varepsilon V^2}{L^2}, \tag{35}$$

$$Q_{\rm max} = \frac{9}{8}\frac{\varepsilon V^2}{L^3}\frac{\pi D^4}{128\eta}. \tag{36}$$

When $u \gg \mu V/L$, the current density is $j \approx \rho u$ which implies that ρ is a constant of x and E is a linear function of x. The final expressions are

$$E(x) = \frac{2Vx}{L^2}, \tag{37}$$

$$j = \frac{2u\varepsilon V}{L^2}, \tag{38}$$

$$\Delta p_{\rm max} = \frac{2\varepsilon V^2}{L^2}, \tag{39}$$

$$Q_{\rm max} = \frac{\varepsilon V^2}{L^3}\frac{\pi D^4}{64\eta}. \tag{40}$$

Here $\Delta p_{\rm max}$ is the maximum pressure at negligible flow-rate but still $u \gg \mu E$.

Let us see the ratio between the delivered hydraulic power and the consumed power. The consumed power is $P_{\rm in} = IV = jSV$ and the maximum energy efficiency for the case $u \ll \mu V/L$ is

$$\text{eff} = \frac{1}{4}\frac{\Delta p_{\rm max} Q_{\rm max}}{IV} = \frac{1}{4}\frac{u_{\rm max}}{\mu V/L} = \frac{9}{1024}\frac{\varepsilon V D^2}{L^2 \eta\mu} \tag{41}$$

where $u_{\max} = 4Q_{\max}/\pi D^2$. The maximum energy efficiency for the case $u \gg \mu V/L$ is

$$\text{eff} = \frac{1}{4}\frac{\Delta p_{\max} Q_{\max}}{IV} = \frac{\frac{1}{2}\Delta p_{\max}\frac{1}{2}Q_{\max}}{S2\varepsilon u V^2/L^2} = \frac{1}{2} \tag{42}$$

where the consumed power when the pump is providing maximum hydraulic power is obtained for $Q = Q_{\max}/2$ and $\Delta p = \Delta p_{\max}/2$.

In a general case, there is some residual conductivity σ and then the current density may be written as $j = \rho(u + \mu E) + \sigma E$. The term σE represents the ordinary bipolar conduction, in which charge carriers are not produced by injection but by dissociation of impurities in the bulk. The consumed power can be approximated by $P_{\text{in}} = jSV = \rho(u + \mu E)SV + \sigma ESV$, where E and ρ are here the mean electric field and mean charge density, respectively, (Crowley et al., 1990). The energy efficiency can be written in an approximate formula that contains the limiting cases

$$\text{eff} \sim \frac{1}{2(1 + \alpha_1 + \alpha_2)} \tag{43}$$

where the coefficients α_1 and α_2 are given by

$$\alpha_1 = \frac{2\mu V/L}{u_{\max}}, \tag{44}$$

$$\alpha_2 = \frac{\sigma E}{\rho u_{\max}} \sim \frac{\sigma L/\varepsilon}{u_{\max}}. \tag{45}$$

The coefficients α_1 and α_2 represent the losses due to electromigration and residual conduction, respectively. Here we have used the estimate $\rho \sim \varepsilon V/L^2$ that comes from Gauss's law. For high efficiency, the ions should go from emitter to collector at the fluid velocity rather than due to electromigration ($\mu V/L \ll u_{\max}$). In addition, the liquid should arrive at the collector in a time $L/u_{\max}$ much shorter than the time the charges take to relax ε/σ. The reciprocal of α_2 has the form of an electric Reynolds number $u\varepsilon/\sigma L$ (Melcher and Taylor, 1969). It represents the ratio between convection current and conduction current.

Some experimental features. Darabi and Wang (2005) working with refrigerant HFE-7100 at 180 V obtained pressures of the order of 300 Pa, flow-rates of the order of 30 mm^3/s and a maximum energy efficiency of around 0.0015. They used an array of co-planar interdigitated saw-tooth/plane electrodes. The saw-tooth electrode (the emitter) produces high electric field at the tips, thus resulting in higher charge injection for a given applied voltage. The array consisted of 200 pairs of emitter/collector electrodes. The gap between emitter and collector was 20 μm and the average electric field between electrodes was of the order of 5×10^6 V/m.

The system studied by Darabi et al. cannot be analysed using the simple expressions we wrote above. For electric fields of the order of 5×10^6 V/m and a dielectric permittivity of $\varepsilon = 7.4\varepsilon_0$, naively we could expect that the generated pressure per electrode pair was of the order of $\varepsilon E^2 \sim 1600$ Pa and because there were 200 stages the total generated pressure could have been 3.2×10^5 Pa. This is three orders of magnitude greater than the pressures measured in these experiments. The main reason for such big difference comes from the difference in geometry. In the experimental system, the microelectrodes are co-planar and the electric force is only applied in a small region of the pump channel, of the order of the size of electrodes and gap. In addition, the force is applied near the walls (where we have the no-slip boundary condition) so that viscous shear stresses are very strong. In fact, Darabi and Rhodes (2006) analysed the previous experiments using numerical computations of the complete model and found good agreement with the experiments.

Kazemi et al. (2009) studied injection micropumps with different designs (planar and 3D microelectrodes) and HFE-7100 as working fluid. The best performance was obtained with arrays of asymmetric 3D electrodes (metallic pillars). The 3-D asymmetric micropump generated a maximum pressure of 2240 Pa at an applied voltage of 900 V. It was also the micropump that consumed less power for a given generated pressure.

A problem when working with injection pumps is that charge injection tends to change the chemical composition of both working liquid and electrodes due to electrochemical reactions. The pump life and run-to-run repeatability can then be affected. According to Zhakin (1998), the choice of the working liquid, the electrode material and the ionizing groups that enter into the electrochemical reactions is an important requirement in order to have reproducible results. According to Darabi and Wang (2005), electrode degradation was observed using the HFE-7100 fluid, however, no noticeable electrode degradation was observed using liquid nitrogen.

2.2 Conduction micropump

This kind of EHD actuation is based upon the Coulomb force on non-equilibrium charged layers adjacent to electrodes that appear in the conduction regime. The representative working fluid is a semi-insulating liquid. For electric fields less than 10^7 V/m, ions from dissociation of impurities or electrolytes are responsible of conduction. Ions generated in the bulk move by electromigration towards the electrodes where they are discharged. Near them, charged layers appear due to ion generation not balanced by recombination. These non-equilibrium charged layers are of opposite sign to that of the adjacent electrode and are called heterocharge layers (Atten and Seyed-Yagoobi, 2003; Seyed-Yagoobi, 2005). The heterocharge layer thickness is typically much greater than the Debye length and this is why we include this kind of EHD flow among those that are originated by electric forces in the liquid bulk. Atten and Seyed-Yagoobi (2003) and Jeong and Seyed-Yagoobi (2002, 2004) reported EHD conduction pumping in the millimeter scale. Pearson and Seyed-Yagoobi (2009) reported recently in Electrostatics Joint Conference a micro-scale EHD conduction pump.

Theoretical analysis. Let us consider the geometry shown in fig. 5: two parallel perforated electrodes immerse in a semi-insulating dielectric liquid. The electrode placed at $x = 0$ is subjected to a potential V and the electrode placed at $x = L$ is grounded.

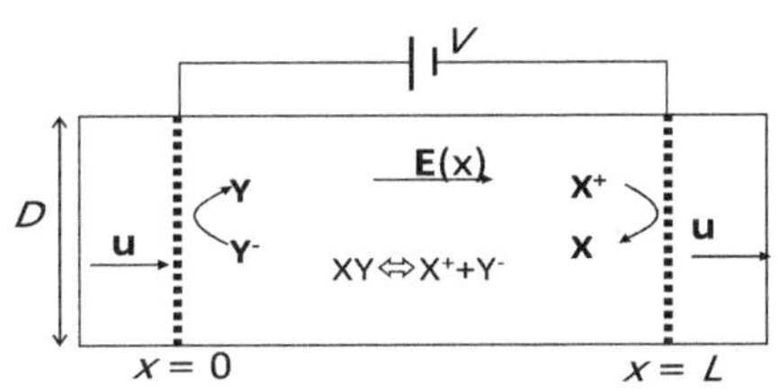

Figure 5. Scheme of conduction pump.

The analysis considers a bipolar conduction model where a neutral species n_0 dissociates reversibly into univalent positive n_+ and negative n_- ions (Atten and Seyed-Yagoobi, 2003; Feng and Seyed-Yagoobi, 2004). There are several simplifications in the model:

- The neutral species is so abundant and the rate of dissociation so small that the concentration of neutral species is considered to be constant.
- Ions are discharged at the electrodes; no charge injection is considered.
- Diffusion and convection of ionic species is negligible.
- A one-dimensional geometry is considered.

The governing 1D-equations in the steady-state are:

$$\frac{\partial \mu_+ n_+ E}{\partial x} = k_d n_0 - k_r n_+ n_- \tag{46}$$

$$\frac{\partial \mu_- n_- E}{\partial x} = -k_d n_0 + k_r n_+ n_- \tag{47}$$

and

$$\frac{\partial E}{\partial x} = e\frac{n_+ - n_-}{\varepsilon} \tag{48}$$

where k_d and k_r denote the dissociation and recombination rates, respectively. At equilibrium $k_d n_0^{\mathrm{eq}} = k_r n_+^{\mathrm{eq}} n_-^{\mathrm{eq}}$ and $n_+^{\mathrm{eq}} = n_-^{\mathrm{eq}} = n^{\mathrm{eq}}$.

The boundary conditions are $n_+ = 0$ at $x = 0$ and $n_- = 0$ at $x = L$, and $\int_0^L E(x)\,dx = V$, with V the potential difference between electrodes. The boundary condition for ions comes from the fact that ion injection is not allowed in the model. In effect, let us consider the positive electrode, negative ions are arriving at it and discharge but no positive ion can come from it (see figure 6). Therefore, at $x = 0$ the flux of positive ions $\mu_+ n_+ E = 0$ which implies that $n_+ = 0$ for finite E. This also implies that near the positive (negative) electrode the majority of ions are negative (positive).

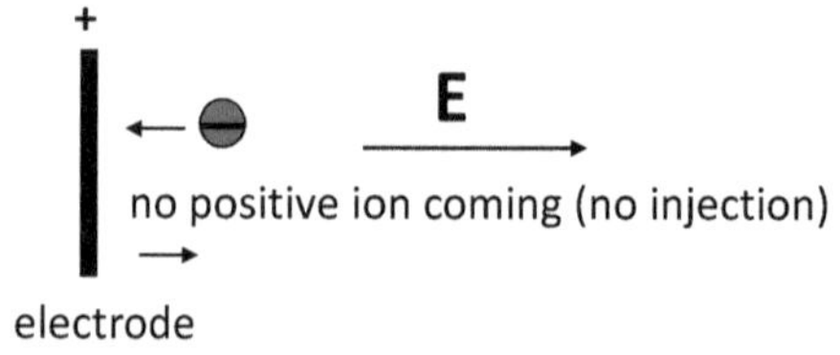

Figure 6. Ion fluxes near positive electrode in conduction regime.

These equations were solved approximately by Thomson and Thomson (1928). The approximate solution is valid under the assumption that the thickness of the heterocharge layer λ is much smaller than the distance between electrodes L (this situation is called the quasi-ohmic regime). This means that the liquid is considered to be electroneutral in the liquid bulk except for a small region near the electrodes, the heterocharge layers. In the bulk the electric field E_b and the concentrations of ionic species $n_+ = n_- = n^{\mathrm{eq}}$ are constant. The approximation consists on neglecting the recombination in the heterocharge layers ($k_r n_+ n_-$ is very small near the electrodes). Let us call λ_+ the thickness of the heterocharge layer adjacent to the positive electrode, the approximate solution for $x \leq \lambda_+$ is

$$\mu_+ n_+ E = k_d n_0 x \qquad \mu_- n_- E = -k_d n_0 x + \text{Const.} \tag{49}$$

This constant is determined from the continuity of electric current along x

$$\mu_- n_- E = (\mu_+ + \mu_-) n^{\mathrm{eq}} E_b - k_d n_0 x \tag{50}$$

where $(\mu_+ + \mu_-)n^{\text{eq}} = \sigma/e$ with σ the bulk conductivity. For $x \geq \lambda_+$, the liquid is assumed to be electroneutral

$$\mu_+ n^{\text{eq}} E_b = k_d n_0 \lambda_+ \qquad \text{at } x = \lambda_+ \,. \tag{51}$$

The thickness λ_+ is then

$$\lambda_+ = \frac{\mu_+ n^{\text{eq}} E_b}{k_d n_0} = \frac{\mu_+ E_b}{k_r n^{\text{eq}}} \tag{52}$$

and, equivalently, for the charge layer adjacent to the negative electrode,

$$\lambda_- = \frac{\mu_- E_b}{k_r n^{\text{eq}}} . \tag{53}$$

The thickness of the charge layer can be seen as the distance that an ion, under the action of the electric field, travels without recombination, taking into account that $t_c = 1/k_r n^{\text{eq}} \approx \varepsilon/\sigma$ is the charge relaxation time (Langevin, 1903).

Gauss's law can now be integrated inside the charge layer, $0 \leq x \leq \lambda_+$. Multiplying the equation by E and substituting the terms En_+ and En_- we get

$$\frac{\varepsilon}{e}\frac{\partial E^2/2}{\partial x} = \frac{k_d n_0 x}{\mu_+} - \frac{(\mu_+ + \mu_-)n^{\text{eq}} E_b - k_d n_0 x}{\mu_-} . \tag{54}$$

The result for the electric field is (Pontiga, 1992)

$$E = E_b \sqrt{1 + (1 - x/\lambda_+)^2 \mu_+/\mu_-} \tag{55}$$

where we have used that $E(\lambda_+) = E_b$ and $n^{\text{eq}}/k_r n_0 = \varepsilon/\sigma$. The value of the electric field in the bulk is close to V/L for $\lambda_+/L \ll 1$ and $\lambda_-/L \ll 1$, typically, when $\varepsilon V/en^{\text{eq}}L^2 \ll 1$. The charge density is

$$\rho = -\frac{\varepsilon E_b}{\lambda_+} \frac{(1 - x/\lambda_+)\mu_+/\mu_-}{\sqrt{1 + (1 - x/\lambda_+)^2 \mu_+/\mu_-}} \tag{56}$$

and we can see that the charge layer sign is opposite to that of adjacent electrode. Therefore, Coulomb force is directed towards the electrode. The generated pressure in the charge layer of thickness λ_+ is (recall eq. 25)

$$\Delta p_g^+ = \frac{\varepsilon}{2}(E_b^2 - E_0^2) = -\frac{\varepsilon}{2} E_b^2 \frac{\mu_+}{\mu_-} . \tag{57}$$

An equivalent expression can be deduced for the other heterocharge layer and we have

$$\Delta p_g^- = \frac{\varepsilon}{2} E_b^2 \frac{\mu_-}{\mu_+} . \tag{58}$$

And the total pressure generated is

$$\Delta p_g = \frac{\varepsilon}{2} E_b^2 \left(\frac{\mu_-}{\mu_+} - \frac{\mu_+}{\mu_-} \right). \tag{59}$$

The difference in mobility between positive and negative ions is responsible for the appearance of a net pressure.

In practice, in order to have a pressure difference even in the case of equal mobilities, the electrodes are made non-symmetric (Atten and Seyed-Yagoobi, 2003; Jeong and Seyed-Yagoobi, 2002; Feng and Seyed-Yagoobi, 2004). Feng and Seyed-Yagoobi (2004) used a perforated disc electrode against a ring electrode (see fig. 7). The Coulomb force generated near the perforated disc electrode is mainly axial while the Coulomb force generated near the ring electrode is mainly radial. The liquid is driven by the axial force while the radial force is balanced by the reaction of the pipe wall. This electrode configuration working as conduction pump always drives the liquid towards the perforated electrode. According to Feng and Seyed-Yagoobi (2004), the pressure generated at zero flow rate for this configuration and $\mu_+ = \mu_-$ is $\Delta p_{\max} = 0.85\varepsilon V^2/L^2$, which is of the order of the expressions previously obtained for only one electrode, eq. (57). For finite flow rate, assuming that the convection current can be neglected, the previous generated pressure leads to a maximum average fluid velocity and flow-rate given by

$$u_{\max} = 0.85\frac{\varepsilon V^2}{L^3}\frac{D^2}{32\eta} \qquad Q_{\max} = 0.85\frac{\varepsilon V^2}{L^3}\frac{\pi D^4}{128\eta}. \tag{60}$$

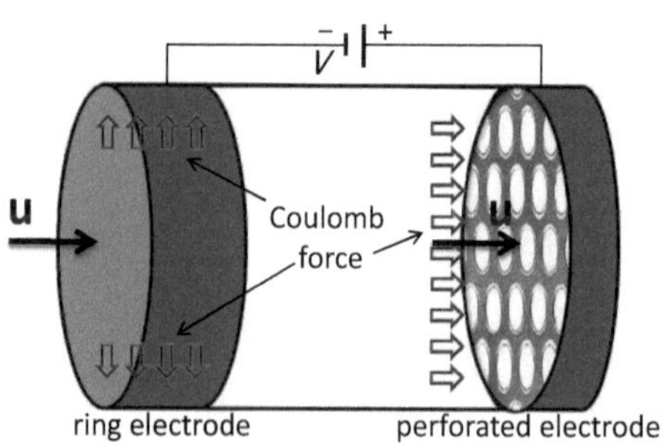

Figure 7. Scheme of conduction pump used in (Feng and Seyed-Yagoobi, 2004).

The consumed power for quasi-ohmic regime ($\lambda \ll L$) can be written as $IV = \sigma V^2 \pi D^2/4L$ and the maximum energy efficiency in the limit of negligible convection current is

$$\text{eff} = \frac{\Delta p_{\max} Q_{\max}}{4IV} \approx 0.2 \frac{u_{\max}\varepsilon}{\sigma L} \approx 0.2 \frac{\varepsilon^2 V^2 D^2}{32 L^4 \eta \sigma}. \tag{61}$$

We can see that energy efficiency is of the order of the ratio between convection and conduction currents (the electric Reynolds number). This is usually very small in microsystems, so that the energy efficiency should be small. In order to increase the energy efficiency, we can increase V, or decrease σ, but there is a limit because these expressions are only valid for $\lambda = \mu\varepsilon V/\sigma L \ll L$. When $\lambda \gg L$, the recombination of ions is also negligible in the bulk and the regimen of saturation of current is reached. Ions are removed on the electrodes at the same rate that are produced in the bulk, which leads to $j = ek_d n_0 L$. In the regime of current saturation, there is a small variation of electric field between electrodes and the generated pressure saturates.

Some experimental features. Feng and Seyed-Yagoobi (2004) working with refrigerant HCFC-123 at 10 kV in a millimeter scale micropump obtained $\Delta p_{\max} \sim 270$ Pa, $Q_{\max} \sim 4000\,\text{mm}^3/s$, $u_{\max} = 5$ cm/s. The power consumption was around 0.25 watts and energy efficiency around 0.001. The estimated thickness of the heterocharge layer was around 100 μm which is much greater than the estimated Debye length $\lambda_D = 1\,\mu$m, as it should be. Pearson and Seyed-Yagoobi (2009) working with refrigerant HCFC-123 at 4 kV for a micrometer scale micropump obtained $\Delta p_{\max} \sim 500$ Pa, and a power consumption around 0.09 watts. The gap between electrodes was 250 μm.

In practice, we can increase the voltage or reduce the dimensions to increase the generated pressure. But then we can reach either the saturation regimen or the regime of charge injection. In some geometries the appearance of charge injection is manifested by the liquid changing direction. For instance, in the point-plane configuration of Atten and Seyed-Yagoobi (2003), the liquid moves towards the point electrode below 17.5 kV by the conduction mechanism and moves from the point to the plane for higher voltages due to the injection mechanism. To avoid charge injection, sharp edges should be avoided.

Because electric fields are not so high as compared to the case of the injection pump, the electrodes and liquids are subjected to much less degradation, which is an advantage.

2.3 Induction micropump

Induction pumping is based on the forces upon charges induced in a liquid by the electric field due to a gradient or discontinuity of the electric conductivity. Typically, an ac travelling-wave electric field attracts or repels these induced charges, leading to fluid flow. Melcher (1966) introduced the concept of EHD induction pumping. He used a semi-insulating liquid and charges were induced at the air/liquid interface. Melcher and Firebaugh (1967) carried out EHD induction pumping acting on induced charges in the liquid bulk. They imposed a conductivity gradient using a temperature gradient and the liquid was then subjected to a travelling-wave potential. Liquid conductivity is generally a function of temperature because ionic mobilities depend on temperature and because the dissociation rate of neutral molecules depends also on temperature. The temperature gradient can be imposed externally or it can also be obtained by Joule heating generated by the imposed electric field. In microsystems, the EHD induction micropumps fabricated by Fuhr and co-workers (Fuhr et al., 1992; Müller et al., 1993; Fuhr et al., 1994) employed Joule heating to establish the temperature gradient. Water saline solutions were used as working fluids, which are very difficult (if not impossible) to pump using EHD induction at the macroscale. More recently, Felten et al. (2006) carried out induction pumping using an externally imposed gradient of temperature in a microdevice. Unidirectional motion can also be obtained with an array of electrodes subjected to a single phase ac signal either with an imposed longitudinal temperature gradient (Green et al., 2000a; González et al., 2006; Stubbe et al., 2007; Holtappels et al., 2009) or using Joule heating and asymmetric electrodes (Wu et al., 2007). In microsystems, most of flows originated by induction pumping are electrothermal flows (Ramos et al., 1998).

Theoretical analysis. An induction pump which does not require a travelling wave signal can be fabricated with two parallel perforated electrodes subjected to both a potential difference and a temperature difference. This is a simplified version of an electrothermal pump using an imposed longitudinal gradient of temperature (Stubbe et al., 2007).

Let us consider the geometry of fig. 8: the electrode at $x = 0$ is subjected to potential V_0 and a temperature T_0 and the electrode at $x = L$ is subjected to potential zero and temperature T_L. We simplify the problem assuming:

- the problem is unidimensional;
- the liquid is quasi-electroneutral; in the bipolar conduction model this means $|n_+ - n_-| \ll n^{\mathrm{eq}}$, which leads to $j = \sigma E$, with $\sigma = e(\mu_+ + \mu_-)n^{\mathrm{eq}}$;
- negligible convection both of charge and temperature;

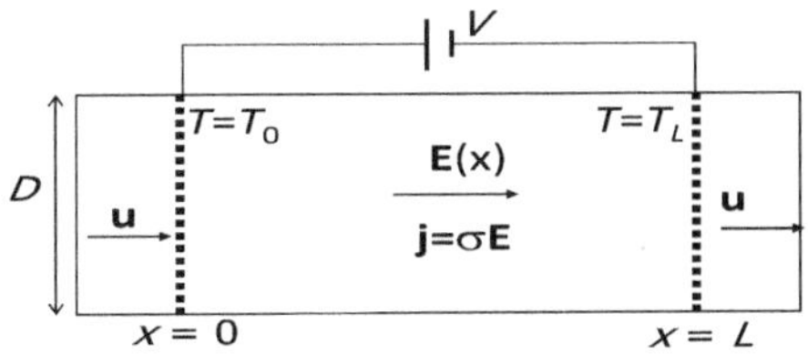

Figure 8. Scheme of induction pump.

- linear variation between T_0 and T_L so that $\sigma(x) = \sigma_1 + ax$, $\varepsilon(x) = \varepsilon_1 + bx$, with $a = (d\sigma/dT)(dT/dx)$ and $b = (d\varepsilon/dT)(dT/dx)$.

Since the electrical current $j = \sigma E$ is a constant of x and $\int_0^L E\,dx = V_0$ we obtain

$$E = \frac{aV_0}{\sigma \ln(1 + aL/\sigma_1)}. \tag{62}$$

The induced charge in the bulk is

$$\rho = \frac{\partial \varepsilon E}{\partial x} = \sigma E \frac{\partial}{\partial x}\left(\frac{\varepsilon}{\sigma}\right) = E(b - a\varepsilon/\sigma). \tag{63}$$

The generated pressure is (recall eq. 25)

$$\Delta p_{\max} = \frac{1}{2}\frac{a^2 V_0^2}{(\ln[1 + aL/\sigma_1])^2}\left(\frac{\varepsilon(L)}{\sigma(L)^2} - \frac{\varepsilon(0)}{\sigma(0)^2}\right). \tag{64}$$

For small increments of temperature so that $\Delta\sigma \ll \sigma$ and $\Delta\varepsilon \ll \varepsilon$, the expression simplifies to

$$\Delta p_{\max} = \frac{1}{2}\varepsilon_1 \frac{V_0^2}{L^2}(2\alpha - \beta)(T_0 - T_L) \tag{65}$$

where $\alpha = (1/\sigma)(d\sigma/dT)$ and $\beta = (1/\varepsilon)(d\varepsilon/dT)$. For water saline solutions, $\alpha = 0.02\,\mathrm{K}^{-1}$ and $\beta = -0.004\,\mathrm{K}^{-1}$ at room temperature. If $T_0 > T_L$ it can be seen that the induced charge is positive and that the electric force and generated pressure are directed from high to low temperature. In this limit of small increment of temperature, the maximum average velocity and flow-rate are

$$u_{\max} = \frac{1}{2}\frac{\varepsilon_1 V^2}{L^3}\frac{D^2}{32\eta}(2\alpha - \beta)(T_0 - T_L) \qquad Q_{\max} = u_{\max}\frac{\pi D^2}{4}. \tag{66}$$

The consumed power is either equal to $P_{\text{in}} = IV$, if the temperature gradient is generated by Joule heating, or to $P_{\text{in}} = IV + \dot{q}$ where $\dot{q}$ is the input of power required to establish the external gradient of temperature. The maximum energy efficiency is then

$$\mathit{eff} = \frac{\Delta p_{\max} Q_{\max}}{4(IV + \dot{q})} \sim \frac{\varepsilon^2 V_0^2 D^2 (\alpha \Delta T)^2}{128 \eta \sigma L^4 (1 + \dot{q}/IV_0)} \approx \frac{1}{4} \frac{(u_{\max}\varepsilon/\sigma L)\alpha \Delta T}{1 + \dot{q}/IV_0} \tag{67}$$

where we have neglected β in front of α as a first approximation since for water-saline solutions $\alpha \gg |\beta|$.

We can see that the maximum pressure and flow-rate are somewhat smaller than for the previous two mechanisms of EHD pumping for the same applied voltage and geometry. In effect, ΔT should be of the order of 10 K or less and the generated pressure is of the order of $\Delta p_g \sim \varepsilon(V^2/L^2)\alpha\Delta T < 0.2\varepsilon(V^2/L^2)$ for water. The energy efficiency is very small, being proportional to the product of two small numbers: the electric Reynolds number and $\alpha\Delta T/(1 + \dot{q}/IV)$.

In practice, in order to avoid Faradaic reactions, ac voltages of high enough frequency are applied. The time-average electric force is different from zero because it is quadratic with voltage. The expressions are the same substituting V_0 by V_{rms} for applied signals with frequencies $\omega \ll \sigma/\varepsilon$.

This geometry induces an electrothermal flow for $\omega \gg \sigma/\varepsilon$, when the liquid behaves as a perfect dielectric. Properly, this case is not induction pumping since no charges are induced; the force actuating here is the dielectric force. For $\omega \gg \sigma/\varepsilon$, it can be seen that the electrothermal flow goes from low to high temperature (contrary to the case with $\omega \ll \sigma/\varepsilon$) and the pressure generated is $\Delta p_{\max} \approx \frac{1}{2}\varepsilon_1\beta(T_0 - T_L)V_{\text{rms}}^2/L^2$.

The analysis of the travelling-wave (TW) experiments is somewhat different. Let us consider the geometry depicted in fig. 9: a straight channel of rectangular cross-section with an array of microelectrodes on the bottom that are subjected to a TW potential $V_0 \cos(\omega t - kx)$, where k is the wavenumber. There is a vertical gradient of temperature that creates vertical gradients of conductivity and permittivity in the liquid. This electric field wave induces a charge wave that lags behind the voltage wave due to the finite time in which charge is induced in the bulk. Maximum longitudinal force is obtained at a frequency around $\omega = \sigma/\varepsilon$ (reciprocal of the charge relaxation time). At much higher frequencies, negligible charge is induced, and at much lower frequencies, there is negligible delay between the induced charge and the TW signal and, therefore, negligible longitudinal force.

Solving Laplace's equation for the potential in the case of $w \gg h \gg 1/k$, and small ΔT, $\Delta\varepsilon$ and $\Delta\sigma$ gives (using phasors)

$$\phi(x, y, t) = V_0 e^{i(\omega t - kx) - ky}. \tag{68}$$

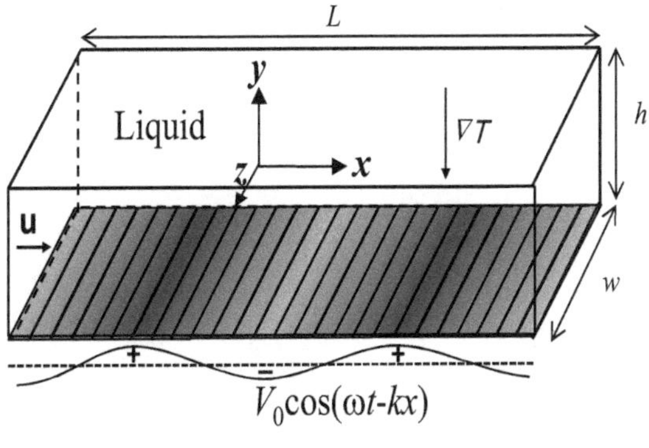

Figure 9. Scheme of TW induction pump.

For small variations in σ and ε the phasor of induced charge is (Ramos et al., 1998)

$$\rho \approx -\frac{\varepsilon\nabla\sigma - \sigma\nabla\varepsilon}{\sigma + i\omega\varepsilon} \cdot \mathbf{E}\,. \tag{69}$$

The time-averaged Coulomb body force is $\langle \mathbf{f} \rangle = \frac{1}{2}\text{Real}[\rho\mathbf{E}^*]$, where $\mathbf{E}^*$ is the complex conjugate of $\mathbf{E}$. The vertical force is compensated by pressure while the longitudinal force generates the fluid flow:

$$\langle f_x \rangle = \frac{1}{2}\varepsilon(\beta - \alpha)\frac{dT}{dy}k^2V_0^2e^{-2ky}\frac{\omega\tau}{1 + (\omega\tau)^2} \tag{70}$$

where $\tau = \varepsilon/\sigma$. This force is maximum at $\omega\tau = 1$. Integrating the longitudinal component of Stokes equation with no applied pressure (that is $0 = \eta\partial^2 u_x/\partial y^2 + \langle f_x \rangle$), we obtain for $\omega\tau = 1$

$$u_x \approx u_0(1 - e^{-2ky} - y/h); \quad u_0 = \frac{\varepsilon(\alpha - \beta)\frac{dT}{dy}V_0^2}{16\eta}. \tag{71}$$

The maximum flow-rate generated is

$$Q_{\text{max}} = \frac{u_0 hw}{2}(1 - (kh)^{-1}) \approx \frac{u_0 hw}{2} \approx \frac{\varepsilon\alpha\frac{dT}{dy}hwV_0^2}{32\eta} \tag{72}$$

where we have neglected β in front of α. To stop this flow the pressure required is

$$\Delta p_{\text{max}} = \frac{12\eta Q_{\text{max}} L}{wh^3} \approx \frac{6\eta u_0 L}{h^2} \approx \frac{3\varepsilon\alpha\frac{dT}{dy}LV_0^2}{8h^2} \tag{73}$$

where L is the total length of the array. The dissipated power is

$$P_{\text{dis}} = \frac{1}{2}\int \sigma|\mathbf{E}|^2 dv = \frac{wL\sigma k V_0^2}{2}. \tag{74}$$

And the maximum energy efficiency is

$$\text{eff} \sim \frac{3\varepsilon^2\alpha^2(\Delta T)^2 V_0^2}{256kh^3\eta\sigma(1+\dot{q}/IV)} = \frac{3}{8}\frac{(u_{\text{max}}\varepsilon k/\sigma)\alpha\Delta T}{k^2h^2(1+\dot{q}/IV)} \tag{75}$$

where $\dot{q}$ is already included in IV when the temperature gradient is generated by Joule heating.

According to these expressions, in order to increase the generated pressure and energy efficiency, the height of the channel can be reduced. If we compare the energy efficiency of this TW induction pump with previous geometry that uses a single ac signal, we can see that the energy efficiency for the TW induction pump is inversely proportional to k^2h^2, which is much greater than one. The fact that the electric force is only applied near the electrodes (in a layer with thickness of the order of k^{-1}) reduces the energy efficiency by the factor of $1/(k^2h^2)$.

Some experimental features. Felten et al. (2006) fabricated a TW induction micropump driven by a four-phase TW potential working at frequencies in the range 1-10 MHz and voltages in the range 4-10 V. They employed water saline solutions of conductivity $\sigma = 0.01$ S/m. The characteristic dimensions of the pump were $h = 50\,\mu$m, $w = 100\,\mu$m, $\lambda = 2\pi/k = 80\,\mu$m, $L = 240\,\mu$m. Using both temperature fields produced by the electric wave itself and by external Peltier elements, they observed velocities near 100 μm/s, for $V_{pp} = 10$ V and $f = 2$ MHz.

Iverson and Garimella (2009) employed a three-phase TW potential working with frequencies in the range 10-300 kHz and voltages around 30 V. The working fluid was saline solution of conductivity $\sigma = 6\times10^{-4}$ S/m. The characteristic dimensions of the pump were $h = 50\,\mu$m, $w = 5$ mm, $\lambda = 2\pi/k = 72\,\mu$m, $L = 10$ mm. A typical velocity was $u = 100\,\mu$m/s, for $V = 30$ V and $f = 122$ kHz using Joule Heating. They also employed external sources of heat to produce the temperature field. Pressure heads of the order of 10 Pa were obtained.

The generated pressure can be increased by increasing the applied voltage. However, this cannot be so high that Joule heating is excessive. An estimation of the temperature rise due to Joule heating is $\Delta T \sim \sigma V_{rms}^2/\kappa$ with κ the thermal conductivity (Ramos et al., 1998). For a voltage of 10 V, typical of induction pumping in microsystems, the increment of temperature may be excessive for $\sigma \geq 1$ S/m (temperature rise of the order or greater than 100 K, so that boiling can be expected). EHD instabilities also limit the maximum voltage, destroying the pattern of σ by convection and disrupting the pumping mechanism (Melcher and Firebaugh, 1967).

Because electric fields are not so high, the electrodes and liquids are not subjected to a degradation comparable to ion injection. Fuhr et al. (1994) using coating on electrodes obtained good reproducibility over several months.

Energy efficiency can be quite low. This, however, may not be a limitation because total consumed power can be low.

2.4 Electroosmotic micropump

Electrical double layers are formed at solid/liquid interfaces because of differences in electrochemical potentials between the solid and liquid phases (Hunter, 1981). The solid surface becomes charged and ions in solution of opposite charge (counterions) are attracted until they screen the electric field created by the surface charge. At equilibrium, the liquid is electroneutral except for a charged layer adjacent to the solid surface. The characteristic thickness of this layer is the Debye length, λ_D, and it is the result of the balance between the electrostatic forces and thermal diffusion. The Debye length is given by

$$\lambda_D = \sqrt{\frac{\varepsilon k_B T}{e^2 \sum_i z_i^2 n_i^\infty}} \tag{76}$$

where z_i and n_i^∞ are, respectively, the valence and number density of ionic species i in solution. For the case of a symmetric electrolyte, the Debye length can be written as $\lambda_D = \sqrt{\varepsilon D_{\text{av}}/\sigma}$, where $D_{\text{av}} = (D_+ + D_-)/2$ is the mean ion diffusion coefficient. As an example of a spontaneous double layer, in the case of silica-based ceramics, like glass, with Si–OH groups at the surface, a fraction of the Si–OH bonds change into Si–O– releasing H^+ when immersed in water. Therefore, we have a negative surface charge balanced by positive counterions.

When an electric field is applied along the surface, the Coulomb force causes the motion of the mobile ions of the double layer dragging the fluid with them. This is called Electroosmosis and it is considered as a superficial phenomenon since the thickness of the Debye layer is very small (1-100 nm in water). Therefore, the electroosmotic (EO) pump actuates upon the electrical double layer that spontaneously develops between a liquid in contact with a solid.

The EO phenomena have been known for 200 years (F.F. Reuss discovered electroosmosis in 1809). Flows generated by EO pumping are used in a range of applications, including soil remediation or contaminant removal from groundwater, and have been used in chemical and biological analysis since a long time. Pretorius et al. (1974) proposed electroosmosis for high-speed chromatography. In miniaturized systems for chemistry and life sciences, a number of important techniques make use of EO pumping: flow injection analysis (Dasgupta and Liu, 1994), on-chip electrophoretic separation (Manz et al., 1994) and on-chip liquid chromatography (Jacobson et al., 1994). Yao and Santiago (2003) presented a detailed description of the history and development of EO pumps that are able to generate high pressure.

Theoretical analysis. Rice and Whitehead (1965) studied theoretically the electrokinetic flow in a cylindrical capillary, which is a simple geometry where EO pumping can be observed (see fig. 10). Consider a cylindrical capillary of radius a, with the wall of the capillary at potential ζ, the zeta potential. The Gouy-Chapman-Stern model describes the electrical double layer as divided into a compact layer and a diffuse layer (Hunter, 1981). The Stern or compact layer is formed of absorbed species onto the wall, while the ions of the Gouy-Chapman diffuse layer are mobile. The zeta potential is the electric potential at the plane separating these two layers.

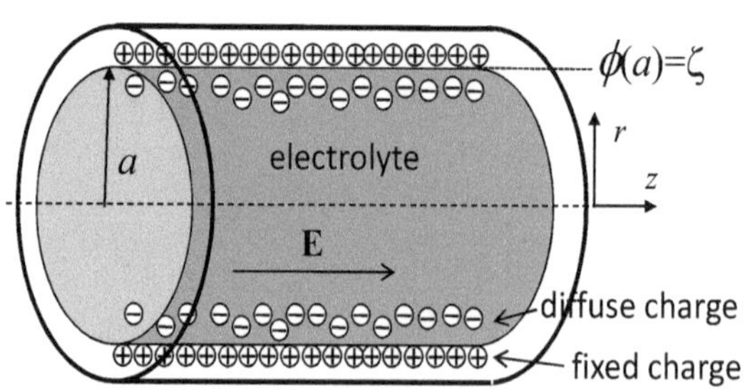

Figure 10. Scheme of double layer and applied electric field for electroosmosis in a capillary.

Inside the capillary there is an electrolyte. For simplicity, we consider a symmetric 1-1 electrolyte with bulk ionic concentration equal to n_∞. The Boltzmann distribution of ions gives

$$n_+ = n_\infty \exp\left[-\frac{e\phi}{k_B T}\right] \qquad n_- = n_\infty \exp\left[\frac{e\phi}{k_B T}\right] \tag{77}$$

and the Poisson-Boltzmann equation is, for this cylindrical geometry,

$$\frac{1}{r}\frac{\partial}{\partial r}\left(r\frac{\partial \phi}{\partial r}\right) = -\frac{\rho(r)}{\varepsilon} = \frac{2en_\infty}{\varepsilon}\sinh\left(\frac{e\phi}{k_B T}\right). \tag{78}$$

In the Debye-Hückel approximation (Hunter, 1981), $e\phi/k_B T \ll 1$, the Poisson-Boltzmann equation becomes

$$\frac{1}{r}\frac{\partial}{\partial r}\left(r\frac{\partial \phi}{\partial r}\right) = \frac{\phi}{\lambda_D^2} \tag{79}$$

where $\lambda_D = \sqrt{\varepsilon k_B T/2e^2 n^\infty}$ is the Debye length. The solution for this equation that holds the boundary condition at $r = a$ and is finite at $r = 0$ is

$$\phi(r) = \zeta\frac{I_0(r/\lambda_D)}{I_0(a/\lambda_D)} \qquad \rho = -\frac{\varepsilon}{\lambda_D^2}\phi \tag{80}$$

where $I_0(x)$ is the zero-order modified Bessel function of the first kind. For values of $a/\lambda_D \gg 1$, the solution is such that potential and charge are only different from zero near the wall.

When an electric field is applied axially the liquid is set into motion. The equation of liquid motion for an applied electric field, E_z, and no applied pressure gradient is

$$\frac{\eta}{r}\frac{\partial}{\partial r}\left(r\frac{\partial u}{\partial r}\right) + E_z\rho = 0. \tag{81}$$

With the boundary condition of no-slip at the capillary wall, $u(a) = 0$, the velocity distribution is

$$u(r) = -\frac{\varepsilon\zeta E_z}{\eta}\left(1 - \frac{I_0(r/\lambda_D)}{I_0(a/\lambda_D)}\right). \tag{82}$$

In the case that the radius is much greater than the Debye length, $a \gg \lambda_D$, the velocity varies from zero at the wall to the asymptotic value $u_s = -\varepsilon\zeta E_z/\eta$ inside a very thin layer of the order of λ_D. The liquid bulk inside the capillary moves as a plug flow with velocity u_s, known as the electroosmotic slip velocity. The slip velocity expression, $u_s = -\varepsilon\zeta E_z/\eta$, is also valid for the case $\zeta > k_B T/e$.

The flow-rate at zero applied pressure is

$$Q_{\max} = \int_0^{D/2} u(r)2\pi r dr = \frac{\pi D^2 \varepsilon\zeta E_z}{4\eta}F(a^*) \tag{83}$$

where $a^* = a/\lambda_D$, $D = 2a$, and

$$F(x) = \left(1 - \frac{2I_1(x)}{xI_0(x)}\right). \tag{84}$$

The pressure difference between the capillary ends required to balance this flow-rate is

$$\Delta p_{\max} = \frac{32\varepsilon\zeta E_z L}{D^2} F(a^*). \tag{85}$$

The function $F(x)$ tends to one for $x \to \infty$, that is for $a \gg \lambda_D$ (see fig. 11).

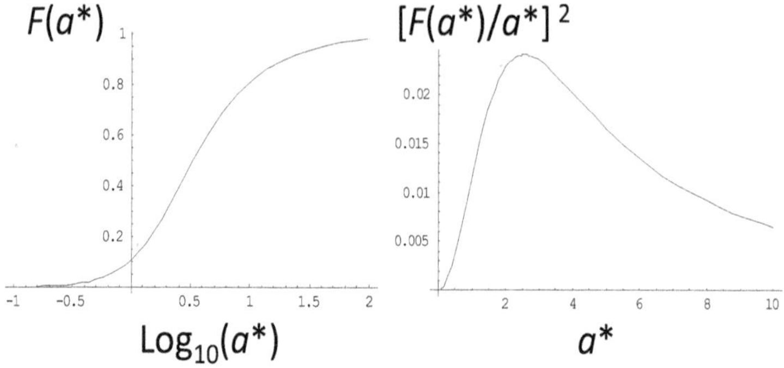

Figure 11. The functions $F(a^*)$ and $(F(a^*)/a^*)^2$.

The input power is IV. The current intensity is the sum of conduction and convection currents

$$I = \frac{\pi D^2 \sigma E_z}{4} + \int_0^{D/2} \rho(r) u(r) 2\pi r dr \tag{86}$$

where $\sigma = e(\mu_+ n_+ + \mu_- n_-) = e(\mu_+ + \mu_-)n_\infty$ everywhere, even inside the double layer, in the approximation $e\phi/k_B T \ll 1$. Neglecting the advection, we can write an expression for the maximum energy efficiency as

$$\text{eff} = \frac{1}{4}\frac{\Delta p_{\max} Q_{\max}}{I V_{\text{app}}} = \frac{V}{V_{\text{app}}} \frac{8\varepsilon^2\zeta^2}{D^2 \eta\sigma} F(a^*)^2 \tag{87}$$

where V_{app} is the applied external voltage and $V = E_z L$ is the effective voltage responsible for the electrokinetic flow. The applied voltage V_{app} includes, in addition to $E_z L$, the potential drop required to start the Faradaic reactions at the electrodes and the junction voltages at inlet and outlet of the capillary. In the limit of very thin diffuse layer, $a^* \gg 1$, the maximum efficiency is eff $= 8\varepsilon^2\zeta^2/D^2\eta\sigma$. In this limit, the efficiency can be increased by decreasing the diameter of the capillary D and by decreasing the ionic concentration of the working fluid (notice that the product $\sigma\eta$ is almost independent on viscosity because of Walden's rule). According to expression

(87), the efficiency as a function of D is maximized when D is of the order of λ_D. In effect, assuming the other parameters are constant, expression (87) is proportional to $F(a^*)^2/(a^*)^2$ which has a maximum for $a^* \approx 2.54$, that is $D \sim 5\lambda_D$. According to Yao and Santiago (2003), maximum efficiency is obtained for $D \sim \lambda_D$ also in the general case when advection current can be important. The model we have described neglects any surface conduction. For $a \gg \lambda_D$, the effects of such surface conduction on the energy efficiency should be negligible as the main contribution to electrical current comes from the bulk current. For $a \ll \lambda_D$, the surface conduction would further increase the consumed power without increasing the hydraulic power. This means that surface conduction should not change the trend that the energy efficiency would decrease for $a \ll \lambda_D$.

In order to increase the generated pressure, the capillary diameter can be reduced, however, this reduces the flow-rate. In practice, EO pumps that generate high pressure make use of porous structures so that each pore acts as a tortuous capillary of small diameter (Yao and Santiago, 2003; Yao et al., 2003; Wang et al., 2006). The porous structure can be seen as a bundle of capillaries; the generated pressure is high because D is small and the flow-rate is high because of parallelization. From the bundle-of-capillaries model, the theoretical maximum pressure of a porous structure is still given by expression (85) with D the characteristic diameter of the pore. The maximum flow rate is given by (Yao and Santiago, 2003)

$$Q_{\max} = -\frac{\psi L^2}{L_e^2} \frac{\varepsilon \zeta A}{\eta} F(a^*) \tag{88}$$

where A is the cross-sectional area of the structure, ψ is its porosity (void volume divided by total volume of the porous medium), L_e is the characteristic length of the tortuous path of the pores and L the length of the porous structure. From these expressions the number $(4\psi AL^2)/(L_e^2 \pi D^2)$ can be seen as an effective number of EO capillaries with diameter D. The energy efficiency of a porous structure will be given approximately by that of one capillary since in the model, both Q and I are proportional to the number of capillaries in parallel and Δp and V are the same for all the capillaries. Therefore, $Q\Delta p/IV$ should be independent of the number of capillaries.

Some experimental features. Yao et al. (2003) employed a porous glass EO micropump with sodium borate buffer as working fluid (volume of micropump around 1300 mm^3). Experiments conducted at 100 V provided: maximum pressure around 10^5 Pa, maximum flow-rate around 500 mm$^3/s$. The maximum energy efficiency was investigated for different pore radius: it was around 0.0025 for $a = 0.5\,\mu$m and 0.0005 for $a = 1.2\,\mu$m. The Debye length was around 13 nm, much smaller than the pore radius and, therefore, the efficiency improved as the pore radius was decreased as expected.

Wang et al. (2006) employed a porous structure made of silica-based monoliths inside a 100 μm diameter capillary (0.5 mm^3 micropump). They conducted experiments with deionized water as working fluid at 6 kV and obtained: maximum pressure around 3×10^5 Pa, maximum flow-rate around $0.05\,\text{mm}^3/\text{s}$. The maximum energy efficiency was around 0.004 with $a = 2\,\mu\text{m}$ and Debye length was $\lambda_D = 0.1\,\mu\text{m}$. They also used acetonitrile as working fluid and they obtained energy efficiency of 0.022. The main difference for the efficiency improvement was due to the smaller conductivity of acetonitrile ($\sigma(\text{water}) \sim 10^{-4}$ S/m and $\sigma(\text{aceton.}) \sim 10^{-6}$ S/m) as expected from theory.

The electrochemical reactions at the electrodes required to maintain the electrical current can be problematic: electrolysis generates gas bubbles and the redox reactions will eventually change the pH. Traditionally, some of these problems are avoided by separating the electrodes from the EO pumping channels with an ion exchange system (Dasgupta and Liu, 1994). The electrodes are placed in reservoirs separated from the channels and the exchange of ions is allowed but not of fluid. The reservoirs are filled with buffers to reduce the changes in pH.

Joule heating can also be problematic. The estimated power dissipated by a capillary is $P_{\text{dis}} \sim \sigma E^2 \pi a^2 L$ and this should be equal to the heat flux through the lateral surface $2\pi a L \kappa \partial T/\partial r$ and the inlet and outlet surfaces $\pi a^2 \kappa \partial T/\partial z$. If we consider that the lateral surface of the cylinder is at constant temperature, the estimated temperature rise is $\Delta T \sim (\sigma V^2 a^2)/(2\kappa L^2)$. The temperature rise can be very high for conductivities greater or around 1 S/m. Temperature increments around 50 K were measured for EO flow in a capillary filled with electrolyte of conductivity 0.2 S/m with an applied electric field of 15 kV/m (Xuan et al., 2004).

EO micropumps usually require high voltage (traditionally in the kilovolt range) and this can be a drawback, especially for portable chips. Takamura et al. (2003) fabricated a low voltage EO pump that produces high pressure (around 25 kPa at 10 V) by using a cascade pump design, which is mainly a set of EO micropumps connected in series (one after the other) which are connected to the same applied dc voltage.

2.5 AC Electroosmotic micropump

In general, there are two ways of charge generation at a solid surface in contact with a liquid: (i) a chemical mechanism, when a difference in electrochemical potentials between phases charges the solid surface and (ii) an electrostatic mechanism, when a solid metal surface gets charged because it is subjected to a potential difference with respect to the bulk electrolyte. The Alternating Current Electroosmotic micropump, also known as Induced Charge Electroosmotic micropump, actuates upon the diffuse charge in the double layer that is induced between an electrolyte in contact with an electrode that is subjected to a potential (Ramos et al., 1999; Ajdari, 2000; Bazant and Squires, 2004). The electrical current coming from the bulk electrolyte charges the electrical double layer at the electrode/electrolyte interface. Meanwhile, the electric field component tangential to the interface produces a force on the induced charge in the diffuse layer that sets the liquid into motion (Green et al., 2000b; González et al., 2000; Green et al., 2002). Typically, unidirectional flow is obtained by arrays of asymmetric pairs of electrodes subjected to a single phase ac signal, known as ACEO pump, (Brown et al., 2001; Ramos et al., 2003; Studer et al., 2004) or by arrays of symmetric electrodes subjected to a travelling-wave signal, known as TWEO pump, (Cahill et al., 2004; Ramos et al., 2005).

Theoretical analysis. Let us analyse the flow generated by an array of electrodes subjected to a travelling-wave potential (see fig. 12). We consider that the number of electrodes in a wave-length is high so that the travelling-wave potential at the level of the electrodes, $z = 0$, is given by $V(x,t) = V_0 \cos(kx - \omega t)$, where ω is the frequency of the applied signal and k is the wave-number.

We are going to apply the linear approximation of the double layer (Debye-Hückel approximation). We also assume that the electrodes are perfectly polarizable, i.e. the applied potential is low enough so that Faradaic currents from the electrodes to the liquid are absent. Above the electrodes, the bulk electrolyte is electro-neutral characterized by its conductivity σ. The electric current follows Ohm's law $\mathbf{j} = \sigma \mathbf{E}$ and the electric potential holds Laplace's equation $\nabla^2 \phi = 0$. The frequency of the applied signal is much smaller than the reciprocal charge relaxation time $\omega \ll \sigma/\varepsilon$, so that the electrode/electrolyte double layer is in quasi-equilibrium.

The ohmic current charges the double-layer capacitance and this is expressed by the following boundary condition (using phasors)

$$\sigma \frac{\partial \phi}{\partial z} = i\omega C_{DL}(\phi - V_0 e^{-i(kx-\omega t)}) \tag{89}$$

where $V_0 e^{-i(kx-\omega t)}$ is the applied potential amplitude, C_{DL} is the double layer capacitance and ϕ is the potential just outside the double layer. The capacitance is given by the series combination of the compact layer capacitance C_s and the Debye layer capacitance C_d: $C_{DL}^{-1} = C_s^{-1} + C_d^{-1}$. The

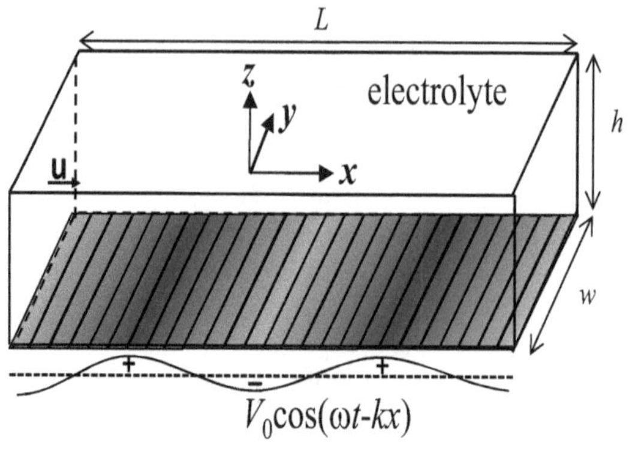

Figure 12. Scheme of TWEO pump.

Debye-Hückel theory gives the value of the diffuse double layer capacitance as $C_d = \varepsilon/\lambda_D$. In complex notation, the potential solution in the bulk is

$$\phi = \frac{i\Omega V_0}{1+i\Omega} e^{-i(kx-\omega t)-kz} \qquad \Omega = \frac{\omega C_{DL}}{k\sigma}. \tag{90}$$

The electroosmotic slip velocity for electrical double layers in quasi-equilibrium on perfectly polarizable electrodes is given by $u_s = -\varepsilon E_x \zeta/\eta$ (Levich, 1962), where ζ is the potential drop across the diffuse double layer and E_x the tangential electric field just outside the diffuse layer. Although in our problem both ζ and E_x are oscillating functions in time, there is a non-zero time-averaged slip velocity. The induced zeta potential can be related to the total double-layer potential drop by $\zeta = C_s(V-\phi)/(C_s+C_d)$. The time-averaged slip velocity is $\langle u_s \rangle = -\frac{1}{2}\varepsilon \mathrm{Re}[E_x \zeta^*]/\eta$ (Cahill et al., 2004; Ramos et al., 2005):

$$\langle u_s \rangle = \Lambda \frac{\varepsilon k V_0^2}{2\eta} \frac{\Omega}{1+\Omega^2} \qquad \text{with} \qquad \Lambda = \frac{C_s}{C_s + C_d}. \tag{91}$$

The TWEO slip velocity has a maximum as a function of frequency at $\Omega = 1$ given by

$$\langle u_s \rangle_{\max} = \Lambda \frac{\varepsilon k V_0^2}{4\eta}. \tag{92}$$

The frequency of maximum velocity is $\omega_{\max} = k\sigma/C_{DL}$, which is the characteristic frequency of an RC circuit, where $R = 1/k\sigma A$ is the characteristic bulk resistance and $C = AC_{DL}$ the double layer capacitance (A, typical surface area).

Taking into account the nonslip velocity condition at $z = h$, the maximum flow-rate generated is

$$Q_{\max} = \Lambda \frac{\varepsilon k V_0^2}{8\eta} wh. \tag{93}$$

The pressure required to stop this flow is

$$\Delta p_{\max} = \frac{12\eta Q_{\max} L}{wh^3} = \Lambda \frac{3\varepsilon k V_0^2 L}{2h^2}. \tag{94}$$

Since in our theoretical model we have neglected charge advection, the mechanical power $\int \rho \mathbf{E} \cdot \mathbf{u} d\tau$ has also been neglected. Therefore, the input power is mainly equal to the consumed power by Joule heating. For $\Omega = 1$ we have

$$P_{\text{in}} \approx \frac{1}{2} \int \sigma |E|^2 \, d\tau = \frac{\sigma}{2} \int_0^\infty k^2 V_0^2 e^{-2kz} \, dz wL = \frac{\sigma k V_0^2 wL}{4}. \tag{95}$$

An estimate of the mechanical power is $\rho E u \lambda_D L w$ and this is much smaller than Joule heating if $(\Lambda u \varepsilon)/(\sigma \lambda_D) \ll 1$. In this limit, the maximum energy efficiency is

$$\text{eff} = \frac{1}{4} \frac{Q_{\max} \Delta p_{\max}}{P_{\text{in}}} = \frac{3\Lambda^2 \varepsilon^2 V_0^2 k}{16\sigma\eta h} = \frac{3\Lambda \, u_{\max} \varepsilon}{2\sigma h}. \tag{96}$$

The energy efficiency should be much smaller than one for negligible charge advection, since it is the product of two small numbers $(\Lambda u \varepsilon)/(\sigma \lambda_D)$ and λ_D/h. In order to increase the generated pressure, or the energy efficiency, we can reduce h, increase kL, or increase the applied voltage. Increasing kL means increasing the total number of electrodes in the array. Decreasing h has the drawback that reduces the flow-rate. If we increase the applied voltage, the threshold for occurrence of Faradaic reactions can be reached. At this point, the double layer starts to leak charge which reduces the expected velocity. Numerical computations show saturation of induced velocity with voltage due to Faradaic reactions (Olesen et al., 2006). Even without the appearance of electrochemical reactions, the induced ζ potential saturates due to the nonlinearity of the double layer capacitance and steric limitations due to the finite size of ions (Bazant et al., 2009).

For both the ACEO pump or the TWEO pump, the linear theory (Ramos et al., 2003, 2005) predicts that the maximum flow-rate is of the form $Q_{\max} = \Gamma wh \Lambda \varepsilon V_0^2/\eta\ell$, where ℓ is the minimum feature size of the electrode structure and Γ is a geometrical nondimensional factor. This factor can be optimized by using 3D electrodes (Bazant and Ben, 2006; García-Sánchez and Ramos, 2008).

Some experimental features. Yang et al. (2009) performed experiments on TWEO and ACEO pumping using the same microelectrode array by changing the electrical connections of gold microelectrodes. The ACEO velocity was around five times smaller than the TWEO velocity for the same applied voltage. The TWEO slip velocity was around 480 μm/s for a TW signal of amplitude 1.5 volts peak to peak and frequency 1 kHz. The working liquid was a saline solution of 100μM KCl in water.

Huang et al. (2010) fabricated an ACEO micropump made of (3D) asymmetric stepped electrodes. The array consisted of 6012 pairs of these electrodes placed in a long serpentine microchannel. Using an ac signal of amplitude 1 Volt rms at a frequency of 1 kHz, the pump was capable of generating a maximum pressure of 1.4 kPa (much higher than previous ACEO pumps) and an effective slip velocity over the electrodes of 1.3 mm/s (which transforms to a maximum flow-rate of 1.7×10^{-3} mm^3/s). The working fluid was distilled water. Under these conditions, the power consumption was 4.8 mW and, therefore, the maximum energy efficiency that is estimated from these values is eff $\sim 10^{-7}$. This is rather small compared to energy efficiencies for distilled water of EO micropumps and to the theoretical value using the effective velocity in equation (96), the latter suggests that there is room for experimental improvement. Since the ACEO pump operates at low voltage and power, it can be integrated in miniaturized portable devices.

As we said previously, we can raise the voltage to increase the pumping but there is the limitation of Faradaic reactions generation. Experimentally, it is observed that the fluid velocity saturates and can even changed direction with voltage (flow reversal) (Studer et al., 2004; García-Sánchez et al., 2006). For the TWEO micropump, the appearance of flow reversal seems to be linked to the threshold voltage for Faradaic reactions (García-Sánchez et al., 2008). For even higher voltages, electrochemically generated bubbles appear. This problem can be reduced by using a dielectric coating on the electrodes, although this reduces the generated slip velocity for a given applied voltage.

The working fluids are typically low ionic strength solutions (< 100 mM), well below physiological salt concentrations. Since the pump should operate at voltages below the ionization potentials, the working liquid is not expected to suffer changes in its electrical properties, which is an advantage.

3 Comparisons

3.1 Liquid electrical conductivity

The ionic strength or electrical conductivity of the working liquid is an important parameter in order to choose a certain EHD micropump. For instance, very insulating liquids ($\sigma \ll 10^{-10}$ S/m) are, very difficult if not impossible, to manipulate by the EHD induction mechanism because this requires a certain level of conductivity. At the other conductivity limit, liquid metals such as mercury can not be actuated by applying electric forces in the bulk. In continuous microfluidics, liquid metals can be actuated by magnetohydrodynamic micropumps (Jang and Lee, 2000). In digital microfluidics, they can be actuated by applying electric forces at their surfaces (Lee et al., 2002), which requires the presence of a dielectric where the electrical energy is stored. Let us analyse the ranges of conductivity that each EHD micropump can actuate.

The ion-drag micropump can actuate very insulating liquids. We do not expect that this mechanism is appropriate for liquids with residual conductivity such that the charge relaxation time (ε/σ) is much smaller than the typical ion transient time $d/\mu E$. If the charge relaxation time is very small compared to the time that the ion takes to travel from the emitter to the collector, the bulk region where there is both free charge and electric force becomes very small. Using typical values for microsystems ($\mu \sim 10^{-7}\,\mathrm{m}^2/(\mathrm{sV})$, $E \sim 10^5$ V/m, $d \sim 10^{-4}$ m, $\varepsilon \sim 7 \times 10^{-10}$ F/m), the estimation says that the conductivity of working fluid should not be much greater than $\sigma = 10^{-7}$ S/m.

The conduction micropump actuates liquids with a certain degree of conductivity, which should not be so small that the saturation regime is reached very soon. If the thickness of the heterocharge layer λ is much greater than the distance between electrodes L, the recombination of ions is negligible in the bulk and the saturation current regime is reached. As we said when we analysed this pump, the generated pressure saturates at this regime. The condition $\lambda \gg L$ says that we do not expect a proper actuation of the conduction pump if conductivity $\sigma \ll \mu\varepsilon V/L^2$. There should be an upper limit of conductivity because for high conductivity the Debye length can be of the same order than the heterocharge layer thickness. In this limit, the diffusion current is not negligible and the approximations that we used to analyse the conduction pump are not applicable. Therefore, the conduction pump mechanism requires $\lambda \gg \lambda_D$, which leads to $\sigma \ll \mu\varepsilon e E^2/k_B T$. This limit does not imply that for a given experiment there will not be EHD pumping beyond a certain conductivity, but it implies that the analysis of the possible EHD actuation should include ion diffusion. Using a value of

$L = 100\,\mu$m, we obtain a range of conductivities between 10^{-10} and 4×10^{-7} S/m for an applied voltage of 10 V, and a range of conductivities between 10^{-9} and 4×10^{-5} S/m for an applied voltage of 100 V.

The induction pump requires the existence of gradients of conductivity to operate. We have analysed the mechanism of induction under the assumption that we only need to consider the conductivity σ of the liquid rather than the full model that considers the conduction due to the different ionic species. This is, we have used in the analysis the Taylor-Melcher leaky dielectric model (Saville, 1997). In order to employ only the classical Ohm's law in the charge conservation equation, the liquid should be quasi-electroneutral. For binary electrolytes, this means that the difference between positive and negative ion densities must be much smaller than the total number density of ions, $|n_+ - n_-| \ll n_+ + n_-$. The estimate of the charge density from Gauss's law $\rho = e(n_+ - n_-) \sim \varepsilon E/L$ leads to

$$\frac{n_+ - n_-}{n_+ + n_-} \sim \frac{\varepsilon E}{eL(n_+ + n_-)} \sim \frac{\mu E \varepsilon}{\sigma L} \ll 1. \tag{97}$$

Therefore, the electrical conductivity has a lower bound given by $\mu\varepsilon E/L$. For $E \sim 10^5$ V/m, $L \sim 10^{-4}$ m, the conductivity should be much greater than 10^{-9} S/m. An upper bound for the conductivity comes from the problems originated by Joule heating. The increment of temperature is estimated from temperature equation to be of the order of $\sigma V_{rms}^2/\kappa$. We expect that boiling temperature is reached when this number is around 100 K. For the typical voltages required to generate significant flow, around 10 V, the condition says that boiling temperature can be reached for conductivities around 1 S/m.

The electroosmotic pumping requires the presence of a double layer at a solid wall. There is no electroosmosis without ions in solution and an interesting question is when the ionic concentration is so small that electrokinetic flow is negligible. If the liquid ionic strength is very small, the Debye length becomes very large and the typical electric field inside the double layer ($\sim k_BT/e\lambda_D$) can be of the same order or smaller than the applied field. In this case, the distribution of ions in the double layer would be very distorted by the applied electric field. Especially in electroosmosis through a porous structure when the electric field lines are bent following the tortuous paths, we expect that induced charge near the walls are of the same order or greater than the intrinsic charge. Therefore, we do not expect that electroosmosis can work properly at very low conductivity. Tentatively, we propose that a condition for proper actuation is $k_BT/e\lambda_D \gg E$, which leads to $\sigma \gg \mu e\varepsilon E^2/k_BT$. This parameter combination has appeared before as an upper bound for the conductivity of working fluid in the conduction pump. If we take $E \sim 10^4$ V/m and $\mu \sim 10^{-7}$ m^2/Vs, the condition implies that σ should be much greater than 10^{-7} S/m for water and greater than 10^{-8} S/m for a liquid with $\varepsilon = 2\varepsilon_0$. Barz et al. (2010) studied the electrokinetic flow for a nonpolar liquid (dodecane) varying the ionic strength (adding OLOA) and, interestingly, they found some limitations at the lowest ionic strengths they investigated. The electric field inside the double layer of the OLOA-dodecane solution with lowest ionic strength was of the order of $\zeta/\lambda_D = 6 \times 10^4$ V/m, somewhat greater than the applied electric field $\sim 10^4$ V/m, which may indicate that they are close to the limit we have proposed for proper actuation. With respect to limitations of electroosmotic pumping at high conductivities, we can see in the literature that classical electrokinetic methods present difficulties in obtaining ζ potentials for ionic concentrations higher than 0.1 M ($\sigma \sim 1$ S/m). The ζ potentials are close to zero at high ionic strengths (Kosmulski and Rosenholm, 2004). When the ionic strength is high, the Debye length becomes very small. A simple model of fixed surface charge q_s leads to a reduction of the ζ-potential with ionic concentration of the form $\zeta = q_s\lambda_D/\varepsilon \sim 1/\sqrt{n_\infty}$. Therefore, to pump a significant flow of an electrolyte with high conductivity, we need to increase the applied voltage, which leads to an increase in Joule heating because of two factors: conductivity and voltage. It seems that there are fundamental and practical limitations for the electroosmotic pumping of liquids with conductivities much greater than 1 S/m.

The ac electroosmotic pumping of electrolytes with conductivities greater than 0.1 S/m has not been reported. This is an upper bound for conductivity much more restrictive than for the dc electroosmotic pump. It seems that there are fundamental limitations around this conductivity value related to ionic Steric effects at high electrolyte concentrations (Bazant et al., 2009) as well as problems with the generation of Faradaic reactions (García-Sánchez et al., 2008). According to the Gouy-Chapman-Stern model, the potential across the metal/electrolyte interface is shared between the Stern layer and the diffuse layer. For a fixed imposed double-layer potential, the increase of conductivity decreases the diffuse-layer potential and, therefore, the slip electroosmotic velocity decreases. In order to generate a given flow-rate while the electrolyte conductivity is increased, we need to raise the applied voltage, which can easily generate Faradaic reactions. With respect to a lower bound of conductivity in ACEO pumping, we can expect a similar condition than for dc electroosmotic pumping, i.e. the electric field inside the double layer should be much greater than the electric field in the bulk. This condition can be written now as $\sigma \gg (\mu e \varepsilon V^2)/(L^2 k_B T)$. For $V \sim 1$ V, $L \sim 2 \times 10^{-5}$m, $\mu \sim 10^{-7}$ m^2/Vs, the conductivity should be greater than $\sigma \sim 10^{-7}$ S/m for water and greater than $\sigma \sim 4 \times 10^{-9}$ S/m for a nonpolar liquid with $\varepsilon = 2\epsilon_0$. For a given experimental setup, EHD pumping below this conductivity may exist although the analysis of the possible electrokinetic flow can not be performed using ACEO theory and the full system of equations is required.

3.2 Mechanical characteristics

For typical values in microsystems, in order to have water flowing with average velocity in the range between 100 μm/s and 10 mm/s inside microchannels with cross-section $100\,\mu\text{m} \times 100\,\mu\text{m}$ (Q between 0.001 to 0.1 mm^3/s) and lengths between 10 and 100 mm, a pressure difference in the range 4 to 4000 Pa is required. Figures 13 and 14 show selected experimental values given in the literature for maximum pressure and flow-rate, respectively. We can see that some EHD micropumps are able to generate pressures and flow-rates inside the ranges that we have proposed.

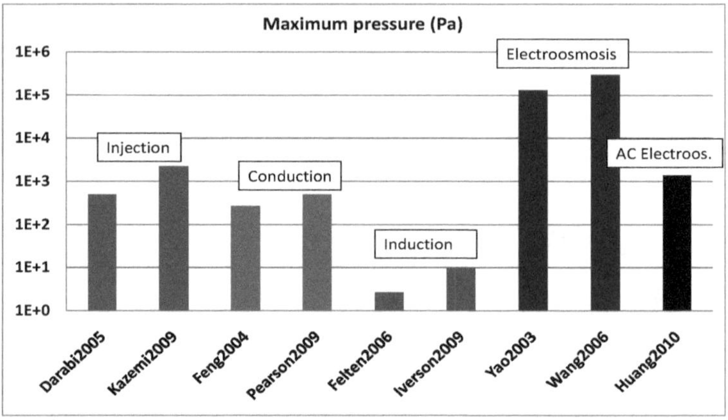

Figure 13. Selected maximum pressure values from different EHD pumping experiments. Some values were obtained from the measured maximum flow-rate multiplied by estimated hydrodynamic resistance.

EO pumps can provide these pressures and flow-rates. They are typically made of porous structures. We recall here that EO pumping in a pipe of small diameter ($\sim 1\mu$m) can generate high pressure, although the generated flow-rate will be small. The porous structures generate high pressure while not decreasing the flow-rate because they can be visualized as a bundle of tortuous narrow capillaries in parallel.

The EHD injection and conduction pumps can generate enough pressure and flow-rate for typical applications, with higher pressure values for the

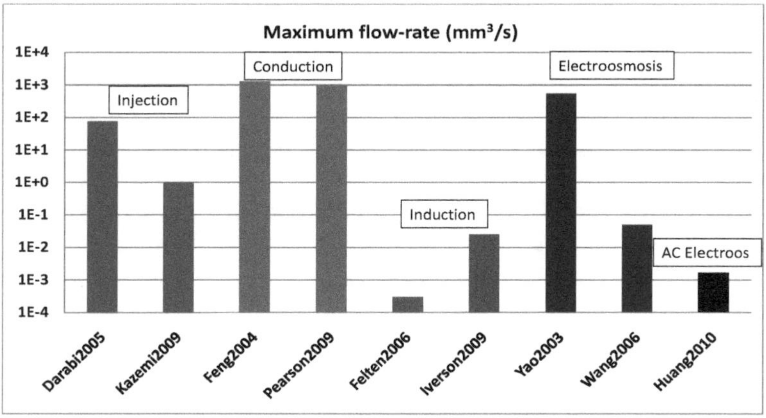

Figure 14. Selected maximum flow-rate values from different EHD pumping experiments. Some values were obtained from the measured maximum pressure divided by estimated hydrodynamic resistance.

injection pump. While for the injection pump maximum pressure is limited by electric breakdown, for the conduction pump maximum pressure is limited by the maximum electric field before ion injection takes place. Thus the conduction pump operates at lower electric field amplitudes generating lower pressures. In the figure the flow-rates for the conduction pumps are greater than for the injection pumps, but this is only due to the fact that the conduction pumps have greater dimensions.

The ACEO pump of Huang et al. (2010) generates high pressure and moderate flow-rate. They managed to increase maximum pressure by two orders of magnitude compared to prior ACEO pumps. This was possible because they used 3D micro-electrodes covering the bottom of a long serpentine channel with small height. To increase the flow-rate, the width of the channel with electrodes could be increased or several devices in parallel could be employed.

The experiments of induction pumping in microsystems have not provided high pressures. Theoretically, it should be possible to increase pressure by reducing the height and increasing the length of the operating part of the device as was done by Huang et al. (2010) for the ACEO micropump.

3.3 Applications

The pumping of refrigerant, insulating liquids has been devised for its use in cryogenic cooling microsystems (Darabi and Wang, 2005; Foroughi et al., 2005; Pearson and Seyed-Yagoobi, 2009). In order to improve the performance of electronics and sensors, cryogenic cooling has become a widely adopted technique because it can increase signal-to-noise ratio of sensors. Therefore, injection and conduction pumps are suitable for the pumping of refrigerant liquids. According to Darabi et al. (2002), other promising application of these EHD micropumps is its use in fuel injection loops.

The induction micropump as well as the dc electroosmotic micropump are suitable for pumping of biofluids for the Lab-on-Chip, since the working fluid conductivity can be as high as 1 S/m. The maximum pressure obtained experimentally by the induction micropump was not high, which limits its applications. This is not the case for the electroosmotic micropump, which can generate high pressure. Its potential applications include the replacement of high-pressure pumps in micro-total-analysis-systems: for drug delivery, sample analysis, separation, and mixing processes. Closed-loop electroosmotic microchannels have also been devised as cooling systems for microelectronics (Jiang et al., 2002).

The ACEO micropump can generate sufficient pressure for certain applications in biomedical or chemical analysis. An important advantage of the ACEO pump compared to other EHD micropumps is that the voltage required for operation is quite low (around 1 volt). This is an important feature for portable devices. However, the working fluid should be a dilute electrolyte (conductivity $\sigma \leq 0.1$ S/m). Huang et al. (2010) used water as a working fluid and, thanks to the high pressure generated, indirect manipulation of biological fluids was demonstrated.

Although generating high pressure can be difficult for certain EHD micropumps, the generation of localized flows in a microdevice does not require high pressure and have many potential applications such as stirring and mixing of analytes (Sigurdson et al., 2005) or diversion of cells (Dittrich and Schwille, 2003). Induction or ACEO pumping with microelectrodes inside the microchannel can be very suitable for the generation of these microflows since they operate with ac voltages, avoiding electrochemical reactions that would deteriorate the electrodes and change the physical properties of the actuated liquid.

4 Concluding remarks

In this chapter we have described different ways of electrical actuation of liquids in the microscale. The main equations of Electrohydrodynamics in microsystems were presented. Basic concepts of electrical conduction in fluids and fluid-mechanical aspects of micropumps were also provided. Then, these concepts were applied to analyse five different strategies of inducing fluid flow in microsystems using electric fields. The intention has been to study the physical mechanisms as well as the mathematical descriptions of electrohydrodynamic pumping in microsystems: from strategies that employ forces in the liquid bulk to those that employ forces in the electrical double layer. It should be noted that there exists other EHD mechanisms to induce fluid flow in microsystems which have not been mentioned in this chapter, such as: electroosmosis of second kind (Mishchuk et al., 2009), EHD actuation on ionic concentration gradients (Morgan et al., 2007), or EHD induced flow by Onsager effect (Ryu et al., 2010).

Each EHD pump has certain characteristics that make it more suitable for certain applications. For example, the electrical properties of the working liquid, especially the electrical conductivity or ionic strength, determine the EHD mechanism that is more suitable for actuation. In the previous section we have analysed the range of conductivity that each micropump can actuate properly. The injection and conduction micropumps seem to be more appropriate for insulating and semi-insulating liquids, while the dc and ac electroosmotic micropumps seem to be more suitable for electrolyte liquids. The induction micropump can be used for liquids in a wide range of ionic concentrations, as far as the liquid have some conductivity.

Other considerations that are important for the choice of an EHD pump are: the voltage amplitude required to operate the pump (a portable device would require low voltage and power); the generation of gases by electrochemical reactions and of heat by the Joule effect (which can deteriorate the microdevices and compromise reproducibility), or the required mechanical properties of the micropump such as its maximum pressure and maximum flow-rate (the electroosmotic micropump seems to meet best the requirement of very high pressure for drug delivery or sample analysis in micro-total-analysis-systems).

Bibliography

S.H. Ahn and Y.K. Kim. Fabrication and experiment of a planar micro ion drag pump. *Sensors and Actuators A: Physical*, 70:1–5, 1998.

A. Ajdari. Pumping liquids using asymmetric electrode arrays. *Phys. Rev. E*, 61:R45–R48, 2000.

P. Atten and J. Seyed-Yagoobi. Electrohydrodynamically induced dielectric liquid flow through pure conduction in point/plane geometry. *IEEE Trans. Dielectr. Electr. Insul.*, 10:27–36, 2003.

A.J. Bard and L.R. Faulkner. *Electrochemical Methods: Fundamentals and Applications, 2nd Edition.* John Wiley and Sons, 2001.

D.P.J. Barz, M.J. Vogel, and P.H. Steen. Determination of the Zeta Potential of Porous Substrates by Droplet Deflection: II. Generation of Electrokinetic Flow in a Nonpolar Liquid. *Langmuir*, 26:3126–3133, 2010.

M.Z. Bazant and Y. Ben. Theoretical prediction of fast 3D AC electro-osmotic pumps. *Lab Chip*, 6:1455–1461, 2006.

M.Z. Bazant and T.M. Squires. Induced-charge electrokinetic phenomena: theory and microfluidic applications. *Phys. Rev. Lett.*, 92:066101, 2004.

M.Z. Bazant, M.S. Kilic, B.D. Storey, and A. Ajdari. Towards an understanding of induced-charge electrokinetics at large applied voltages in concentrated solutions. *Advances in Colloid and Interface Science*, 152: 48–88, 2009.

A.B.D. Brown, C.G. Smith, and A.R. Rennie. Pumping of water with ac electric fields applied to asymmetric pairs of microelectrodes. *Phys. Rev. E*, 63:016305, 2001.

B.P. Cahill, L.J. Heyderman, J. Gobrecht, and A. Stemmer. Electro-osmotic streaming on application of traveling-wave electric fields. *Phys. Rev. E*, 70:036305, 2004.

A. Castellanos. Basic concepts and equations in electrohydrodynamics. In A. Castellanos, editor, *Electrohydrodynamics*. Springer-Verlag, 1998.

J.M. Crowley, G.S. Wright, and J.C. Chato. Selecting a working fluid to increase the efficiency and flow rate of an EHD pump. *IEEE Trans. Ind. Appl.*, 26:42–49, 1990.

J. Darabi and C. Rhodes. CFD modeling of an ion-drag micropump. *Sensors and Actuators A*, 127:94–103, 2006.

J. Darabi and H. Wang. Development of an electrohydrodynamic injection micropump and its potential application in pumping fluids in cryogenic cooling systems. *J. Microelectromech. Syst.*, 14:747–755, 2005.

J. Darabi, M. Rada, M. Ohadi, and J. Lawler. Design, fabrication, and testing of an electrohydrodynamic ion-drag micropump. *J. Microelectromech. Syst.*, 11:684–690, 2002.

P.K. Dasgupta and S. Liu. Electroosmosis: a reliable fluid propulsion system for flow injection analysis. *Anal. Chem.*, 66:1792–1798, 1994.

P.S. Dittrich and P. Schwille. An Integrated Microfluidic System for Reaction, High-Sensitivity Detection, and Sorting of Fluorescent Cells and Particles. *Anal. Chem.*, 75:5767–5774, 2003.

M. Felten, P. Geggier, M. Jäger, and C. Duschl. Controlling electrohydrodynamic pumping in microchannels through defined temperature fields. *Phys. Fluids*, 18:051707, 2006.

Y. Feng and J. Seyed-Yagoobi. Understanding of electrohydrodynamic conduction pumping phenomenon. *Phys. Fluids*, 16:2432–2441, 2004.

P. Foroughi, V. Benetis, M. Ohadi, Y. Zhao, and J. Lawler. Design, testing and optimization of a micropump for cryogenic spot cooling applications. In *Proceedings of Semiconductor Thermal Measurement and Management Symposium*, pages 335–340, 2005.

G. Fuhr, R. Hagedorn, T. Müller, W. Benecke, and B. Wagner. Microfabricated electrohydrodynamic (EHD) pumps for liquids of higher conductivity. *J. Microelectromech. Syst.*, 1:27–36, 1992.

G. Fuhr, T. Schnelle, and B. Wagner. Travelling wave-driven microfabricated electrohydrodynamic pumps for liquids. *J. Micromech. Microeng.*, 4:217–226, 1994.

P. García-Sánchez and A. Ramos. The effect of electrode height on the performance of travelling-wave electroosmotic micropumps. *Microfluidics Nanofluidics*, 5:307–312, 2008.

P. García-Sánchez, A. Ramos, N.G. Green, and H. Morgan. Experiments on AC electrokinetic pumping of liquids using arrays of microelectrodes. *IEEE Trans. Dielectr. Electr. Insul.*, 13:670–677, 2006.

P. García-Sánchez, A. Ramos, N.G. Green, and H. Morgan. Traveling-Wave Electrokinetic Micropumps: Velocity, Electrical Current, and Impedance Measurements. *Langmuir*, 24:9361–9369, 2008.

A. González, A. Ramos, N.G. Green, A. Castellanos, and H. Morgan. Fluid flow induced by non-uniform AC electric fields in electrolytes on microelectrodes II: A linear double-layer analysis. *Phys. Rev. E*, 61:4019–4028, 2000.

A. González, A. Ramos, A. Castellanos, N.G. Green, and H. Morgan. Electrothermal flows generated by alternating and rotating electric fields in microsystems. *J. Fluid Mech.*, 564:415–433, 2006.

N.G. Green, A. Ramos, A. González, A. Castellanos, and H. Morgan. Electric field induced fluid flow on microelectrodes: the effect of illumination. *J. Phys. D: Appl. Phys.*, 33:L13–L17, 2000a.

N.G. Green, A. Ramos, A. González, H. Morgan, and A. Castellanos. Fluid flow induced by non-uniform AC electric fields in electrolytes on microelectrodes I: Experimental measurements. *Phys. Rev. E*, 61:4011–4018, 2000b.

N.G. Green, A. Ramos, A. González, H. Morgan, and A. Castellanos. Fluid flow induced by non-uniform AC electric fields in electrolytes on microelectrodes III: Observation of streamlines and numerical simulations. *Phys. Rev. E*, 66:026305, 2002.

H.A. Haus and J.R. Melcher. *Electromagnetic Fields and Energy.* Prentice Hall, 1989.

M. Holtappels, M. Stubbe, and J. Gimsa. ac-field-induced fluid pumping in microsystems with asymmetric temperature gradients. *Phys. Rev. E*, 79:026309, 2009.

C.C. Huang, M.Z. Bazant, and Todd Thorsen. Ultrafast high-pressure AC electro-osmotic pumps for portable biomedical microfluidics. *Lab Chip*, 10:80–85, 2010.

R.J. Hunter. *Zeta Potential in Colloid Science.* Academic Press, 1981.

B.D. Iverson and S.V. Garimella. Experimental characterization of induction electrohydrodynamics for integrated microchannel pumping. *J. Micromech. Microeng.*, 19:055015, 2009.

S.C. Jacobson, R. Hergenroder, L.B. Koutny, and J.M. Ramsey. Open-channel electrochromatography on a microchip. *Anal. Chem.*, 66:2369–2373, 1994.

J. Jang and S.S. Lee. Theoretical and experimental study of mhd (magnetohydrodynamic) micropump. *Sensors Actuators*, 80:84–89, 2000.

S. Jeong and J. Seyed-Yagoobi. Experimental study of electrohydrodynamic pumping through conduction phenomenon. *J. Electrostatics*, 56:123–133, 2002.

S.I. Jeong and J. Seyed-Yagoobi. Innovative electrode designs for electrohydrodynamic conduction pumping. *IEEE Trans. Ind. Appl.*, 40:900–904, 2004.

L.N. Jiang, J. Mikkelsen, J.M. Koo, D. Huber, S.H. Yao, L. Zhang, P. Zhou, J.G. Maveety, R. Prasher, J.G. Santiago, T.W. Kenny, and K.E. Goodson. Closed-loop electroosmotic microchannel cooling system for VLSI circuits. *IEEE Transactions on Components and Packaging Technologies*, 25:347–355, 2002.

P.Z. Kazemi, P.R. Selvaganapathy, and C.Y. Ching. Electrohydrodynamic micropumps with asymmetric electrode geometries for microscale electronics cooling. *IEEE Trans. Dielectr. Electr. Insul.*, 16:483–488, 2009.

M. Kosmulski and J.B. Rosenholm. High ionic strength electrokinetics. *Advances in Colloid and Interface Science*, 112:93–107, 2004.

P. Langevin. Recombinaison et mobilites des ions dans les gaz. *Annales de Chimie et de Physique*, 28:433, 1903.

D.J. Laser and J.G. Santiago. A review of micropumps. *J. Micromech. Microeng.*, 14:R35–R36, 2004.

J. Lee, H. Moon, J. Fowler, T. Schoellhammer, and C.J. Kim. Electrowetting and electrowetting-on-dielectric for microscale liquid handling. *Sensors Actuators A*, 95:259–268, 2002.

V.G. Levich. *Physicochemical Hydrodynamics.* Prentice-Hall, 1962.

A. Manz, C.S. Effenhauser, N. Burggraf, D.J. Harrison, K. Seiler, and K. Fluri. Electroosmotic pumping and electrophoretic separations for miniaturized chemical analysis systems. *J. Micromech. Microeng.*, 4: 257–265, 1994.

J.R. Melcher. Traveling-wave induced electroconvection. *Phys. Fluids*, 9: 1548–1555, 1966.

J.R. Melcher and M.S. Firebaugh. Travelling-wave bulk electroconvection induced across a temperature gradient. *Phys. Fluids*, 10:1178–1185, 1967.

J.R. Melcher and G.I. Taylor. Electrohydrodynamics: a review of the role of interfacial shear stresses. *Ann. Rev. Fluid Mech.*, 1:111–146, 1969.

N.A. Mishchuk, T. Heldal, T. Volden, J. Auerswald, and H. Knapp. Micropump based on electroosmosis of the second kind. *Electrophoresis*, 30:3499–3506, 2009.

H. Morgan, N.G. Green, A. Ramos, and P. Garcia-Sanchez. Control of two-phase flow in a microfluidic system using ac electric fields. *Appl. Phys. Lett.*, 91:254107, 2007.

T. Müller, W.M. Arnold, T. Schnelle, R. Hagedorn, G. Fuhr, and U. Zimmermann. A traveling-wave micropump for aqueous solutions: Comparison of 1 g and μg results. *Electrophoresis*, 14:764–772, 1993.

L.H. Olesen, H. Bruus, and A. Ajdari. AC electrokinetic micropumps: The effect of geometrical confinement, Faradaic current injection, and nonlinear surface capacitance. *Phys. Rev. E*, 73:056313, 2006.

M.R. Pearson and J. Seyed-Yagoobi. Experimental Study of EHD Conduction Pumping at the Micro-scale. In *2009 Electrostatics Joint Conference*, 2009.

W.F. Pickard. Ion Drag Pumping. I. Theory. *J. Appl. Phys.*, 34:246—250, 1963a.

W.F. Pickard. Ion Drag Pumping. II. Experiment. *J. Appl. Phys.*, 34: 251–258, 1963b.

F. Pontiga. *Sobre la Estabilidad de una Capa de Líquido Sometida a un Campo Eléctrico y un Gradiente Térmico.* PhD thesis, Universidad de Sevilla, 1992.

V. Pretorius, B.J. Hopkins, and J.D. Schieke. Electro-osmosis: A new concept for high-speed liquid chromatography. *J. Chromatography*, 99:23–30, 1974.

A. Ramos, H. Morgan, N.G. Green, and A. Castellanos. Ac electrokinetics: a review of forces in microelectrode structures. *J. Phys. D: Appl. Phys.*, 31:2338–2353, 1998.

A. Ramos, H. Morgan, N.G. Green, and A. Castellanos. AC electric-field-induced fluid flow in microelectrodes. *J. Colloid and Interface Science*, 217:420–422, 1999.

A. Ramos, A. González, A. Castellanos, N.G. Green, and H. Morgan. Pumping of liquids with ac voltages applied to asymmetric pairs of microelectrodes. *Phys. Rev. E*, 67:056302, 2003.

A. Ramos, H. Morgan, N.G. Green, A. González, and A. Castellanos. Pumping of liquids with traveling-wave electroosmosis. *J. Appl. Phys.*, 97: 084906, 2005.

C.L. Rice and R. Whitehead. Electrokinetic flow in a narrow cylindrical capillary. *J. Phys. Chem.*, 69:4017–4024, 1965.

A. Richter and H. Sandmaier. An electrohydrodynamic micropump. In *IEEE Proceedings of Micro Electro Mechanical Systems, 1990*, pages 99–104, 1990.

A. Richter, A. Plettner, K.A. Hofmann, and H. Sandmaier. Electrohydrodynamic pumping and flow measurement. In *IEEE Proceedings of Micro Electro Mechanical Systems, 1991*, pages 271–276, 1991.

J.C. Ryu, H.J. Park, J.K. Park, and K.H. Kang. New Electrohydrodynamic Flow Caused by the Onsager Effect. *Phys. Rev. Lett.*, 104:104502, 2010.

D.A. Saville. Electrohydrodynamics: The Taylor-Melcher leaky dielectric model. *Annu. Rev. Fluid Mech.*, 29:27–64, 1997.

W.F. Schmidt. Conduction mechanisms in liquids. In R. Bartnikas, editor, *Electrical Insulating Liquids. Engineering Dielectrics, Vol III.* ASTM International, 1994.

J. Seyed-Yagoobi. Electrohydrodynamic pumping of dielectric liquids. *J. Electrostatics*, 63:861–869, 2005.

M. Sigurdson, D. Wang, and C.D. Meinhart. Electrothermal stirring for heterogeneous immunoassays. *Lab Chip*, 5:1366–1373, 2005.

H.A. Stone, A.D. Stroock, and A. Ajdari. Engineering flows in small devices: Microfluidics toward a Lab-on-a-Chip. *Annu. Rev. Fluid Mech.*, 36:381–411, 2004.

J.A. Stratton. *Electromagnetic Theory.* McGraw Hill, 1941.

M. Stubbe, M. Holtappels, and J. Gimsa. A new working principle for ac electro-hydrodynamic on-chip micro-pumps. *J. Phys. D: Appl. Phys.*, 40:6850–6856, 2007.

V. Studer, A. Pepin, Y. Chen, and A. Ajdari. An integrated AC electrokinetic pump in a microfluidic loop for fast and tunable flow control. *Analyst*, 129:944–949, 2004.

O.M. Stuetzer. Instability of Certain Electrohydrodynamic Systems. *Phys. Fluids*, 2:642–648, 1959.

O.M. Stuetzer. Ion Drag Pumps. *J. Appl. Phys.*, 31:136–146, 1960.

Y. Takamura, H. Onoda, H. Inokuchi, S. Adachi, A. Oki, and A. Horiike. Low-voltage electroosmosis pump for stand-alone microfluidics devices. *Electrophoresis*, 24:185–192, 2003.

J.J Thomson and G.P. Thomson. *Conduction of Electricity Through Gases, 3rd ed.* University Press, 1928.

P. Wang, Z.L. Chen, and H.C. Chang. A new electro-osmotic pump based on silica monoliths. *Sensors Actuators B*, 113:500–509, 2006.

J. Wu, M. Lian, and K. Yang. Micropumping of biofluids by alternating current electrothermal effects. *Appl. Phys. Lett.*, 90:234103, 2007.

X. Xuan, B. Xu, D. Sinton, and D. Li. Electroosmotic flow with Joule heating effects. *Lab Chip*, 4:230–236, 2004.

H. Yang, H. Jiang, A. Ramos, and P. García Sánchez. AC electrokinetic pumping on symmetric electrode arrays. *Microfluid Nanofluid*, 7:767–772, 2009.

S.H. Yao and J.G. Santiago. Porous glass electroosmotic pumps: theory. *J. Colloid Interface Sci.*, 268:133–142, 2003.

S.H. Yao, D.E. Hertzog, S.L. Zeng, J.C. Mikkelsen, and J.G. Santiago. Porous glass electroosmotic pumps: design and experiments. *J. Colloid Interface Sci.*, 268:143–153, 2003.

A.I. Zhakin. Conduction phenomena in dielectric liquids. In A. Castellanos, editor, *Electrohydrodynamics*. Springer-Verlag, 1998.

Electrohydrodynamic Stability

Chuan-Hua Chen*
Department of Mechanical Engineering and Materials Science,
Duke University, Durham, NC, USA

Abstract Stability of electrohydrodynamic flows is essential to a variety of applications ranging from electrokinetic assays to electrospray ionization. In this series of lecture notes, a few basic concepts of electrohydrodynamic stability are illustrated using two model problems, electrokinetic mixing flow and electrohydrodynamic cone-jet, respectively wall-bounded and free surface flow. After a review of the governing equations, spatiotemporal analysis of the two example problems is presented using linearized bulk- or surface-coupled models. The operating regimes for these flows are discussed within the framework of electrohydrodynamic stability.

1 Introduction

Electrohydrodynamic transport phenomena are fundamental to a variety of engineering applications such as electrokinetic assays, electrospray ionization, electro-coalescence and mixing, electrostatic printing and spinning. Although unstable flow is desired in certain applications (e.g. in mixing), a stable flow is typically the preferred state (e.g. in assays and ionization). In either case, the demarcation between the stable and unstable states is of practical importance. The theme of these lecture notes is to develop a systematic methodology in identifying such stability boundaries of electrohydrodynamic flows. The topics have been selected mainly for their pedagogical value in illustrating the basic concepts in continuum electromechanics and hydrodynamic stability, often motivated by practical applications and backed by experimental observations. Although the governing equations and analysis methodology should be of general applicability, no attempt

*I am indebted to the organizer, Dr. A. Ramos, who graciously helped on delivering the videotaped lectures as a result of visa complications that prevented me from traveling to CISM. My Ph.D. adviser J. G. Santiago helped on the initial outline of the lecture notes. Drs. M. P. Brenner and A. M. Ganan-Calvo provided helpful correspondence about their research. This work was funded in part by an NSF CAREER Award.

has been made to comprehensively review the rapidly expanding field of electrohydrodynamic stability. Extensive coverage of electrohydrodynamics and the associated flow stability can be found in Melcher and Taylor (1969); Melcher (1981); Saville (1997); Fernandez de la Mora (2007); Chang and Yeo (2010). In addition, an educational film developed by Melcher (1974) offers many intuitive insights on electrohydrodynamics.

2 Basics of Electrohydrodynamics

Electrohydrodynamics deals with the interaction of electric and flow fields where the Ohmic model is frequently an excellent approximation (Melcher and Taylor, 1969; Saville, 1997). In this section, we first present an intuitive derivation of the Ohmic model and clearly identify the assumptions behind. We then offer some physical insights on the Maxwell stress tensor which plays the crucial role of coupling the electrostatics and hydrodynamics. Finally, we discuss the governing equations for both surface- and bulk-coupled models. The surface-coupled model takes essentially the same form as in Melcher and Taylor (1969) and Saville (1997), while the bulk-coupled model takes roots in Levich (1962) and Melcher (1981).

2.1 Ohmic model

Electrohydrodynamic systems can usually be approximated as electroquasistatic (Saville, 1997). In the absence of external magnetic fields, magnetic effects can be ignored completely. The electrostatic field is solenoidal,

$$\nabla \times \mathbf{E} = 0. \tag{1}$$

The electric field ($\mathbf{E}$) obeys the Gauss's law, which for electrically linear medium reduces to,

$$\nabla \cdot \varepsilon \mathbf{E} = \rho_f, \tag{2}$$

where ε is the permittivity, and ρ_f is the free charge density. The free charge density is related to the current ($\mathbf{i}$) by the charge conservation equation,

$$\frac{\partial \rho_f}{\partial t} + \nabla \cdot \mathbf{i} = 0. \tag{3}$$

In the Ohmic model, or the so-called leaky dielectric model (Saville, 1997), an Ohmic constitutive law of the conduction current is assumed (Melcher and Taylor, 1969),

$$\mathbf{i} = \mathbf{i}_C + \mathbf{i}_O = \rho_f \mathbf{v} + \sigma \mathbf{E}, \tag{4}$$

where $\mathbf{i}_C = \rho_f \mathbf{v}$ is the convection current, $\mathbf{i}_O = \sigma\mathbf{E}$ is the Ohmic current, $\mathbf{v}$ is the fluid velocity, and σ is the electrical conductivity. The Ohmic model can be derived from the electro-diffusive transport of individual charged ions (Melcher, 1981; Saville, 1997; Levich, 1962). We assume for algebraic simplicity a monovalent binary electrolyte which is fully dissociated with constant properties (Chen et al., 2005). For derivations involving multivalent electrolyte and chemical reaction, see Levich (1962) and Saville (1997), respectively. In the Ohmic model, bulk quantities of conductivity and charge density are tracked instead of individual ions. The charge density (ρ_f) and electric conductivity (σ) are related to the ionic concentrations through

$$\rho_f = F(c_+ - c_-), \tag{5}$$

$$\sigma = F^2(c_+ m_+ + c_- m_-), \tag{6}$$

where F is the Faraday constant, $c_\pm$ is the cationic/anionic molar concentration, and $m_\pm$ is the ionic mobility (in $\mathrm{mol\,N^{-1}\,m\,s^{-1}}$).

The key simplifying assumption in the derivation is electro-neutrality, which can be assumed in the limit of (Chen et al., 2005)

$$\Theta_1 = \frac{F m_+ \rho_f}{\sigma} = \frac{c_+ - c_-}{c_+ + \frac{m_-}{m_+} c_-} \ll 1, \tag{7}$$

where Θ_1 represents the ratio of cationic and anionic concentration difference (which contributes to the charge density) to the total concentration of ions (which contributes to the electrical conductivity). Applying the Gauss's law,

$$\Theta_1 = \frac{F m_+ \rho_f}{\sigma} \sim \frac{F m_+ \nabla \cdot \varepsilon \mathbf{E}}{\sigma} \sim \frac{\varepsilon/\sigma}{L_r / m_+ F E} \sim \frac{\tau_e}{\tau_r}, \tag{8}$$

where $\tau_e = \varepsilon/\sigma$ is the charge relaxation time, $m_+ FE$ is by definition the electro-migration velocity of the cation, L_r is a reference length scale over which the electric field varies (typically the smallest length scale of the electrohydrodynamic system), and τ_r is the time scale to travel L_r by electro-migration. Therefore, electro-neutrality is an excellent approximation for most electrolyte solutions (typically with $\sigma > 10^{-4}$ S/m) with fast charge relaxation (typically with $\tau_e < 10\ \mu$s), because the charge relaxation time is much shorter than the electrohydrodynamic time scale of interest (typically with $\tau_r > 1$ ms). When the electrolyte solution is approximately neutral,

$$c_+ \simeq c_- = c; \quad |c_+ - c_-| \ll c, \tag{9}$$

where c is the reduced ionic concentration (Levich, 1962). Under electro-neutrality, the conductivity is proportional to this reduced concentration by

$$\sigma = F^2(m_+ + m_-)c. \tag{10}$$

To derive the governing equation for conductivity, we start from the Nernst-Planck equations for ionic species (Levich, 1962),

$$\frac{\partial c_+}{\partial t} + \nabla \cdot c_+\mathbf{v} = D_+\nabla^2 c_+ - m_+F\nabla \cdot c_+\mathbf{E}, \tag{11}$$

$$\frac{\partial c_-}{\partial t} + \nabla \cdot c_-\mathbf{v} = D_-\nabla^2 c_- + m_-F\nabla \cdot c_-\mathbf{E}, \tag{12}$$

where $D_\pm$ is the ionic diffusivity. The diffusivity and mobility is related by Einstein's relation $D_\pm = RTm_\pm$ where R is the universal gas constant and T is the absolute temperature. Subtracting Eqs. 11 and 12 and noting the electro-neutrality condition,

$$(D_+ + D_-)F\nabla \cdot c\mathbf{E} \simeq RT(D_+ - D_-)\nabla^2 c, \tag{13}$$

where the equality holds to the leading order of ionic concentrations. Substituting Eq. 13 to Eq. 11 and noting Eq. 10, the electro-diffusion equation becomes

$$\frac{\partial \sigma}{\partial t} + \nabla \cdot \sigma\mathbf{v} = D_{eff}\nabla^2\sigma, \tag{14}$$

where D_{eff} is an effective diffusivity,

$$D_{eff} = \frac{2D_+D_-}{D_+ + D_-}. \tag{15}$$

To derive the equation for charge density, we subtract Eqs. 11 and 12 again in an exact manner,

$$\frac{\partial(c_+ - c_-)}{\partial t} + \nabla\cdot(c_+ - c_-)\mathbf{v} = \nabla^2(D_+c_+ - D_-c_-) - F\nabla\cdot(m_+c_+ + m_-c_-)\mathbf{E}, \tag{16}$$

or in terms of bulk quantities,

$$\frac{\partial \rho_f}{\partial t} + \nabla \cdot (\rho_f\mathbf{v} + \mathbf{i}_D + \sigma\mathbf{E}) = 0, \tag{17}$$

where the diffusive current $\mathbf{i}_D = -F\nabla(D_+c_+ - D_-c_-) \simeq -(D_+ - D_-)F\nabla c$ (Levich, 1962). Eq. 17 reduces to the the charge conservation equation in the Ohmic regime,

$$\frac{\partial \rho_f}{\partial t} + \nabla \cdot \rho_f\mathbf{v} = -\nabla \cdot \sigma\mathbf{E}, \tag{18}$$

when the diffusive current can be neglected, i.e.

$$\Theta_2 = \left|\frac{\mathbf{i}_D}{\mathbf{i}_O}\right| \ll 1. \tag{19}$$

In electrohydrodynamic systems, the diffusive current is usually much smaller than the Ohmic conduction current. The ratio of the two currents scales as,

$$\Theta_2 = \left|\frac{\mathbf{i}_D}{\mathbf{i}_O}\right| \sim \frac{(D_+ - D_-)F\nabla c}{(m_+ + m_-)F^2 cE} \sim \frac{D_+ - D_-}{D_+ + D_-}\frac{RT}{F}\frac{\nabla c}{Ec} \sim \frac{RT/F}{EL_r} \sim \frac{\Phi_T}{\Phi_r}, \tag{20}$$

where $\Phi_T = RT/F$ (25 mV at room temperature) is the thermal voltage driving the diffusive current, and Φ_r is the reference voltage drop along a concentration gradient. Since the applied electric field is typically high in an electrohydrodynamic system, the diffusive current can be safely neglected (i.e. Eq. 19 is valid) for most practical cases. For example, with a field of 10^5 V/m, $\Theta_2 \ll 1$ for a diffusive interface as thin as 1 μm.

The Ohmic model consists of the conservation equations for conductivity (Eq. 14) and charge density (Eq. 18). Physically, the material derivative of conductivity and charge density is balanced by the divergence of an effective diffusive flux and Ohmic current flux, respectively. The underlying assumptions in the Ohmic model are instantaneous charge relaxation (Eq. 7) and negligible diffusive current (Eq. 19).[1] Both assumptions hold for most practical electrohydrodynamic systems driven by direct-current (DC) fields, where the time scale of interest is typically above 1 ms and the length scale of interest is typically above 1 μm. With rapid charge relaxation, the electrolyte solution is approximately electro-neutral in the bulk and the cations and anions are almost always paired together, as the difference in cationic and ionic concentrations is very small compared to the background concentration of electrolytes. Consequently, conductivity becomes a conserved material property with an effective diffusivity averaging the cationic and anionic properties. Note that the Ohmic model does not work inside electric double layer, where the net charged layer has a typical thickness of well below 1 μm.

2.2 Maxwell stress

The electrostatics and hydrodynamics are coupled together through the Maxwell stress tensor. In vacuum, the Coulombic force density exerted on

[1] Although we have assumed constant properties ($D_\pm$ and $m_\pm$) in the derivation, the Ohmic model consisting of Eqs. 14 and 18 are generally believed to hold for cases with non-constant properties, e.g. due to temperature gradients.

free charges can be rearranged noting the solenoidal nature of the electrostatic field (Panofsky and Phillips, 1962),

$$\begin{aligned}\mathbf{f}_0^e &= \rho_f \mathbf{E} = (\nabla \cdot \varepsilon_0 \mathbf{E})\mathbf{E} \\ &= \nabla \cdot \left(\varepsilon_0 \mathbf{E}\mathbf{E} - \frac{1}{2}\varepsilon_0 E^2 \mathbf{I} \right) = \nabla \cdot \mathbf{T}_0^e, \end{aligned} \tag{21}$$

where ε_0 is the permittivity of vacuum, $\mathbf{I}$ is the identity matrix, and $\mathbf{T}_0^e$ is the Maxwell stress (or electric stress) tensor in vacuum.

The derivation of Maxwell stress tensor for a dielectric medium is rather complicated (Melcher, 1981; Panofsky and Phillips, 1962). Here, we summarize salient points of the Maxwell stress tensor and refer the readers to Melcher (1981) for details. The electrical force density can be derived using either the Kelvin approach or the Korteweg-Helmholtz approach. The Kelvin force density is useful for appreciating the underlying microscopic electromechanics, while the Korteweg-Helmholtz force density is more useful for predicting the consequences of electromechanical coupling (Melcher, 1974). The Kelvin force density is the sum of the Coulombic force exerted on free charges and the polarization force exerted on the dipoles (Melcher, 1981),

$$\begin{aligned}\mathbf{f}_K^e &= \rho_f \mathbf{E} + \mathbf{P} \cdot \nabla \mathbf{E} = (\nabla \cdot \varepsilon \mathbf{E})\,\mathbf{E} + (\varepsilon - \varepsilon_0)\,\mathbf{E} \cdot \nabla \mathbf{E} \\ &= \nabla \cdot \left(\varepsilon \mathbf{E}\mathbf{E} - \frac{1}{2}\varepsilon_0 E^2 \mathbf{I} \right) = \nabla \cdot \mathbf{T}_K^e, \end{aligned} \tag{22}$$

where $\mathbf{P}$ is the polarization density, and $\mathbf{T}_K^e$ is the Maxwell stress tensor corresponding to the Kelvin force density. The key concept due to Kelvin is that the polarization force is exerted on the dipoles ($\mathbf{P}$), not on individual polarization charges ($-\nabla \cdot \mathbf{P}$) (Melcher, 1981). The Korteweg-Helmholtz force density stems from thermodynamic principles. For an electrically linear medium with polarization dependent on mass density (ρ) and temperature (T) alone, the force density can be shown to be,

$$\begin{aligned}\mathbf{f}_{KH}^e &= \rho_f \mathbf{E} - \frac{1}{2} E^2 \nabla \varepsilon + \nabla \left[\frac{1}{2} \rho \left(\frac{\partial \varepsilon}{\partial \rho} \right)_T E^2 \right] \\ &= \nabla \cdot \left[\varepsilon \mathbf{E}\mathbf{E} - \frac{1}{2}\varepsilon E^2 \mathbf{I} + \frac{1}{2} \rho \left(\frac{\partial \varepsilon}{\partial \rho} \right)_T E^2 \mathbf{I} \right] = \nabla \cdot \mathbf{T}_{KH}^e, \end{aligned} \tag{23}$$

where the last term is the electrostriction force density associated with volumetric change in the material; See Melcher (1981, Sec. 3.7) for a detailed derivation. The Kelvin and Korteweg-Helmholtz force densities are different by the gradient of a scalar,

$$\mathbf{f}_K^e - \mathbf{f}_{KH}^e = \nabla \left[\frac{1}{2}(\varepsilon - \varepsilon_0)E^2 - \frac{1}{2}\rho \left(\frac{\partial \varepsilon}{\partial \rho} \right)_T E^2 \right], \tag{24}$$

which can be absorbed into a lumped pressure. The difference represents the interaction between dipoles, which is omitted in the Kelvin derivation. For incompressible flow, where pressure becomes a "left-over" variable, any two force densities differing by the gradient of a scalar pressure will give rise to the same incompressible deformation (Melcher, 1981). Both force densities, if used consistently, will yield the same answer as far as incompressible mechanical deformation is concerned. Because the actual electric force distributions of the two approaches are ususally very different, we stress that the same force density should be used consistently; See an example in Melcher (1981, Sec. 8.3).

Hereon, we shall restrict our discussions to the practically important case of *electrically linear, incompressible* dielectric medium, and *consistently* adopt the Korteweg-Helmholtz force density,

$$\begin{aligned} \mathbf{f}^e &= \rho_f \mathbf{E} - \frac{1}{2} E^2 \nabla \varepsilon \\ &= \nabla \cdot \left(\varepsilon \mathbf{E}\mathbf{E} - \frac{1}{2} \varepsilon E^2 \mathbf{I} \right) = \nabla \cdot \mathbf{T}^e. \end{aligned} \tag{25}$$

When the permittivity is that of vacuum, the Maxwell stress tensor $\mathbf{T}^e$ in Eq. 25 reduces to $\mathbf{T}_0^e$ in Eq. 21. This similarity enables us to take advantage of the "bisect rule" graphically shown in Figure 1 (Panofsky and Phillips, 1962, Sec. 6-5). The bisect rule is useful in graphically identifying the direction of electric stress when the field direction is known. A few example applications of the bisect rule are shown in Figure 2.

From the Korteweg-Helmholtz force density in Eq. 25 and the corresponding bisect rule in Figure 2, it is apparent why a leaky dielectric is necessary to support any tangential electric stress at electrostatic interfaces (Melcher and Taylor, 1969). For a perfect conductor such as the charged drop in (a), the Coulombic force is always along the electric field which is perpendicular to the interface. For a perfect dielectric such as a dielectric jet in (b), the polarization force is always along the permittivity gradient which is again perpendicular to the interface. For a leaky dielectric with finite, non-zero conductivity and permittivity, a tangential shear stress can develop such as in the electrified cone in (c), where the electric field at 45° to the surface normal gives rise to a shear force along the conical surface.

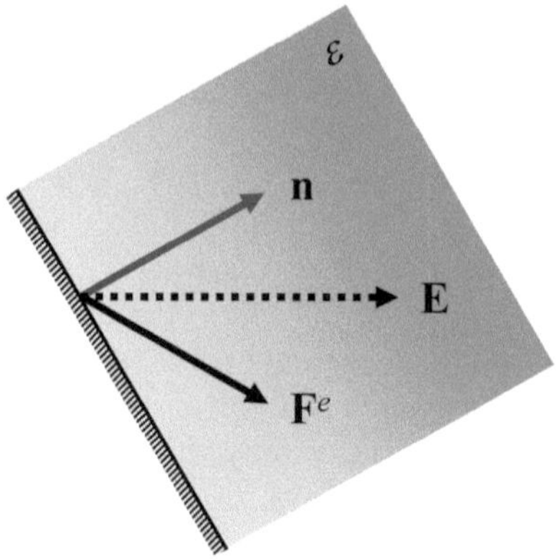

Figure 1. The bisect rule: the electric field (**E**) bisects the angel between the normal to the surface (**n**) and the direction of the resultant force ($\mathbf{F}^e$) acting on the surface (dS). In a dielectric medium with a local permittivity of ε, the magnitude of the electric force is $F^e = T^e dS = \frac{1}{2}\varepsilon E^2 dS$. After Panofsky and Phillips (1962).

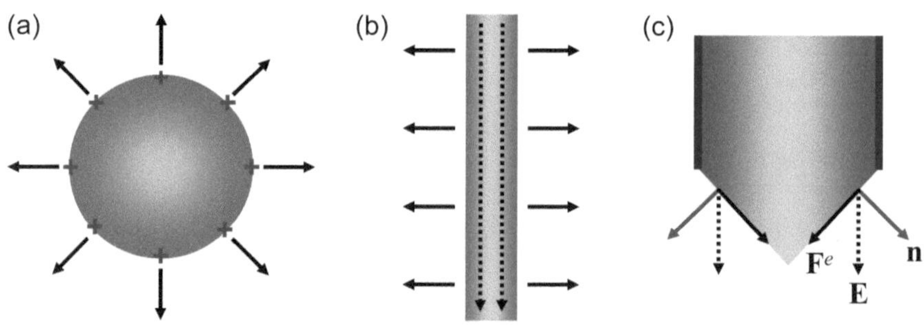

Figure 2. Example applications of the bisect rule: (a) on a charged conducting drop, the surface normal is along the direction of the electric field, hence the Coulombic repulsion is also along the direction of the electric field; (b) on a dielectric liquid jet, the surface normal is perpendicular to the electric field, hence the polarization force is perpendicular to the electric field and pointing outward; (c) on a leaky dielectric cone, the surface normal is at 45° to the electric field, hence the (total) electric force is along the tangential direction of the cone which is also at 45° to the electric field.

2.3 Governing equations

To summarize, the electrohydrodynamic leaky dielectric model consists of the following equations:

$$\nabla \times \mathbf{E} = 0, \tag{26a}$$

$$\nabla \cdot \varepsilon \mathbf{E} = \rho_f, \tag{26b}$$

$$\frac{\partial \rho_f}{\partial t} + \nabla \cdot \rho_f \mathbf{v} = -\nabla \cdot \sigma \mathbf{E}. \tag{26c}$$

$$\nabla \cdot \mathbf{v} = 0, \tag{26d}$$

$$\rho \frac{\partial \mathbf{v}}{\partial t} + \rho \mathbf{v} \cdot \nabla \mathbf{v} = -\nabla p + \mu \nabla^2 \mathbf{v} + \rho_f \mathbf{E} - \frac{1}{2} E^2 \nabla \varepsilon, \tag{26e}$$

where p is the pressure and μ is the dynamic viscosity. The electric body forces in the momentum equation (Eq. 26e) include both a Coulombic and a polarization component. These equations are valid for incompressible, electrically linear leaky dielectrics with rapid charge relaxation and negligible diffusive current, for example, aqueous electrolyte solutions. The set of governing equations (26) is closed if the distribution of material properties (ε, σ, μ) is either given or modeled with additional equations.

We have employed the electro-neutrality assumption in the derivation of the Ohmic model, but kept the charge density in the governing equations. Although the material derivative of ρ_f may often be neglected in the charge conservation equation (Eq. 26c), it will not be appropriate to neglect the Coulombic body force term $\rho_f \mathbf{E}$ in the momentum equation (Eq. 26e) because this force *drives* the electrohydrodynamic flow. Combining the charge conservation equation (Eq. 26c) and the Gauss's law (Eq. 26b) (Melcher, 1974),

$$\frac{D\rho_f}{Dt} = \frac{\rho_f}{\tau_e} - \sigma \mathbf{E} \cdot \left(\frac{\nabla \sigma}{\sigma} - \frac{\nabla \varepsilon}{\varepsilon} \right). \tag{27}$$

In regions of uniform conductivity and permittivity, the net free charge decays with the charge relaxation time for an observer following a particle of fixed identity (Melcher and Taylor, 1969):

$$\rho_f = \rho_{f,0} e^{-t/\tau_e}. \tag{28}$$

Therefore, unless an element of material having uniform properties can be traced along a particle line to a source of net charge, it supports no net charge (Melcher and Taylor, 1969). In inhomogeneous material, however, free charge density can be generated by an electric field component along the gradients of conductivity and/or permittivity.

Systems having nonuniform properties can be modeled either as a bulk region with continuously varying properties or as adjoining regions with piecewise uniform properties (Melcher, 1981; Hoburg and Melcher, 1976). Since interesting electrohydrodynamics often happen at material interfaces, we shall discuss two approaches to close the problem depending on the interfacial sharpness (Hoburg and Melcher, 1976).

In the bulk-coupled model for a diffusive interface, additional equations of the materials properties must be added. For example, in aqueous electrolytes the permittivity is essentially constant but the conductivity variation can be significant; in this case, the "conservation" equation of conductivity from the Ohmic model can be used,

$$\frac{\partial \sigma}{\partial t} + \nabla \cdot \sigma \mathbf{v} = D_{eff} \nabla^2 \sigma. \tag{29}$$

In the surface-coupled model for a sharp interface, the material properties are usually piecewise constant on either side of the interface; However, jump conditions are needed to relate the interfacial and bulk properties. Except for the empirical no-slip condition (Eq. 30e), the following jump conditions can be obtained by integrating the differential equations across the interface (Melcher and Taylor, 1969; Melcher, 1981; Saville, 1997; Leal, 2007),

$$\mathbf{n} \times \|\mathbf{E}\| = 0, \tag{30a}$$

$$\mathbf{n} \cdot \|\varepsilon \mathbf{E}\| = q_s, \tag{30b}$$

$$\frac{\partial q_s}{\partial t} + \nabla_s \cdot (q_s \mathbf{v}) = -\mathbf{n} \cdot \|\sigma \mathbf{E}\|, \tag{30c}$$

$$\mathbf{n} \cdot \|\mathbf{v}\| = 0, \tag{30d}$$

$$\mathbf{n} \times \|\mathbf{v}\| = 0, \tag{30e}$$

$$\mathbf{n}\,\|p\| = \mathbf{n} \cdot \|\mathbf{T}^m + \mathbf{T}^e\| + \nabla_s \gamma - \gamma \mathbf{n}(\nabla_s \cdot \mathbf{n}), \tag{30f}$$

where $\|\,.\,\|$ denotes the jump in a variable across the interface, $\mathbf{n}$ denotes the outward normal vector, subscript s denotes surface quantities, q_s is the surface charge density, $\nabla_s = (\mathbf{I} - \mathbf{nn}) \cdot \nabla$ is the surface gradient operator, and γ is the surface tension. In the stress balance (Eq. 30f), the surface gradient of surface tension gives rise to a tangential force while surface tension on a curved surface leads to a normal force (Leal, 2007, p. 78). The viscous and electric stress tensors are

$$\mathbf{T}^m = \mu(\nabla \mathbf{v} + \nabla \mathbf{v}^T); \qquad \mathbf{T}^e = \varepsilon \mathbf{E}\mathbf{E} - \frac{1}{2} \varepsilon E^2 \mathbf{I}. \tag{31}$$

On a surface with an outward normal $\mathbf{n}$ and orthogonal tangential vectors $\mathbf{t}_1$ and $\mathbf{t}_2$, the normal and tangential Maxwell stress components are

$$\mathbf{n} \cdot \|\mathbf{T}^e \cdot \mathbf{n}\| = \frac{1}{2}\|\varepsilon(\mathbf{E} \cdot \mathbf{n})^2 - \varepsilon(\mathbf{E} \cdot \mathbf{t}_1)^2 - \varepsilon(\mathbf{E} \cdot \mathbf{t}_2)^2\|, \tag{32a}$$

$$\mathbf{t}_i \cdot \|\mathbf{T}^e \cdot \mathbf{n}\| = \|\varepsilon\mathbf{E} \cdot \mathbf{n}\|(\mathbf{E} \cdot \mathbf{t}_i). \tag{32b}$$

From Eq. 32b, we note again that neither a perfect dielectric ($q_s = 0$) nor a perfect conductor ($\mathbf{E}_t = 0$) can support a tangential Maxwell stress (Chang and Yeo, 2010).

The above set of jump conditions (Eq. 30) implies that a sharp material interface can support a surface charge density, but not a surface mass density (Melcher, 1981, p. 7.8).[2] The jump condition for surface charge density (Eq. 30c) can be further expanded as

$$\frac{\partial q_s}{\partial t} + \nabla_s \cdot (q_s \mathbf{v}_s) + q_s \mathbf{n} \cdot \mathbf{v}(\nabla_s \cdot \mathbf{n}) = -\mathbf{n} \cdot \|\sigma \mathbf{E}\|, \tag{33}$$

where the surface velocity $\mathbf{v}_s = (\mathbf{I} - \mathbf{nn}) \cdot \mathbf{v}$, and the surface curvature $\kappa = \nabla_s \cdot \mathbf{n} = 1/R_1 + 1/R_2$ where R_1 and R_2 are the principal radii of curvature of the surface, e.g. $\kappa = 2/R$ for a sphere and $1/R$ for a cylinder. The term involving the surface curvature accounts for the variation of q_s due to the dilation of the surface (Leal, 2007, p. 93). Note that unlike the surface charge convection, the bulk charge convection does not enter into the balance of surface charge density because bulk free charges never reach the interface by convection (Melcher, 1981, p. 2.18). The jump condition for surface charge density (Eq. 30c) assumes negligible surface diffusion current and surface Ohmic current, in other words, the surface convection current is assumed to dominate. Although the negligence of the surface diffusion current appears to be consistent with negligible bulk diffusion current, the negligence of the surface conduction current is not entirely justifiable from first principles. However, the simplified jump condition in the form of Eq. 30c is often sufficient to model electrohydrodynamic phenomena, see for example the literature reviewed in Saville (1997) and Zeng and Korsmeyer (2004).

2.4 Model problems of electrohydrodynamic stability

We shall now apply the governing equations of electrohydrodynamics to two model problems, electrokinetic mixing flow and electrohydrodynamic

[2]Electrical double layer is not considered here because the Ohmic model does not apply within the double layer. The inclusion of double layer may also disrupt the continuation of tangential electric fields (Melcher, 1981, p. 2.16).

cone-jet (Figure 3). Both problems are motivated by a broad range of practical applications, typically with a working fluid of relatively high conductivity ($\gtrsim 10^{-4}$ S/m) where the Ohmic model holds. Since deionized water has a conductivity of approximately 10^{-4} S/m when equilibrated with carbon dioxide in the atmosphere, the "high-conductivity" regime encompasses all practical aqueous solutions (see also discussions below around Eq. 49).

The model problems are chosen to represent two extreme scenarios. In the electrokinetic mixing flow discussed here, two *miscible* working fluids are driven by an electric field approximately *parallel* to the material interface. In contrast, in the electrohydrodynamic cone-jet, two *immiscible* working fluids are stressed by an electric field with a component *perpendicular* to the interface. We will start with the mixing flow problem where the base state is well defined, and then take on the more difficult case of cone-jet flow where a clear-cut description of the base state is a challenge in itself.

Because both model problems are open flow systems, it is useful to study the instability in the spatial frame work in addition to the more conventional temporal frame work. A system that is unstable in a temporal framework can be either convectively or absolutely unstable in the spatial framework (Huerre and Rossi, 1998; Schmid and Henningson, 2001). The onset condition of convective instability is the same as that of a temporal instability in which a global disturbance grows in time (Schmid and Henningson, 2001). Consider a disturbance introduced at a localized point in space: if it grows only downstream, the system is convectively unstable; if the disturbance grows both downstream and upstream, the system is absolutely unstable. Physically, a convectively unstable system is a noise amplifier in which a disturbance at the origin is amplified downstream, while an absolutely unstable system is an intrinsic oscillator in which the downstream propagating wave oscillates simultaneously with an upstream propagating waves (Huerre and Rossi, 1998).

3 Electrokinetic Mixing Flow

Micro total analysis systems (μTAS) aim to integrate multiple assaying functions including sample pretreatment, mixing and separation on a microfabricated chip (Manz et al., 1990; Stone et al., 2004). Electrokinetics is often the method of choice for reagent transport and manipulation in μTAS (Stone et al., 2004; Chang and Yeo, 2010). As devices gain complexity, robust control of electrokinetic processes with heterogeneous samples becomes critical. One important regime is on-chip biochemical assays with high conductivity gradients, which might occur intentionally as in sample stacking processes, or unavoidably as in multi-dimensional assays. Such conductivity

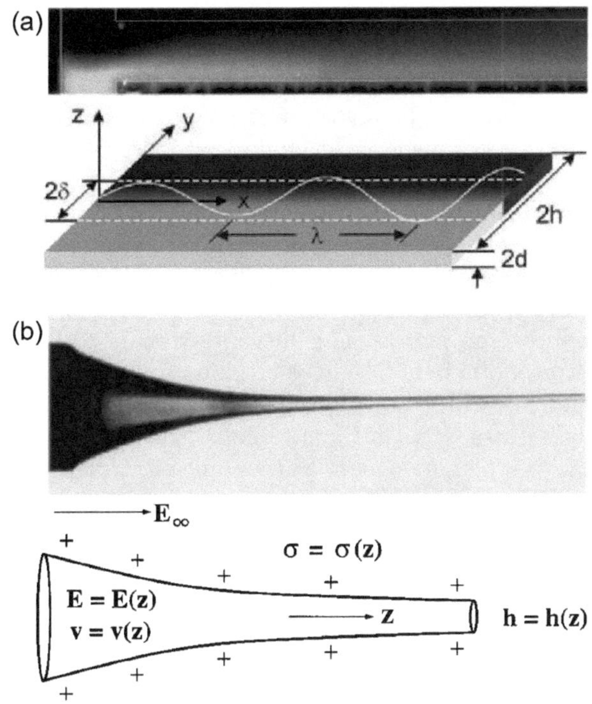

Figure 3. Model problems of electrohydrodynamic stability: (a) Electrokinetic mixing flow of two miscible streams at a T-junction. The two fluids have the same properties except for different electrical conductivities. Although the conductivity gradient leads to instability at high electric fields, the base state is the well-known diffusive solution when the material interface is approximately parallel to the applied electric field. (b) Electrohydrodynamic cone-jet with a liquid jet issued from the Taylor cone on an electrified nozzle. The inner fluid and the surrounding air have dramatically different electrical properties, with the liquid modeled as a leaky dielectric while the air as an insulator. The base state is strongly dependent on the applied electric field which has a significant component perpendicular to the material interface. A complete mechanistic understanding of the base state is not yet available. Figure (a) reprinted with permission from Chen et al. (2005), © 2005 Cambridge University Press. Figure (b) reprinted with permission from Hohman et al. (2001a,b), © 2001 American Institute of Physics.

gradients may lead to instabilities under high electric fields (Figure 4), with anecdotal evidences widespread in the μTAS literature (Chen et al., 2005). Although instability is undesirable for robust electrokinetic assay, it is useful for rapid mixing at low Reynolds number (Oddy et al., 2001). In either case, the instability mechanism must be understood before enhancement or suppression can be engineered.

As shown below, these "electrokinetic instabilities" (Lin et al., 2004; Chen et al., 2005) are fundamentally electrohydrodynamic, in which electroosmotic flow mainly acts as a convecting medium (Chen et al., 2005). The instability waves in Figure 4 clearly originate at the liquid interface with gradients of material properties, not on charged solid walls. We will also show that electroosmotic bulk flow in electrokinetic systems leads to convective and absolute instability; see Figure 4a and Figure 4b, respectively. The absolute instability sets in when the internally generated electroviscous velocity disturbances are high enough to overcome electroosmotic convection. Both the electroviscous and electroosmotic velocities result from balancing electric body forces and viscous stresses. However, the electroviscous velocity is due to the accumulated net charge density in the *bulk* (Melcher, 1981), while the electroosmotic velocity is due to the net charge within the electric double layer at the *boundary* (Chang and Yeo, 2010).

According to Eq. 27, electromechanical coupling may arise from a gradient of permittivity and/or conductivity along the electric field (Melcher, 1974). However, permittivity gradient is negligible because a dilute solution of electrolyte does not significantly alter the permittivity of pure water (Chen et al., 2005). As shown in a series of papers published by Melcher and coworkers, a sharp interface separating regions of identical properties except disparate conductivities is linearly stable (Melcher and Schwartz, 1968); however, a diffuse interface under the same condition gives rise to unstable electromechanical coupling (Hoburg and Melcher, 1976). Compared to the surface-coupled model where the electromechanical coupling is through interfacial stresses, the bulk-coupled model implies coupling through volumetric forces distributed across the diffusive interface. The diffusive interface model is particularly relevant to microsystems with small length scales, e.g. a diffusion length comparable to the channel width is visible in the unperturbed base state (Figure 3a). Consistent with the bulk-coupled model, the diffusive term should be kept in the conductivity conservation Eq. 29 (Baygents and Baldessari, 1998). Without the diffusive term, the linearized equations will be unconditionally unstable as in Hoburg and Melcher (1976), which contradicts with an experimentally measured threshold electric field below which the flow is stable (Chen et al., 2005).

The following discussions on electrokinetic mixing flow will closely follow

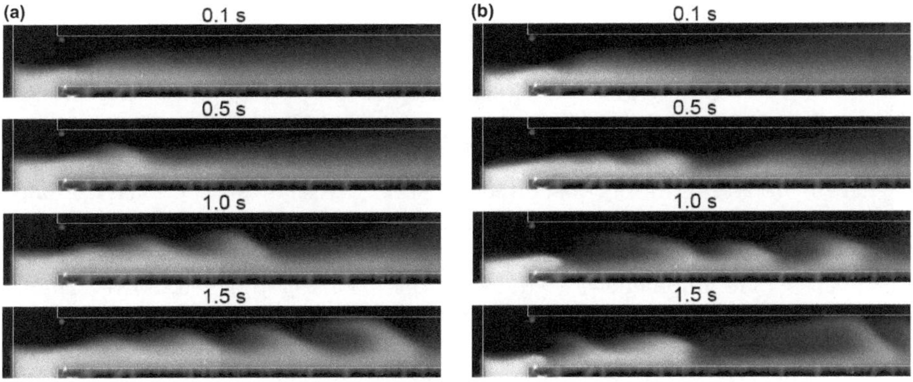

Figure 4. Instability in electrokinetic mixing flow of two miscible streams of aqueous electrolytes: (a) convective instability at a nominal external field of 1.0 kV/cm; and (b) absolute instability at 2.0 kV/cm in (b). The channel has a width of 155 μm and a depth (into the paper) of 11 μm; the conductivity ratio is $\eta = 10$ with the dyed side having the higher conductivity of $\sigma = 7.7 \times 10^{-2}$ S/m; the permittivity of both streams is $\varepsilon = 6.9 \times 10^{-10}$ C/V·m. Reprinted with permission from Chen et al. (2005), © 2005 Cambridge University Press.

Chen et al. (2005). A review of recent development in the area of electrokinetic instability can be found in Lin (2009).

3.1 Linearized thin-layer equations

The governing equations for the bulk-coupled model (Chen et al., 2005) are essentially the same as Eqs. 26 and 29. Because of the rapid charge relaxation in aqueous electrolytes, the charge conservation equations (Eq. 26c) is simplified to the continuity of Ohmic current,

$$\nabla \cdot \sigma \nabla \phi = 0, \tag{34}$$

where the quasi-electrostatic field is related the potential by $\mathbf{E} = -\nabla \phi$. The polarization term drops out of Eq. 26e because of negligible permittivity gradients.

With thin Debye length, the physics of the electric double layer is assumed to influence the instability dynamics *only* in that the double layer determines an electroosmotic velocity very close to the microchannel wall

(Chen et al., 2005). The boundary conditions at the solid walls are therefore,

$$\mathbf{n} \cdot \nabla\sigma = 0, \tag{35a}$$

$$\mathbf{n} \cdot \nabla\phi = 0, \tag{35b}$$

$$\mathbf{v} = -\varepsilon\zeta\nabla\phi/\mu, \tag{35c}$$

where $\mathbf{n}$ denotes wall-normal direction, and ζ is the zeta potential of the electrical double layer. Boundary conditions (35a) and (35b) are consequences of non-penetrating walls. The electroosmotic velocity at the wall is is given by the Smoluchowski equation (35c) (Chang and Yeo, 2010). Here, we assume constant zeta potential for simplicity. In reality, zeta potential is related to ionic concentration and therefore changes with the electrical conductivity. However, the dependence of zeta potential on local conductivity has been shown to be not essential to the instability mechanism (Chen et al., 2005).

The base state is assumed parallel, in particular, the conductivity distribution assumes a diffusive profile with a constat diffusion length δ (Figure 3a),

$$\sigma_0 = \sigma_L + \frac{\sigma_H - \sigma_L}{2}\operatorname{erfc}\left(\frac{y}{\delta}\right), \tag{36}$$

where σ_H and σ_L are respectively the high and low conductivities of the streams prior to mixing. The base electroosmotic velocity profile is assumed uniform, $U_{eo} = \varepsilon\zeta_r E_a/\mu$, where ζ_r is the reference zeta potential and E_a is the applied electric field which is assumed uniform. By the parallel base state assumption, the linear stability analysis is greatly simplified and no boundary conditions are needed in the streamwise (x-direction). The validity of the parallel base state assumption is discussed in details in Chen et al. (2005).

As mentioned earlier, two velocity scales are necessary to properly scale the governing equations in electrokinetic mixing flow. The electroosmotic velocity (U_{eo}) is the imposed velocity scale due to the net charge in the electric double layer (close to the wall), and is introduced as part of the boundary conditions. The electroviscous velocity (U_{ev}) is the internal scale for velocity disturbances due to net charge accumulation in the diffusive conductivity interface (in the bulk).

Motivated by the high-aspect ratio experimental system, we shall focus on the asymptotic thin-layer limit where $d/h \ll 1$, i.e. the channel depth ($2d$) is much thinner than the horizontal length scales which are all assumed to be on the same order of the channel width ($2h$). In the thin-layer limit, conductivity (σ) and potential (ϕ) are both independent of the z, but velocity ($\mathbf{v}$) has a z-dependence. The difference in z-dependence reflects the fact that the walls at $z = \pm d$ prohibit mass or electric fluxes, but accommodates momentum fluxes. Therefore, the velocity profile takes the form of,

$$\mathbf{v} = \mathbf{v}_0 + \tilde{\mathbf{v}}_H = \mathbf{v}_0 + \bar{\mathbf{v}}_{ev}(x, y, t)g(z) + \bar{\mathbf{v}}_{eo}(x, y, t), \tag{37}$$

where the subscript 0 denotes the base state, the subscript H denotes the horizontal direction, g(z) is a parabolic function $g(z) = \frac{3}{2}(1 - z^2/d^2)$, $\bar{\mathbf{v}}_{ev}$ and $\bar{\mathbf{v}}_{eo}$ are the electroviscous and electroosmotic velocity perturbation with the over bar denoting a depth-averaged quantity, and the electroosmotic velocity perturbation is given by

$$\bar{\mathbf{v}}_{eo} = -U_{eo}\left(\frac{\partial_x \bar{\phi}}{E_a}\boldsymbol{e}_x + \frac{\partial_y \bar{\phi}}{E_a}\boldsymbol{e}_y\right). \tag{38}$$

After depth averaging, the linearized conductivity conservation and momentum equations take the form of

$$\frac{\partial \bar{\sigma}}{\partial t} + \mathbf{v}_0 \cdot \nabla_H \bar{\sigma} + \bar{\mathbf{v}} \cdot \nabla_H \sigma_0 = {\nabla_H}^2 \bar{\sigma}, \tag{39}$$

$$\rho\frac{\partial \bar{\mathbf{v}}}{\partial t} + \rho(\mathbf{v}_0 \cdot \nabla_H)\bar{\mathbf{v}} + \rho(\bar{\mathbf{v}} \cdot \nabla_H)\mathbf{v}_0 = -\nabla_H \bar{p} + \mu {\nabla_H}^2 \bar{\mathbf{v}} - \frac{3\mu}{d^2}(\bar{\mathbf{v}} - \bar{\mathbf{v}}_{eo}) + \bar{\rho}_f \boldsymbol{E}_0, \tag{40}$$

where $\boldsymbol{E}_0 = E_a \boldsymbol{e}_x$. The form of mass continuity and current continuity equations remains essentially unchanged. See Chen et al. (2005) for an asymptotic derivation of the thin-layer equations.

Balancing the viscous and electric forces in the linearized momentum Eq. 40, the electroviscous velocity can be shown to scale as (Chen et al., 2005),

$$U_{ev} \sim \frac{\bar{\rho}_f E_a}{\mu/d^2} \sim \frac{(\eta - 1)^2}{(\eta + 1)^2}\frac{d^2}{\delta}\frac{\varepsilon E_a^2}{\mu}. \tag{41}$$

Note δ is the diffusion half length across which conductivity varies, and is therefore the relevant length scale for charge density. However, d is the relevant length scale for wall-bounded viscous transport. The terms involving $\eta = \sigma_H/\sigma_L$ account for the conductivity gradient. This definition of U_{ev} strives to account for all the relevant physical parameters involved in the electroviscous balance of the problem. Fundamentally, Eq. 41 is the same as the simpler version in Hoburg and Melcher (1976) and Melcher (1981), both containing the electroviscous time scale of $\mu/\varepsilon E_a^2$. Note that U_{ev} scales as E_a^2 while U_{eo} scales as E_a, because the electroviscous velocity results from induced charge density which itself scales as E_a.

In modeling electrokinetic instabilities, it is important to note that U_{eo} is the velocity scale for the base electroosmotic flow, while U_{ev} is the internally generated velocity scale to properly scale the perturbation velocity. With proper nondimensionalization, two governing parameters naturally appears in the conductivity conservation Eq. 39,

$$\frac{1}{R_v}\left(\partial_T\bar{\Sigma} + U_0\partial_X\bar{\Sigma}\right) + (\partial_Y\Sigma_0)\bar{V} = \frac{1}{Ra_e}\nabla_H^2\bar{\Sigma}, \tag{42}$$

where capital letters denote nondimensional variables (except E which does not appear in the nondimensional equations), i.e. a general field variable Λ is decomposed in nondimensional form as $\Lambda = \Lambda_0 + \bar{\Lambda}$. The electroviscous-to-electroosmotic velocity ratio R_v is defined as

$$R_v = \frac{U_{ev}}{U_{eo}} = \frac{(\eta-1)^2}{(\eta+1)^2}\frac{E_a d^2}{\zeta_r \delta}. \tag{43}$$

The electric Rayleigh number Ra_e is defined as

$$Ra_e = \frac{U_{ev}h}{D_{eff}} = \frac{(\eta-1)^2}{(\eta+1)^2}\frac{h}{\delta}\frac{\varepsilon E_a^2 d^2}{\mu D_{eff}}, \tag{44}$$

which measures the relative importance of dynamic (electric body) forces to dissipative forces by molecular and viscous diffusion.

3.2 Results of linear stability analysis

The instability mechanism is evident from the most unstable eigenmodes of a representative convective instability shown in Figure 5 (Hoburg and Melcher, 1976; Chen et al., 2005). Figure 5a: The eigenmode of conductivity perturbation has a cellular pattern and alternates in sign in the x-direction. Figure 5b: Such conductivity perturbation will change the electric potential distribution due to current continuity (see also Figure 5e). Figure 5c: The perturbed electric field will produce a charge density perturbation by Gauss's Law, and leads to electric body forces. Figure 5d: The electric body forces produce cellular fluid motion through the Navier-Stokes equations. This fluid motion further alters the conductivity field through the convection-diffusion equation of conductivity, and the positive feedback leads to instability. The similarity of perturbation patterns of the stream function ($\bar{\Psi}$) and y-direction gradient of electric potential ($\bar{\Phi}$), shown by Figure 5d and Figure 5e, confirms a useful relationship in the absence of electroosmotic flow (Chen et al., 2005),

$$\bar{\Psi} \propto \partial_Y\bar{\Phi}. \tag{45}$$

The analytical expression (Eq. 45) directly relates the electric and velocity fields, which enables us to explain the instability mechanism in a

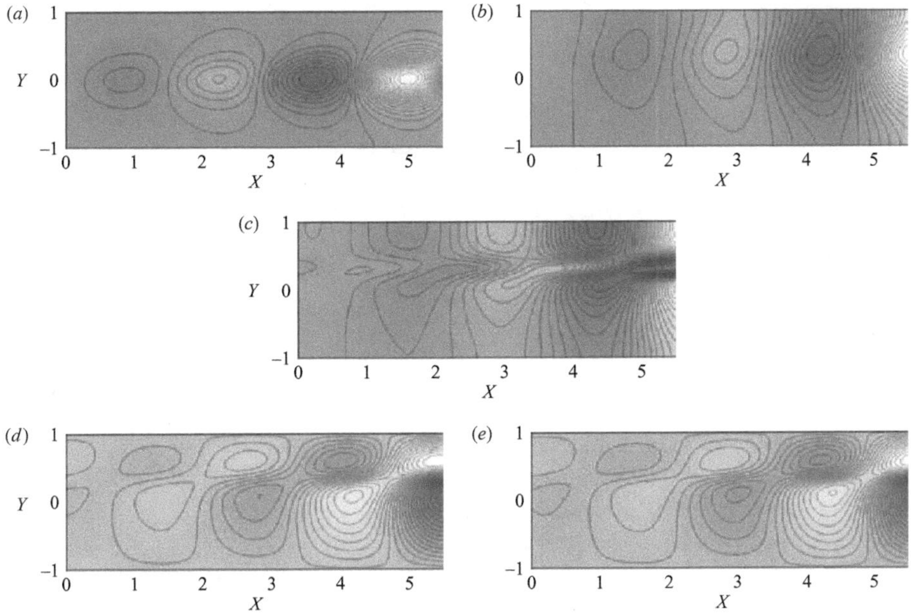

Figure 5. Electrokinetic instability mechanism shown by the most unstable eigenmodes. Dark gray is negative and light gray is positive, contour plots are overlayed: (a) $\bar{\Sigma}$, conductivity perturbation; (b) $\bar{\Phi}$, potential perturbation; (c) $\nabla_H^2\bar{\Phi}$, which is proportional to negative charge density; (d) $\bar{\Psi}$, streamfunction perturbation; (e) $\partial_Y\bar{\Phi}$, electric field perturbation. Conductivity perturbation alternates electric potential distribution and induces bulk charge accumulation, which in turn results in electric body forces and promotes cellular fluid motion. The cellular flow further perturbs the conductivity field and this positive feedback leads to instability. Reprinted with permission from Chen et al. (2005), © 2005 Cambridge University Press.

more intuitive way (Figure 6) (Chen et al., 2005). When the initially flat interface is perturbed, the perturbed conductivity field will alter the electric field. The local electric field is strengthened where lower conductivity replaces higher conductivity (region II), and vice versa (region I). The electric field perturbation is strongest at the conductivity interface (dot-dashed line) and decays away from it. According to Eq. 45, the velocity perturbation further stretches the interface upward for regions I and downward for regions II. This tendency for the perturbed interface to be further stretched is competing with molecular diffusion. When the electroviscous process is faster than molecular diffusion, i.e.

$$\frac{h}{U_{ev}} < \frac{h^2}{D_{eff}} \quad \Rightarrow \quad Ra_e > 1, \tag{46}$$

the perturbation at interface will grow and lead to instability.

In the presence of electroosmotic flow, the unstable perturbations grow as they are convected downstream, which leads to convective instability if the disturbance is not too strong. In the regime of convective instability, growth rate is finite at any downstream location and the regions upstream of the initial disturbance are largely unaffected. However, when the internal electroviscous process is faster than the external electroosmotic convection, i.e.

$$\frac{h}{U_{ev}} < \frac{h}{U_{eo}} \quad \Rightarrow \quad R_v > 1, \tag{47}$$

the flow may become absolutely unstable because the electroosmotic flow can not carry the disturbances downstream fast enough. Therefore, in the regime of absolute instability, the disturbance grows in time (t) at the origin and the upstream flow is perturbed.

The above heuristic arguments are confirmed by the linear stability analysis (Chen et al., 2005). All seven nondimensional parameters in the problem are systematically varied, some over three orders of magnitude. The critical conditions for convective and absolute instabilities are plotted on a $Ra_e - R_v$ phase diagram (Figure 7). It is obvious from the figure that the onset of convective instability collapses around $Ra_{e,cr} \simeq 10$, while the onset of absolute instability collapses around $R_{v,cr} \simeq 4$. When the system properties are fixed, the electric Rayleigh number and the electroviscous-to-electroosmotic velocity ratio vary along a fixed curve $Ra_e \propto R_v^2$. As electric field is increased, the system first become convectively unstable when Rayleigh number exceeds $Ra_{e,cr}$, and then absolutely unstable when the velocity ratio exceeds $R_{v,cr}$.

Despite the linearization with depth averaging and parallel base state, the above model agrees reasonably well with experiments (Chen et al., 2005).

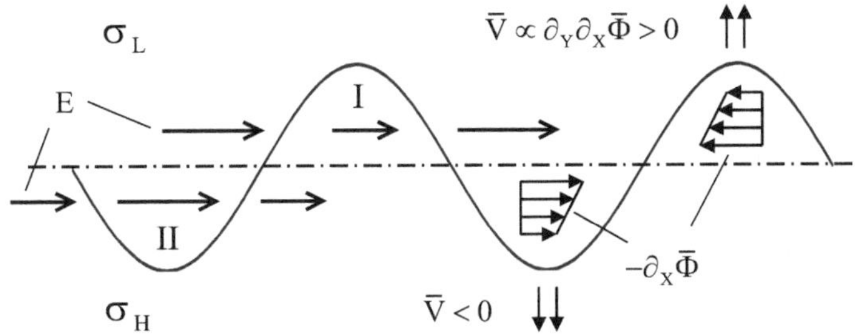

Figure 6. Schematic of electrohydrodynamic instability mechanism. When the conductivity interface is perturbed, the electric field is perturbed due to alternation of conductivity. The perturbed electric field is strongest at the conductivity interface (dot-dashed line) and decays away from it, this gradient in electric field results in vertical velocities which further stretch the interface. Reprinted with permission from Chen et al. (2005), © 2005 Cambridge University Press.

The model captures trends of the instability as a function of increasing field strength, and predicts the observable quantities such as spatial growth rates within a factor of three. Take the critical electric field for example, the nominal threshold for onset of (convective) instability was 0.5 ± 0.1 kV/cm in experiments, and is 0.14 kV/cm from the linear stability analysis; The onset of absolute instability was about 1.5 kV/cm in experiments (Figure 4), and is 0.65 kV/cm from the analysis. With slight modifications, the model also predicts the onset of convective instability in a different geometry where three liquid streams mix at a cross-junction (Posner and Santiago, 2006).

The model presented here offers a useful framework to address the stability issues of electrokinetic mixing flow. We emphasize again that the electrokinetic flow instability resulting from electrical conductivity gradients is essentially an electrohydrodynamic instability convected by the electroosmotic flow. To suppress instability in electrokinetic assays, one needs to minimize Ra_e, the ratio of dynamic electric body forces to dissipative forces due to molecular and viscous diffusion, which controls the onset of instability. To enhance instability for microfluidic mixing, one needs to maximize R_v, the ratio of internally generated electroviscous velocity to bulk electroosmotic velocity, which controls the absolute verses convective nature of instability.

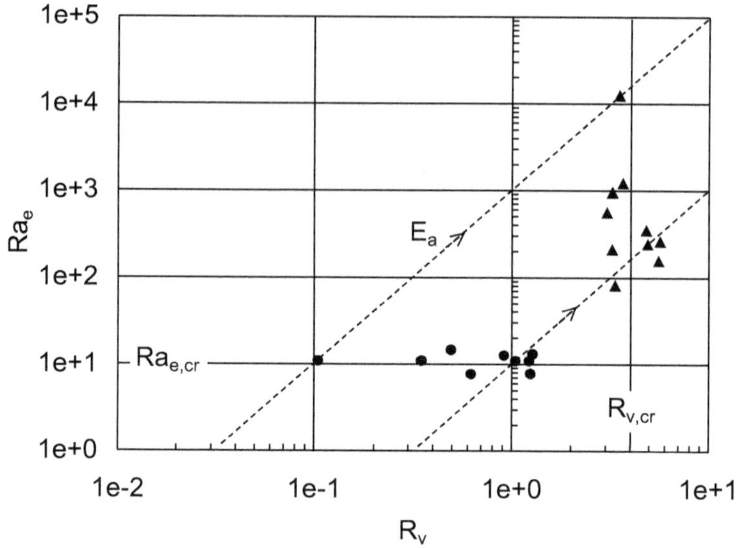

Figure 7. Phase diagram for convective and absolute instability. The critical conditions are plotted with circles representing onset of convective instability and triangles absolute instability. At onset of convective instability, the critical Rayleigh number collapses around $Ra_{e,cr} \simeq 10$. At onset of absolute instability, the critical velocity ratio collapses around $R_{v,cr} \simeq 4$. Note that at a given condition, Ra_e/R_v^2 is a constant so Ra_e and R_v can not be independently varied. Reprinted with permission from Chen et al. (2005), © 2005 Cambridge University Press.

4 Electrohydrodynamic cone-jet

Electrohydrodynamic cone-jet transition is a unique phenomenon that permits the production of a tiny liquid jet from a much larger nozzle (Cloupeau and Prunet-Foch, 1989; Fernandez de la Mora, 2007). The large drawdown ratio enables a variety of techniques including electrospraying, electrospinning and electrostatic printing (Figure 8). Electrospraying has a broad spectrum of applications, most notably crop and paint spraying and electrospray ionization (Bailey, 1988; Kebarle and Verkerk, 2009). Electrospinning is mainly used in producing nanoscale fibers and miniaturized encapsulation (Reneker et al., 2007; Barrero and Loscertales, 2007). Electrostatic printing is useful for high-resolution production of biomolecular microarrays and electronic circuits (Yogi et al., 2001; Park et al., 2007). The stability of electrohydrodynamic cone-jet is of paramount importance to its practical applications. Electrospraying and electrospinning hinges on the *downstream* destabilization of a steady cone-jet to generate fine droplets or thin fibers (Fernandez de la Mora and Loscertales, 1994; Hohman et al., 2001a), while electroprinting relies on transient cone-jets to deploy drops on demand (Chen et al., 2006a,b).

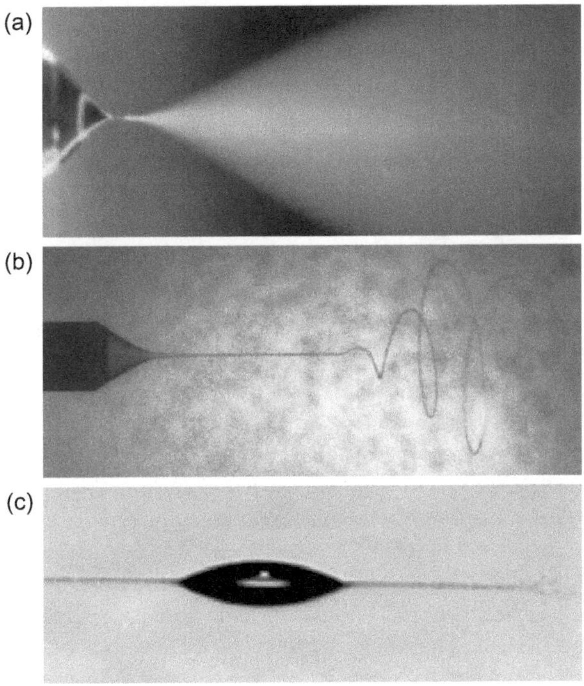

Figure 8. Electrohydrodynamic cone-jets in the context of (a) electrospraying, (b) electrospinning, and (c) electrostatic explosion. Unlike the steady cone-jet in (a) and (b), the cone-jet in (c) is inherently transient, similar to the transient jetting process in electrostatic printing. Figures (a) and (b) reprinted with permission from Barrero and Loscertales (2007), © 2007 Annual Reviews. Figure (c) reprinted with permission from Duft et al. (2003), © 2003 Nature Publishing Group.

In contrast to its wide-ranging applications, a complete mechanistic understanding of the cone-jet dynamics remains elusive. In addition to the mathematical complexities in analyzing multi-physical free surface flow, experimental measurements are also difficult for the flow fields in a free jet, the diameter of which is often on the verge of optical resolution. For a summary of the state-of-the-art and a discussion of unresolved issues, see Fernandez de la Mora (2007); Barrero and Loscertales (2007); Ganan-Calvo and Montanero (2009). The incomplete understanding of the steady cone-jet significantly complicates the stability analysis which needs a base state to start with.

In this section, we will first discuss the operating regime of a steady cone-jet, and then discuss a few selected papers on the cone-jet stability. The dynamics of the electrified jet will be compared to its uncharged counterpart (Eggers and Villermaux, 2008). Because there is a well-defined interfacial boundary between the liquid jet and its immiscible surrounding fluid (e.g. air), a surface-coupled model will be adopted as in the study of conventional uncharged jets.

4.1 Operating diagram of a steady cone-jet

Steady cone-jet is the most useful of the many possible functioning modes of an electrified meniscus (Cloupeau and Prunet-Foch, 1989; Fernandez de la Mora, 2007). For a given working fluid, the conditions leading to steady cone-jet are best summarized by the stability island in the $E-Q$ operating diagram (Figure 9), which is obtained by independently varying the externally applied parameters of electric field (E) and flow rate (Q). The detailed shape of the stability island varies among different working fluids and different electrode configuration. The stability boundary also exhibits a fairly pronounced hysteresis (Cloupeau and Prunet-Foch, 1989; Fernandez de la Mora, 2007). Fortunately, the minimum flow rate (Q_m) and the associated electric field (E_m) in Figure 9 can be rationalized as follows.

When a liquid meniscus is electrified, the Maxwell stress leads to interfacial deformation. At a sufficiently high electric field, the meniscus takes the shape of a Taylor cone (Taylor, 1964). Balancing the electric stress and surface tension, the critical electric field scales as (Taylor, 1964; Smith, 1986)

$$E_m \sim \sqrt{\frac{\gamma}{\varepsilon_0 a}}, \tag{48}$$

where a is a characteristic length scale such as the radius of the nozzle to which the meniscus attaches.[3] For a perfectly conducting liquid, the cone surface is equipotential with a theoretical half angle of 49.3° (Taylor, 1964). See Fernandez de la Mora (2007) for a discussion of other possible conical configurations.

[3]Some operating diagrams use voltage as the control variable. For the commonly used

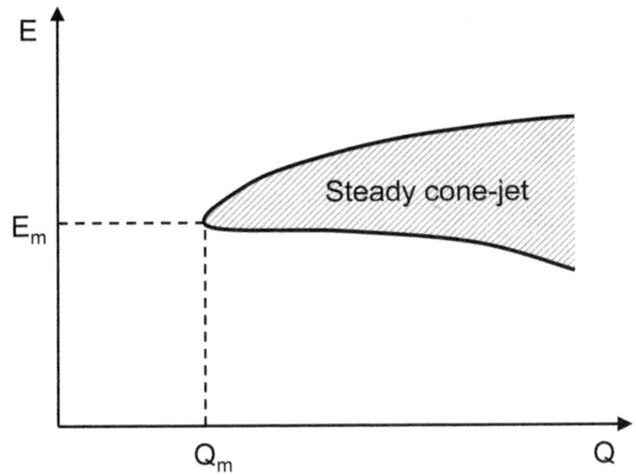

Figure 9. Operating regime of steady cone-jet: At the minimum flow rate Q_m, the applied electric field should be E_m to produce a stead cone-jet; Above Q_m, a steady cone-jet can be produced in a limited range of electric field around E_m; Below Q_m, a steady cone-jet is not possible but pulsating cone-jet results around E_m.

For a relatively conducting liquid, the minimum flow rate does not depend on viscosity (the jet has an approximately flat velocity profile); On dimensional ground Q_m scales as (Fernandez de la Mora and Loscertales, 1994; Fernandez de la Mora, 1996),

$$Q_m \sim \frac{\gamma \varepsilon}{\rho \sigma}. \tag{49}$$

This scaling law is empirically known to work in the "high-conductivity" limit with $\sigma \gtrsim 10^{-4}$ S/m, but a rigorous theoretical justification is still missing (Fernandez de la Mora, 2007). Higher conductivities usually lead to finer sprays, but for conductivities higher than 1 S/m, direct ion evaporation may be possible (Fernandez de la Mora, 2007). Many electrospray working fluids (including most aqueous solutions) are within this intermediate conductivity regime between 10^{-4} and 1 S/m.

---

needle-to-plate configuration, the critical voltage can be approximated as

$$\Phi_m \approx E_m a \sqrt{\frac{\cos\theta}{2}} \ln\left(\frac{4l}{a}\right),$$

where θ is the conical half angle. This expression is valid when the electrode separation (l) is much larger than the needle radius (a) (Smith, 1986; Kebarle and Verkerk, 2009).

The dynamics outside of the steady cone-jet regime is very poorly understood, see for example Cloupeau and Prunet-Foch (1994); Grace and Marijnissen (1994). With the caveats of being overly simplified and *incomplete*, a sketch of the operating modes in the $E - Q$ diagram is offered as follows:

- $Q < Q_m$: If the flow rate is smaller than the theoretical minimum, a stable jet is not possible and a variety of pulsating modes may result.
  - $E \simeq E_m$: *Pulsating cone-jet*, which results from the imbalance of supplied and jetted flow rates. Many electrostatic printing and "nanoelectrospray" (Wilm and Mann, 1994) systems fall in this pulsating regime as a result of the increased drag in miniaturized nozzles. The pulsation mechanism will be discussed below.
- $Q \gtrsim Q_m$: If the flow rate is intermediate for which a steady cone-jet is possible, the mode of operation depends strongly on the electric field:
  - $E \ll E_m$: *Dripping*, which is analogous to hydrodynamic dripping (Eggers and Villermaux, 2008).
  - $E \lesssim E_m$: *Pulsating cone-jet*, which sometimes resembles the aforementioned case when $Q \lesssim Q_m, E \simeq E_m$. This transition region is very complex and can be subdivided further into bursting, pulsating and astable modes, where the astable mode is not periodic at all (Marginean et al., 2007).
  - $E \simeq E_m$: *Cone-jet*, which is the most useful mode (Fernandez de la Mora, 2007) and the benchmark that all other operating modes are compared against. Electrospraying and electrospinning systems typically employ a stable cone-jet, which eventually destabilizes downstream in the *varicose* or *kink* (whipping) mode, respectively. The instability mechanism will be discussed below.
  - $E \gg E_m$: *Multi-jet*, which occurs when the rapidly accumulated surface charges must be redistributed through multiple jets. There are many other possible modes in addition to the multi-jet one (Cloupeau and Prunet-Foch, 1994), but these highly unstable modes are rarely used in practice.
- $Q \gg Q_m$: If the flow rate is much larger than the theoretical minimum, a hydrodynamic jet results instead of a cone-jet; this hydrodynamic regime is usually not of practical interest (Grace and Marijnissen, 1994).

4.2 Long-wavelength model

Consistent with the surface-coupled model adopted below, the working fluid is assumed to be a leaky dielectric such that (i) the liquid is sufficiently conducting to carry electric charges only on its surface and any bulk charge will instantaneously relax to the interface; (ii) the liquid is sufficiently dielectric to support a tangential electric field (Melcher and Taylor, 1969; Melcher and Warren, 1971; Ganan-Calvo, 1997a; Hohman et al., 2001a; Feng, 2002). Note that the assumption of instantaneous charge relaxation refers to the *local* relaxation of bulk charge density (Ganan-Calvo, 1997a; Hohman et al., 2001a). The external fluid (air) is assumed to have no effect on the jet except to provide a uniform external pressure. As we have stressed earlier, only a leaky dielectric can support tangential shear stress, which pulls charge and mass toward the conical tip and is the root cause for the eventual jet issuance (Hayati et al., 1986); cf. Figure 2c. We shall restrict our discussions to Newtonian fluids; for non-Newtonian effects, see for example Reneker et al. (2007); Feng (2002).

In the long-wavelength limit, the electrified jet is asymptotically governed by a set of one-dimensional equations (Hohman et al., 2001a; Feng, 2002). In the axisymmetric case,

$$\partial_t(\pi h^2) + \partial_z(\pi h^2 v) = 0, \tag{50a}$$

$$\partial_t(2\pi h q_s) + \partial_z(2\pi h q_s v + \pi h^2 \sigma E) = 0, \tag{50b}$$

$$\rho(\partial_t v + v\partial_z v) = -\partial_z p_{tot} + \rho g + \frac{2}{h} q_s E + \frac{3}{h^2}\mu\partial_z(h^2\partial_z v), \tag{50c}$$

with

$$p_{tot} = \gamma\kappa - \frac{1}{2}(\varepsilon - \varepsilon_0)E^2 - \frac{q_s^2}{2\varepsilon_0},$$

and

$$E = E_\infty + \ln\xi \left[\frac{1}{\varepsilon_0}\partial_z(q_s h) - \frac{\varepsilon - \varepsilon_0}{2\varepsilon_0}\partial_z^2(Eh^2)\right], \tag{50d}$$

where $h(z)$ is the jet radius, v and E are the velocity and electric field in the axial z-direction, E_∞ is the external field in the absence of the jet, ξ is the local aspect ratio assumed to be small; for the linearized system, ξ is proportional to the wave number (Hohman et al., 2001a). Eq. 50a is the conservation of mass. Eq. 50b is the conservation of current, where the jet current consists of both surface convection and bulk conduction. The momentum Eq. 50c incorporates the same hydrodynamic terms as nonelectrical jets (Eggers and Villermaux, 2008) and a few additional electrical terms: the $q_s E$ term is related to the tangential electrical force (cf. Figure 2c), $(\varepsilon - \varepsilon_0)E^2/2$ is the polarization force under an external field (cf. Figure 2b), and $q_s^2/2\varepsilon_0$ is the radial repulsion of surface charge (cf. Figure 2a). Eq. 50d approximates the electrostatic field inducted by the surface charge by an effective line charge along the axis; for alternative approximations, see e.g. Ganan-Calvo (1997a). The key notion is that the self-inducted field modifies the imposed electric field, which complicates the coupled dynamics between electric and flow fields. Note that Eq. 50 is asymptotically valid for a slender jet in the long-wavelength limit with an additional requirement unique to electrified jet: the tangential electric stress must be much smaller than the radial viscous stress, $q_s E \ll \mu v/h$ (Hohman et al., 2001a).

The long-wavelength limit represents a major simplification to the governing equations while keeping the essential physics. The governing equations suitable for a whipping jet with curved centerline have also been derived in Hohman et al. (2001a). Among other complications, the bending introduces nonaxisymmetric distortions which necessitates the introduction of dipole in addition to monopole charge density. In the interest of space, we will base our discussions on the axisymmetric equations (Eq. 50), which contains the majority of the physics; however, we will cite results from the full governing equations applicable to both varicose and whipping instabilities.

Before discussing the electrified jet, it is instructive to review the stability physics of a purely mechanical jet. In a nonelectrical jet, surface tension is destabilizing and drives the well-known varicose instability first analyzed by Rayleigh (1879). The instability mechanism is that the perturbed jet gives rise to a lower overall surface energy when the wavelength is longer than the perimeter of the jet. All azimuthal modes are stable as surface corrugation always carries a penalty in surface energy. Both longitudinal stretching and viscous stresses slow down the instability and increases the most amplified wavelength. More details on nonelectrical jet can be found in Eggers and Villermaux (2008).

In an electrified jet, three additional terms are in action at the interface:[4] the normal polarization force due to the permittivity gradient at the liquid/air interface, the normal Coulombic force due to *self*-repulsion of the free surface charge, and the tangential electric stress on the free surface charge. Their roles in electrohydrodynamic stability are summarized as follows:

- The polarization forces is stabilizing in a current-carrying jet. A local constriction of the jet increases the electric field (by current conservation) which increases the outward polarization force to resist the constriction (Melcher and Warren, 1971).
- The self repulsion of surface charges stabilizes the nonelectrical varicose mode by reducing the effective surface tension, but at the same time promotes an electrical whipping mode by Coulombic repulsion

[4]The Korteweg-Helmholtz force concept is adopted, as usual, which nicely confines the polarization forces on the interfaces where there is a nonzero $\nabla\varepsilon$.

(Hohman et al., 2001a). Contrary to surface energy, electrostatic energy shrinks with expanding surface area. Nonaxisymmetric surface corrugation is therefore energetically favorable as far as electrostatic energy is concerned.

- The tangential electric stress provides a stabilizing longitudinal stretching by accelerating the jet. However, the same tangential electric stress can also destabilize the jet by promoting out-of-phase oscillation between surface charge rearrangement and the fluid response (Hohman et al., 2001a); see also discussions of Figure 12 below.

Both tangential shear stress and self repulsion can be important in promoting instability, depending on the relative magnitude of the imposed tangential electric field and the normal field induced by surface charge (Hohman et al., 2001a); see also discussions around Eq. 56 below.

The linear stability analysis by Hohman et al. (2001a) showed three instability modes: the axisymmetric Raleigh mode (the conventional varicose instability modified by the electric field), the axisymmetric conducting mode (caused by the redistribution of surface charges under an external field), and the whipping conducting mode (cf. Figure 8b). The latter two are called "conducting modes" because they only exist with a finite conductivity. The demarcation between the varicose and kink modes follows the following scaling,

$$q_{s,cr} \sim \sqrt{\frac{2\varepsilon_0 \gamma}{h}}, \tag{51}$$

where the critical surface charge density ($q_{s,cr}$) is typically an order of magnitude smaller than the Rayleigh limit for electrostatic explosion (Rayleigh, 1882). The Rayleigh limit ($\sqrt{2\varepsilon_0\gamma/h}$ for a jet with a radius of h) represents the competition between electrostatic repulsion and surface tension.

To quantitative compare the simplified 1D model with experimental results, the axial distribution of the jet diameter and surface charge density were calculated by fitting the experimentally measured profile of a stable cone-jet as it thins down from the nozzle (Hohman et al., 2001b); for alternatively approaches, see the discussions in Feng (2002). The operating diagram of an electrified jet of aqueous polymer is shown in Figure 10. Remarkably, the experimentally measured operating diagram of steady cone-jet is qualitatively reproduced by the stability analysis. The lower stability boundary demarcates the varicose instability and the steady jet, and the upper one demarcates the steady jet and the whipping instability. The linear stability analysis showed that both viscosity and surface charge are important to capture the long-wavelength and whipping features of the instability, respectively (Hohman et al., 2001a). These observations are consistent with the role of viscosity in mechanical jet instability and the fact that self-repulsive surface charge tends to maximize surface area. The model also captured scaling trends predicted by Eq. 49, with a higher conductivity leading to lower minimum flow rate (Hohman et al., 2001b). However, it should be noted that the theoretically predicted minimum flow rate based on the linear stability analysis seems to be a few orders of magnitude higher than the experimentally measured one (Figure 10).

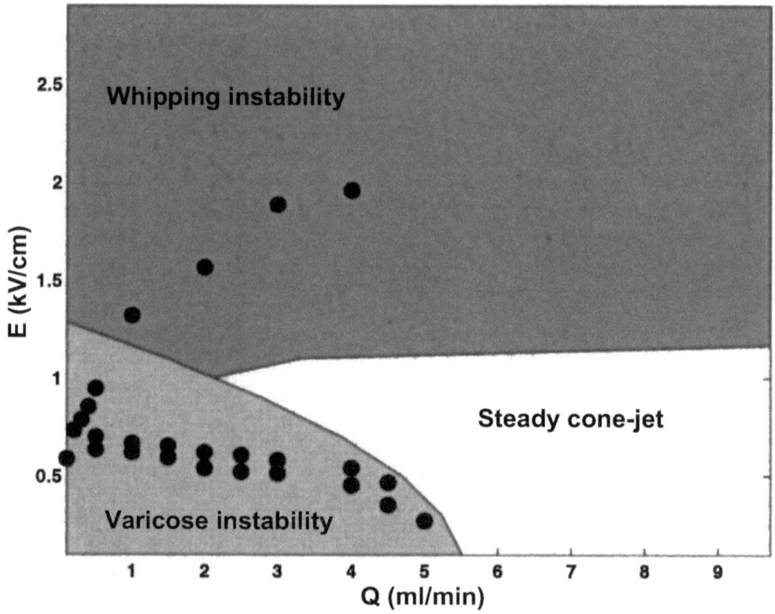

Figure 10. Prediction of varicose and whipping instabilities of an electrohydrodynamic jet using the one-dimensional model. Experimental measurements denoted by dots are overlayed on the numerical prediction with the white region denoting stable jet, light gray varicose instability, and dark gray whipping instability. The working fluid is aqueous solution of polyethylene oxide (assumed Newtonian), with $\sigma = 1.2 \times 10^{-2}$ S/m, $\varepsilon = 3.8 \times 10^{-10}$ C/V·m, $\mu = 1.7$ Pa·s, $\rho = 1.2 \times 10^3$ kg/m^3, and $\gamma = 6.4 \times 10^{-2}$ N/m. Reprinted with permission from Hohman et al. (2001b), © 2001 American Institute of Physics.

4.3 Convective and absolute instability

The convective and absolute instability transition in a charge liquid jet is studied by Lopez-Herrera et al. (2010) using a two-dimensional model of leaky dielectric jets assuming axisymmetric perturbations and negligible gravity and tangential electric field.[5] The assumption of negligible tangential electric field is relevant on a *developed* electrohydrodynamic jet where the electric current is dominated by surface convection instead of bulk conduction, and the tangential field is (typically) much smaller than the radial one induced by the free surface charge (Ganan-Calvo, 1997a,b). In this limit, the governing electrical parameters are the relative permittivity β, the relaxation parameter α, and the charge parameter χ,

$$\beta = \frac{\varepsilon}{\varepsilon_0}, \quad \alpha = \sqrt{\frac{\rho h^3 \sigma^2}{\gamma \epsilon^2}}, \quad \chi = \frac{h q_s^2}{\gamma \varepsilon_0}, \tag{52}$$

where α compares the capillary-inertial time with the charge relaxation time and χ compares the electrostatic pressure to capillary pressure. Close to the onset conditions for a stable cone-jet, calculations in Lopez-Herrera et al. (2010) showed that most cone-jet systems are moderately charged with a charge parameter χ below 1 ($\chi = 2$ at the Rayleigh limit given by Eq. 51).

For cone-jets with moderate charge parameter (χ), the absolute to convective instability (dripping to jetting) transition is found to be insensitive to electrical parameters; Instead, the transition on moderately charged jets is mainly governed by the mechanical parameters (Lopez-Herrera et al., 2010). The Weber number (We) is related to the Reynolds (Re) and Capillary (Ca) numbers by

$$We = \frac{\rho U^2 h}{\gamma} = \frac{\rho U h}{\mu} \frac{\mu U}{\gamma} = Re \cdot Ca, \tag{53}$$

The Weber number be viewed as the ratio of the imposed flow velocity (U) to the intrinsic capillary-inertial velocity ($\sqrt{\gamma/\rho h}$), similar to $1/R_v$ which is equal to the imposed electroosmotic flow to the intrinsic electroviscous velocity. Note that the Reynolds and Capillary numbers are related by the

[5] Within the limit of negligible tangential electric field, Lopez-Herrera et al. (2005) compared the more general two-dimensional model against the one-dimensional long-wavelength description, and showed that the long-wavelength limit breaks down for low-viscosity, low-conductivity/permittivity liquid. However, the long-wavelength model is usually valid for aqueous solutions due to their large conductivity and permittivity. See Lopez-Herrera et al. (2005) for details.

Ohnesorge number,

$$Oh = \frac{\mu}{\sqrt{\rho\gamma h}} = \sqrt{\frac{Ca}{Re}}, \tag{54}$$

which is not a function of velocity. Heuristically, with increasing convection velocity U, the flow transits from absolute to convective instability as the external convection process outcompetes the intrinsic capillary waves, i.e.

$$\frac{h}{U} < \frac{h}{\sqrt{\gamma/\rho h}} \quad \Rightarrow \quad We > 1. \tag{55}$$

The heuristical argument of the critical Weber number is confirmed by the spatiotemporal analysis in Figure 11, where the theoretical results are plotted along with experimental values of Reynolds and Capillary numbers for dozens of stable cone-jet configurations in the context of both electrospraying and electrospinning. The experimental conditions for stable cone-jets more or less follow a line of constant Weber number. Note that this hydrodynamically dominated criterion of stability transition breaks down in the limit of low Reynolds number, where the stability behavior is strongly sensitive to the electrical parameters (Lopez-Herrera et al., 2010).

The assumption of negligible tangential electric field is invalid close to the "neck" region of the cone-jet, where the dominant charge transport mechanism transits from bulk conduction to surface convection (Ganan-Calvo, 1997b; Lopez-Herrera et al., 2010). In the neck region, the tangential electrical stress is crucial to the dynamics of the cone-jet.

The essential role of tangential electric stress is illustrated by Figure 12 (Melcher and Warren, 1971); see also the film by Melcher (1974). The polarity of the free charge on the jet surface is controlled by a stack of wall electrodes (horizontal lines in the figure). The induced surface charge takes positive or negative polarity, depending on the relative potential between the wall electrodes and their corresponding location on the jet. The resulting tangential electric stress is either along (Figure 12a) or opposite to (12b) the jet flow direction, leading to "supercritical" and "subcritical" flow, respectively. The 1D model employed by Melcher (1974) is essentially the same as Eq. 50, with the viscous stresses and surface charge convection neglected and Eq. 50d replaced by a much simpler induction relationship. The neglection of viscous stresses reduces the system into a set of first order wave equations with upstream and downstream wave velocities a_- and a_+ *in the convective frame* traveling with the jet. In a supercritical flow represented by Figure 12a, the jet velocity exceeds the upstream wave velocity ($v > a_-$) so that downstream disturbances can not propagate upstream; In a subcritical flow ($v < a_-$) represented by Figure 12b, disturbances can

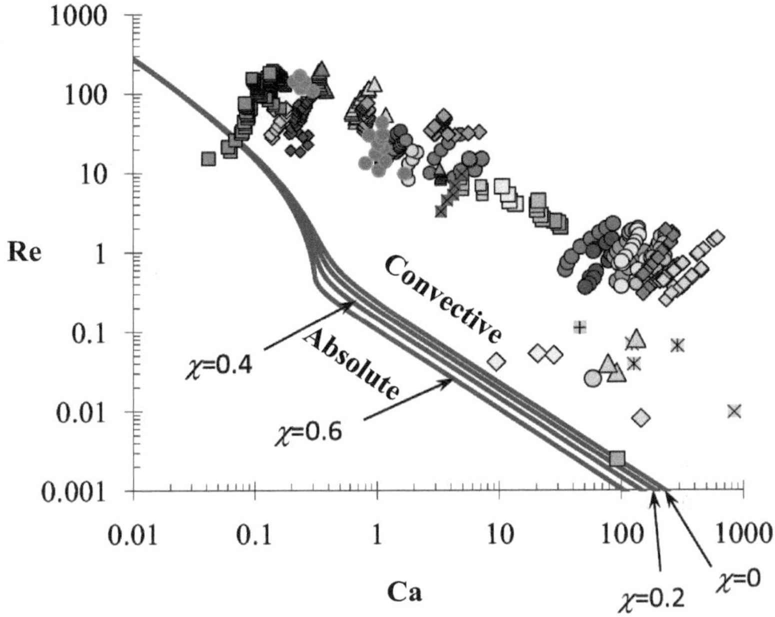

Figure 11. Absolute to convective instability transition for cone-jets. Solid curves correspond to the theoretical predictions of the critical condition as a function of the charge parameter χ (Eq. 52), while discrete data points are the Reynolds and Capillary numbers estimated from cone-jets reported in both electrospraying and electrospinning literature. Reprinted with permission from Lopez-Herrera et al. (2010), © 2010 American Institute of Physics.

propagate in both directions so the flow is dependent on both upstream and downstream conditions (Melcher and Warren, 1971). In this sense, the subcritical/supercritical jet flow is analogous to subsonic/supersonic flow of compressible gas.

The supercritical concept was originally developed to explain the unusually stable electrohydrodynamic jets with much longer breakup length compared to their nonelectrical counterparts (Melcher and Warren, 1971). Ganan-Calvo (1997a) extended this line of work to include the convection of surface charge and more importantly, self-induction of electric fields from the free surface charges (using a approach different from Eq. 50d). In Melcher (1974), the entire jet is either subcritical or supercritical when the surface charge is induced by wall electrodes. Ganan-Calvo (1997a) showed that a

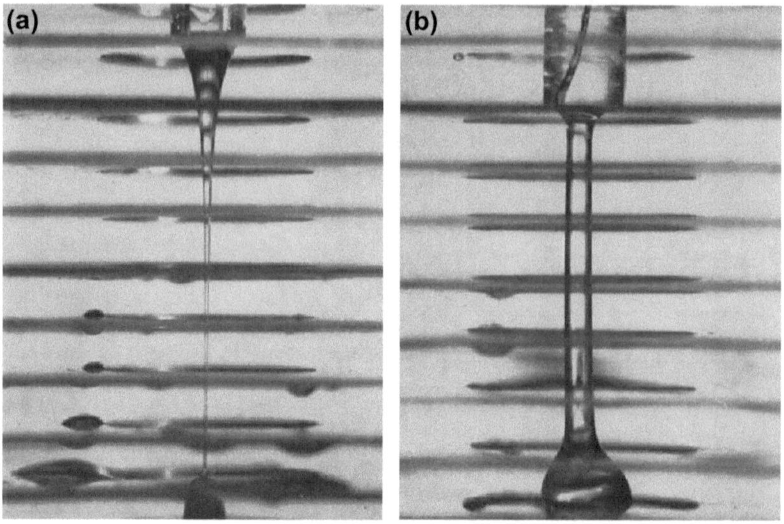

Figure 12. Reversal of the tangential electric stress on the jet leads to the transition from (a) supercritical to (b) subcritical flow. Voltage is applied between a nozzle at the top and a counter electrode at the bottom. Horizontal lines are the wall electrodes with a linear voltage dependence that is separately conrolled. In (a), dripping glycerin is accelerated into a jet by the applied electric field; In (b), the polarity of surface charge on the jet is reversed by changing the wall potentials relative to the potential on the jet; the reversed tangential electric stress leads to a subcritical flow that approaches the critical ("sonic") point at the bottom electrode. Reprinted with permission from Melcher and Warren (1971), © 1971 Cambridge University Press.

self-inductive cone-jet transits from the subcritical to supercritical regime at a critical point along the cone-jet. Downstream of the critical point, the supercritical jet eventually develops a convective instability. The supercritical portion of the jet between the critical point and the point of instability provides a shield to prevent perturbations (e.g. from jet breakup) from propagating upstream, giving rise to the unusual stability of electrohydrodynamic jets. Lopez-Herrera et al. (2010) confirmed that the jet issued from a Taylor cone is supercritical at the region prior to breakup.

In a global analysis of the cone-jet, the relative importance of the radial to tangential electric stress,

$$\vartheta = \frac{T_{rr}^e}{T_{rz}^e} \sim \frac{q_s^2/\varepsilon_0}{q_s E_\infty} = \frac{q_s/\varepsilon_0}{E_\infty}, \tag{56}$$

is a function of location along the developing cone-jet. Close to the cone, tangential electric field dominates with a small surface charge density ($\vartheta \ll 1$); Along the jet, radial electric field eventually dominates with increasing surface charge density downstream ($\vartheta \gg 1$). The surface charge density is dependent on the detailed cone-jet dynamics. In the form of Eq. 50 with negligible gravity, q_s is governed by the imposed electric field and flow rate ($\frac{E_\infty}{E_m}, \frac{Q}{Q_m}$) and the associated geometrical ($\frac{z}{a}, \xi$), mechanical (We, Oh), and electrical (α, β, χ) parameters. Because of the complex, global nature of the problem, no controlling parameters (like Ra_e in Eq. 44) have emerged with appropriate scaling of the surface charge density in a developing cone-jet.

One possibility to reduce the complexity is to operate exactly at the minimum flow rate $Q = Q_m$; accordingly, it is necessary for $E = E_m$ to obtain a steady cone-jet. Unfortunately, except for the empirical scaling Eq. 49, the mechanism determining Q_m is still unknown (Fernandez de la Mora, 2007).

4.4 Pulsating cone-jet

The crucial role of the minimum flow rate (Q_m) is apparent on the operating diagram of steady cone-jets (Figure 9). In the previous section, we showed that a minimum flow rate (a critical Weber number) is necessary for the dripping to jetting transition (Eq. 55) *if* the electrohydrodynamic jet behaves analogously to its hydrodynamic counterpart. However, when $Q < Q_m$ but $E \simeq E_m$, pulsating cone-jet sets in from a supported meniscus which does not have a clear hydrodynamic analogue.

Pulsating cone-jets are also observed on charged drops (e.g. Figure 8c); see Fernandez de la Mora (2007) for a comprehensive review. For an inviscid charged drop in air, the lowest frequency of free drop oscillation (f_d) is given by Rayleigh (1882)

$$f_d = \frac{4}{\pi}\sqrt{\frac{\gamma}{\rho d^3}\left(1 - \frac{q^2}{q_R^2}\right)} \propto f_c, \tag{57}$$

where d is the diameter of the drop, q is the *total* charge on the surface, and

$$q_R = \pi\sqrt{8\varepsilon_0 \gamma d^3}, \tag{58}$$

is the Rayleigh limit for the maximum electrostatic charge allowed on a static drop. Without any surface charge ($q = 0$), the frequency is simply governed by the capillary-inertial process with a characteristic frequency of $f_c = \sqrt{\gamma/\rho d^3}$. At the Rayleigh limit, electrostatic repulsion reduces the effective surface tension to zero, leading to electrostatic fission of the drop. The electrostatic explosion can proceed through a fine fission mode *via* a cone-jet (Figure 8c and Figure 13b), or a rough fission mode into a few drops of comparable size (Fernandez de la Mora, 2007).

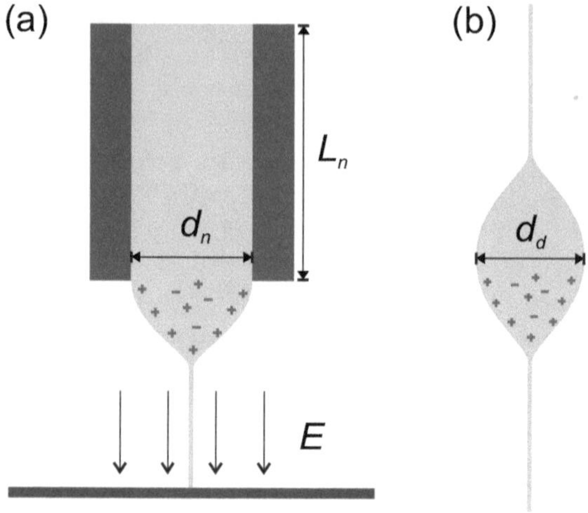

Figure 13. Analogy between pulsating cone-jets from a supported meniscus and an isolated drop: (a) The cone-jet transition under an external electric field between a nozzle and a plate; (b) The cone-jet transition when a droplet experiences electrostatic fission.

Fernandez de la Mora (1996) argued that the cone-jet pulsation on a supported meniscus is analogous to that on an isolated drop, as both result from the redistribution of the excessive electrostatic charge to a larger surface area (Figure 13).[6] When the time scale for cone-jet formation is short compared to the duration of the cone-jet, which is the case for high conductivity liquid with rapid charge relaxation, the transient cone-jet is quasi-steady and is assumed to behave similarly to a steady cone-jet on a supported meniscus (Fernandez de la Mora, 1996). Since pulsation takes place when the supply rate of liquid to the cone is less than the loss rate through the jet (Juraschek and Rollgen, 1998), a pulsating cone-jet will

[6]There are also many differences between steady and transient cone-jets, especially for liquids with high viscosity and low conductivity; see Fernandez de la Mora (2007) for details.

likely observe the minimum flow rate scaling (Fernandez de la Mora, 1996; Chen et al., 2006b). When the transient jet is on, the flow rate, jet diameter and electric current scale as (Fernandez de la Mora and Loscertales, 1994; Fernandez de la Mora, 1996; Barrero and Loscertales, 2007)

$$Q_j \sim \frac{\gamma\varepsilon}{\rho\sigma}, \tag{59a}$$

$$d_j \sim \left(\frac{\gamma\varepsilon^2}{\rho\sigma^2}\right)^{1/3}, \tag{59b}$$

$$i_j \sim g(\varepsilon)\left(\frac{\varepsilon_0\gamma^2}{\rho}\right)^{1/2}, \tag{59c}$$

where $g(\varepsilon)$ accounts for the effects of liquid dielectric constant; For aqueous solutions, $g \approx 18$ (Fernandez de la Mora and Loscertales, 1994). The lifetime of a transient cone-jet on an exploding drop can be obtained by integrating $dt = -dq/i$ (Fernandez de la Mora, 1996),

$$\Delta t_j \sim \frac{\Delta q}{i_j} \sim \frac{\Delta q}{q_R}\frac{q_R}{i_j} \sim \frac{2\sqrt{2}\pi\Delta q}{g(\varepsilon)q_R}\sqrt{\frac{\rho d^3}{\gamma}} \sim \sqrt{\frac{\rho d^3}{\gamma}} = \tau_c, \tag{60}$$

where Δt_j is the time scale for a drop with surface charge approaching the Rayleigh limit (q_R) to emit enough charge (Δq) to reach a new electrostatic equilibrium. It is generally agreed that the mass loss due to electrostatic fission is negligible (order of 1%), but the charge loss is substantial (order of 10%); see Fernandez de la Mora (1996). For water, the pre-factor in the scaling of jet duration is close to 1 (Chen et al., 2006b). Interestingly, the jetting duration (Eq. 60) reduces to the capillary-inertial time scale ($\tau_c = 1/f_c$) which also sets the oscillation frequency of an uncharged drop (Eq. 57). If the pulsating cone-jet on a supported meniscus is analogous to that on an exploding drop, the above scaling laws apply to both situations with d taken as either the drop diameter (d_d) or the *anchoring* diameter of the Taylor cone (e.g. the inner diameter of a non-wetting nozzle, d_n).

We are now in a position to reconcile two different models for the pulsating cone-jets by Marginean et al. (2006) and Chen et al. (2006b). At the onset of cone-jet pulsation, Marginean et al. (2006) observed that the pulsation frequency on a support meniscus closely follows Eq. 57 with d equal to the anchoring diameter d_n. However, the free oscillation frequency is independent of the supplied flow rate Q, and this independence contradicts with the observations in Chen et al. (2006b) as well as empirical evidences that the emergence of cone-jet pulsation strongly depends on the flow rate (see Section 4.1).

Chen et al. (2006b) proposed a different model for the pulsation frequency, motivated by the hypothesis that cone-jet pulsations result when a transient jet discharges liquid mass at a rate (Q_j) higher than the supplied flow rate (Q). For each pulsation with a duration of Δt_j, the volume of liquid extracted scales as $Q_j \Delta t_j$. By simple balance of mass flow supplied to the cone and emitted by the jet, the pulsation frequency scales as (Chen et al., 2006b)

$$f_j \sim \frac{Q}{Q_j \Delta t_j} \sim \frac{Q}{Q_m} f_c. \tag{61}$$

In other words, cone-jet pulsations result when the mass flow is limited ("choked") by upstream conditions. Note that for high conductivity fluid, the limiting factor leading to pulsation is the mass flow, not free charge, because surface charges can be generated at a rate governed the nearly instantaneous charge relaxation process. The pulsation model Eq. 61 has been confirmed by Chen et al. (2006a,b) for cone-jet pulsations with a stable conical base (see insets in Figure 14).

Xu and Chen recently showed that both models are correct within their applicable regime (Xu, 2010). In Figure 14, both laws have been identified in the same system using flow rate as the only controlling parameter. As the flow rate increases, the frequency plateaus at relatively high flow rates. Microscopic imaging indicated that the oscillation modes at low and high flow rates are fundamentally different. At low flow rates, the Taylor cone only deforms at the conical apex ("mass flow choking" regime); at high flow rates, the entire Taylor cone deforms significantly ("conical oscillation" regime). The demarcation between the two regimes is related to the minimum flow rate (Xu, 2010); For the conditions in Figure 14, $Q_m \approx 50$ μL/h according to Eq. 49.

- $Q < Q_m$: The mass flow is choked upstream, and the pulsation frequency scales as $f \sim (Q/Q_m) f_c$ (Eq. 61). The linear dependence of the pulsation frequency on the flow rate is confirmed at small flow rates (Figure 14). The mass flow choking regime is unambiguously shown by cases with self-induced flow rate (Xu, 2010), which scales inversely with the length of the slender nozzle (Chen et al., 2006b). The nozzle length (L_n) is an upstream condition that will not change the conical oscillation frequency (Eq. 57).
- $Q \gtrsim Q_m$: The conical base oscillates when ample liquid is supplied, and the frequency scales as $f \sim f_c$ (Eq. 57). The $f_c \sim d^{-\frac{3}{2}}$ scaling is verified by Marginean et al. (2006) and confirmed by Xu (2010). The diameter should be taken as that anchoring the conical base. For low viscosity liquid, the anchoring diameter can become ambiguous and a

non-wettable nozzle (such as teflon) should be used to avoid exciting the conical oscillation regime even at flow rates below Q_m.

Note that the charge relaxation time (τ_e) is 3 μs in Figure 14, an order of magnitude smaller than the 30 μs jetting duration (Δt_j) predicted by Eq. (60) and confirmed by experiments. Hence, the quasi-steady assumption holds for the transient cone-jet discussed here.

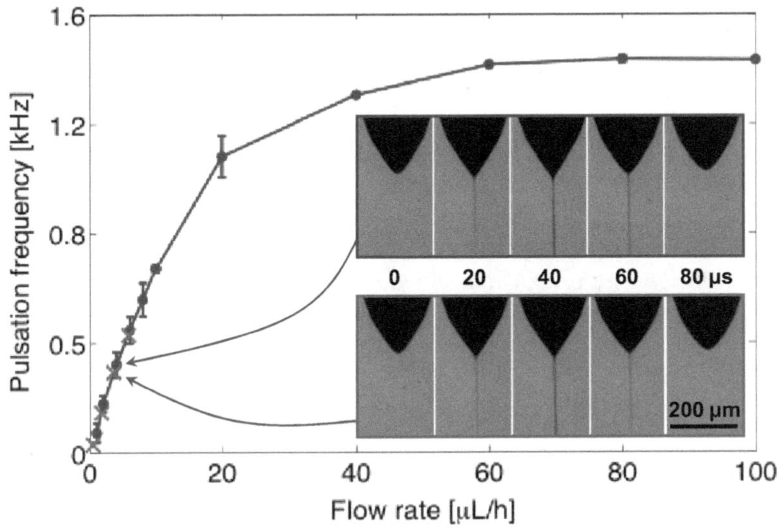

Figure 14. Pulsation frequency as a function of the imposed (•) and induced (×) flow rate. The imposed flow rate is controlled by a syringe pump, while the electric-stress-induced flow rate is varied by the nozzle length. The glass nozzle has an inner diameter of 100 μm and an outer diameter of 160 μm; here the outer diameter is the anchoring diameter. Inset shows one duty cycle of pulsation for the imposed (top) and induced (bottom) flow rate. The applied voltage of 1.8 kV is slightly lower than the threshold ($\simeq$ 2.0 kV) for a steady cone-jet. The working fluid is doped ethylene glycol with $\sigma = 1.0 \times 10^{-3}$ S/m, $\varepsilon = 3.3 \times 10^{-10}$ C/V·m, $\mu = 1.6 \times 10^{-2}$ Pa·s, $\rho = 1.1 \times 10^3$ kg/m^3, and $\gamma = 4.5 \times 10^{-2}$ N/m.

The mechanistic understanding of pulsating cone-jets has practical implications for both electrohydrodynamic printing and miniaturized electrospray. In both applications, slender nozzles with a diameter from 100 μm down to 1 μm are routinely used. The viscous drag on the slender nozzle limits the flow rate and leads to the regime of mass flow choking (Chen et al., 2006b). The intrinsic pulsations in the choking regime will influence the performance of both electroprinting and nanoelectrospray. For example, the highest speed of electrohydrodynamic printing is limited by the intrinsic pulsation frequency (Chen et al., 2006a). The advantage of nanoelectrospray over conventional electrospray diminishes with increasing flow rates (Schmidt et al., 2003); the critical flow rate is around the minimum flow rate (Eq. 49), suggesting a possible role of cone-jet pulsations in nanoelectrospray.

5 Concluding Remarks

In this chapter, electrokinetic mixing flow and electrohydrodynamic cone-jet are reviewed as two model problems of electrohydrodynamic stability. In both cases, knowledge of the stability boundary is of practical interest for either enhancing or suppressing the instability. Both problems are studied using the Ohmic model for leaky dielectrics with the assumptions of instantaneous local charge relaxation and negligible diffusive current, which hold for relatively conducting working fluids such as aqueous solutions.

The stability of electrokinetic mixing flow is relatively simple, owing in part to the well-understood base state of a diffusive interface. Using the bulk-coupled model which accounts for the electric body forces induced by conductivity gradients, a large set of relevant experimental parameters can be conveniently reduced to two controlling parameters: the electric Rayleigh number relating the dynamic to dissipative processes which governs the onset of instability, and the electroviscous-to-electroosmotic velocity ratio which governs the onset of absolute instability.

The stability of electrohydrodynamic cone-jet is significantly more difficult. Much of the difficulty is associated with the lack of a fundamental understanding of the base state, i.e. the mechanisms leading to the formation of the Taylor cone-jet. Surface-coupled model is usually employed to relate the electric and flow fields through the Maxwell stress, which now consists of contributions from both free and bound charge due to conductivity and permittivity gradients, respectively. Compared to its nonelectrical counterpart, the physics of electrohydrodynamic jets is enriched but complicated by the electric field. The most useful guideline is still the electric field–flow rate operating diagram. Strategies to rationalize the operating diagram in the framework of electrohydrodynamic stability have only been partially successful, e.g. within specific parametrical regimes of pulsating, spraying, or spinning cone-jets. Further research is required to understand key issues such as the origin of the minimum flow rate and the role of tangential electric stress before a unified theoretical picture is possible.

Bibliography

A. G. Bailey. *Electrostatic Spraying of Liquids.* Research Studies Press, 1988.

A. Barrero and I. G. Loscertales. Micro- and nanoparticles via capillary flows. *Annu. Rev. Fluid Mech.*, 39:89–106, 2007.

J. C. Baygents and F. Baldessari. Electrohydrodynamic instability in a thin fluid layer with an electrical conductivity gradient. *Phys. Fluids*, 10:301–311, 1998.

H. C. Chang and L. Y. Yeo. *Electrokinetically-Driven Microfluidics and Nanofluidics.* Cambridge University Press, 2010.

C. H. Chen, H. Lin, S. K. Lele, and J. G. Santiago. Convective and absolute electrokinetic instability with conductivity gradients. *J. Fluid Mech.*, 524:263–303, 2005.

C. H. Chen, D. A. Saville, and I. A. Aksay. Electrohydrodynamic 'drop-and-place' particle deployment. *Appl. Phys. Lett.*, 88:154104, 2006a.

C. H. Chen, D. A. Saville, and I. A. Aksay. Scaling laws for pulsed electrohydrodynamic drop formation. *Appl. Phys. Lett.*, 89:124103, 2006b.

M. Cloupeau and B. Prunet-Foch. Electrostatic spraying of liquids in cone-jet mode. *J. Electrostat.*, 22:135–159, 1989.

M. Cloupeau and B. Prunet-Foch. Electrohydrodynamic spraying functioning modes: a critical review. *J. Aerosol Sci.*, 25:1021–1036, 1994.

D. Duft, T. Achtzehn, R. Muller, B. A. Huber, and T. Leisner. Coulomb fission: Rayleigh jets from levitated microdroplets. *Nature*, 421:128, 2003.

J. Eggers and E. Villermaux. Physics of liquid jets. *Rep. Prog. Phys.*, 71: 036601, 2008.

J. J. Feng. The stretching of an electrified non-Newtonian jet: a model for electrospinning. *Phys. Fluids*, 14:3912–3926, 2002.

J. Fernandez de la Mora. On the outcome of the coulombic fission of a charged isolated drop. *J. Colloid Interface Sci.*, 178:209–218, 1996.

J. Fernandez de la Mora. The fluid dynamics of Taylor cones. *Annu. Rev. Fluid Mech.*, 39:217–243, 2007.

J. Fernandez de la Mora and I. G. Loscertales. The current emitted by highly conducting Taylor cones. *J. Fluid Mech.*, 260:155–184, 1994.

A. M. Ganan-Calvo. On the theory of electrohydrodynamically driven capillary jets. *J. Fluid Mech.*, 335:165–188, 1997a.

A. M. Ganan-Calvo. Cone-jet analytical extension of Taylor's electrostatic solution and the asymptotic universal scaling laws in electrospraying. *Phys. Rev. Lett.*, 79:217–220, 1997b.

A. M. Ganan-Calvo and J. M. Montanero. Revision of capillary cone-jet physics: electrospray and flow focusing. *Phys. Rev. E*, 79:066305, 2009.

J. M. Grace and J. C. M. Marijnissen. A review of liquid atomization by electrical means. *J. Aerosol Sci.*, 25:1005–1019, 1994.

I. Hayati, A. I. Bailey, and Th. F. Tadros. Mechanism of stable jet formation in electrohydrodynamic atomization. *Nature*, 319:41–43, 1986.

J. F. Hoburg and J. R. Melcher. Internal electrohydrodynamic instability and mixing of fluids with orthogonal field and conductivity gradients. *J. Fluid Mech.*, 73:333–351, 1976.

M. Hohman, M. Shin, G. C. Rutledge, and M. P. Brenner. Electrospinning and electrically forced jets. I. Stability theory. *Phys. Fluids*, 13:2201–2220, 2001a.

M. Hohman, M. Shin, G. C. Rutledge, and M. P. Brenner. Electrospinning and electrically forced jets. II. Applications. *Phys. Fluids*, 13:2221–2236, 2001b.

P. Huerre and M. Rossi. Hydrodynamic instabilities in open flows. In C. Godreche and P. Manneville, editors, *Hydrodynamics and Nonlinear Instabilities*. Cambridge University Press, 1998.

R. Juraschek and F. W. Rollgen. Pulsation phenomena during electrospray ionization. *Int. J. Mass. Spectrom*, 177:1–15, 1998.

P. Kebarle and U. H. Verkerk. Electrospray: from ions in solution to ions in the gas phase, what we know now. *Mass Spectrom. Rev.*, 28:898–917, 2009.

L. G. Leal. *Advanced Transport Phenomena*. Cambridge University Press, 2007.

W. G. Levich. *Physicochemical Hydrodynamics*. Prentice-Hall, 1962.

H. Lin. Electrokinetic instability in microchannel flows: a review. *Mech. Res. Comm.*, 36:33–38, 2009.

H. Lin, B. D. Storey, M. H. Oddy, C. H. Chen, and J.G. Santiago. Instability of electrokinetic microchannel flows with conductivity gradients. *Phys. Fluids*, 16:1922–1935, 2004.

J. M. Lopez-Herrera, A. M. Ganan-Calvo, and M. A. Herrada. Linear stability analysis of axisymmetric perturbations in imperfectly conducting liquid jets. *Phys. Fluids*, 17:034106, 2005.

J. M. Lopez-Herrera, P. Riesco-Chueca, and A. M. Ganan-Calvo. Absolute to convective instability transition in charged liquid jets. *Phys. Fluids*, 22:062002, 2010.

A. Manz, N. Graber, and H. M. Widmer. Miniaturized total chemical-analysis systems - a novel concept for chemical sensing. *Sensors and Actuators-B*, 1:244–248, 1990.

I. Marginean, P. Nemes, L. Parvin, and A. Vertes. How much charge is there on a pulsating Taylor cone? *Appl. Phys. Lett.*, 89:064104, 2006.

I. Marginean, P. Nemes, and A. Vertes. Astable regime in electrosprays. *Phys. Rev. E*, 76:026320, 2007.

J. R. Melcher. Electric fields and moving media. *IEEE Trans. Educ.*, E-17: 100–110, 1974. This article describes the 16 mm film by J. R. Melcher, *Electric Fields and Moving Media*, produced by the Education Development Center, 39 Chapel Street, Newton, MA.

J. R. Melcher. *Continuum Electromechanics.* MIT Press, 1981.

J. R. Melcher and W. J. Schwartz. Interfacial relaxation overstability in a tangential electric field. *Phys. Fluids*, 11:2604–2616, 1968.

J. R. Melcher and G. I. Taylor. Electrohydrodynamics: a review of the role of interfacial shear stress. *Annu. Rev. Fluid Mech.*, 1:111–146, 1969.

J. R. Melcher and E. P. Warren. Electrohydrodynamics of a current-carrying semi-insulating jet. *J. Fluid Mech.*, 47:127–143, 1971.

M. H. Oddy, J. G. Santiago, and J. C. Mikkelsen. Electrokinetic instability micromixing. *Anal. Chem.*, 73:5822–5832, 2001.

W. K. H. Panofsky and M. Phillips. *Classical Electricity and Magnetism.* Addison-Wesley, 2nd edition, 1962. Dover, 2005.

J. U. Park, M. Hardy, S. J. Kang, K. Barton, K. Adair, D. K. Mukhopadhyay, C. Y. Lee, M. S. Strano, A. G. Alleyne, J. G. Georgiadis, P. M. Ferreira, and J.A. Rogers. High-resolution electrohydrodynamic jet printing. *Nat. Mater.*, 6:782–789, 2007.

J. D. Posner and J. G. Santiago. Convective instability of electrokinetic flows in a cross-shaped microchannel. *J. Fluid Mech.*, 555:1–42, 2006.

Lord Rayleigh. On the instability of jets. *Proc. London Math. Soc.*, 10: 4–13, 1879.

Lord Rayleigh. On the equilibrium of liquid conducting masses charged with electricity. *Philos. Mag.*, 14:184–186, 1882.

D. H. Reneker, A. L. Yarin, E. Zussman, and H. Xu. Electrospinning of nanofibers from polymer solutions and melts. *Adv. Appl. Mech.*, 41: 43–195, 2007.

D. A. Saville. Electrohydrodynamics: the Taylor-Melcher leaky dielectric model. *Annu. Rev. Fluid Mech.*, 29:27–64, 1997.

P. J. Schmid and D. S. Henningson. *Stability and Transition in Shear Flows.* Springer, 2001.

A. Schmidt, M. Karas, and T. Dulcks. Effect of different solution flow rates on analyte ion signals in nano-ESI MS, or: when does ESI turn into nano-ESI? *J. Am. Soc. Mass Spectrom.*, 14:492–500, 2003.

D. P. H. Smith. The electrohydodynamic atomization of liquids. *IEEE Trans. Indus. Appl.*, IA-22:527–535, 1986.

H. A. Stone, A. D. Stroock, , and A. Ajdari. Engineering flows in small devices: microfluidics toward a lab-on-a-chip. *Annu. Rev. Fluid Mech.*, 36:381–411, 2004.

G. I. Taylor. Disintegration of water drops in an electric field. *Proc. R. Soc. London A*, 280:383–397, 1964.

M. S. Wilm and M. Mann. Electrospray and Taylor-cone theory, Dole's beam of macromolecules at last? *Int. J. Mass Spectrom.*, 136:167–180, 1994.

S. Xu. Operating regimes of self-regulated electrohydrodynamic cone-jets. Master's thesis, Duke University, 2010.

O. Yogi, T. Kawakami, M. Yamauchi, J.Y. Ye, and M. Ishikawa. On-demand droplet spotter for preparing pico- to femtoliter droplets on surfaces. *Anal. Chem.*, 73:1896–1902, 2001.

J. Zeng and T. Korsmeyer. Principles of droplet electrohydrodynamics for lab-on-a-chip. *Lab Chip*, 4:265–277, 2004.

Induced-Charge Electrokinetic Phenomena

Martin Z. Bazant [*]

[*] Departments of Chemical Engineering and Mathematics,
Massachusetts Institute of Technology, Cambridge, MA 02139 USA

Abstract This chapter provides an introduction to a certain class of nonlinear electrokinetic phenomena, where the applied electric field acts on its own induced-charge in an electrolytic solution near a polarizable surface. Many applications are discussed, such as colloidal particle dynamics (induced-charge electrophoresis) and microfluidic mixing and pumping (induced-charge electro-osmosis), while emphasizing the basic physics of each phenomenon. A Standard Model for these situations is introduced and analyzed in simple cases. Similarities and differences are noted with other electrokinetic phenomena, such as classical linear (fixed-charge) electrokinetics in electrolytes and electrohydrodynamics in leaky dielectrics.

1 Introduction

Electrokinetic phenomena (electrically driven fluid flow and particle motion) in liquid electrolytes have been studied for well over a century in colloid science (Hunter, 2001; Lyklema, 1995; Anderson, 1989), but much recent interest in this classical subject has been triggered by the development of microfluidics (Stone et al., 2004; Squires and Quake, 2005; Laser and Santiago, 2004; Squires, 2009; Schoch et al., 2008). In electrolytes, electrokinetic phenomena are associated with thin electric double layers on charged surfaces, and as a result they have favorable scaling with miniaturization, with increasing surface to volume ratio. Electrokinetic phenomena also offer other unique advantages for "lab-on-a-chip" systems, such as low hydrodynamic dispersion, no moving parts, electrical actuation and sensing, and easy integration with microelectronics. Beyond microfluidics, it is becoming increasingly recognized that electrokinetic phenomena can play an important role in the dynamics of electrified interfaces in other fields such as biology (e.g. vesicle motion, membrane fluctuations, electroporation) and electrochemistry (e.g. porous electrode charging, desalination dynamics, dendritic growth).

Until recently, almost all studies of electrokinetic phenomena have assumed linear response in the applied voltage, based on the hypothesis of fixed surface charge (or fixed "zeta potential" relative to the bulk solution). This assumption is reasonable for most insulating or dielectric surfaces, but not for metallic or ion-conducting surfaces. For applications in microfluidic systems, linear electrokinetic phenomena have a number of possible drawbacks: Direct current (DC) must be passed to sustain electric fields; it is difficult to produce vortices for mixing; and large voltages must be applied along centimeter or greater distances to achieve the necessary field strengths, giving little direct control over local fields and flows within microchannels. Related to these issues, there are also well known drawbacks of linear electrokinetic phenomena in colloid science, e.g. that electrophoresis cannot separate particles with fixed, uniform zeta potential by size or shape in free solution. This is the reason that electrophoretic separation of DNA or other large molecules is usually done in a gel, rather in free solution, to take advantage of entropic effects of trapping rather than differences in electrophoretic mobility.

As shown in Figure 1, much richer dynamics are possible with nonlinear electrokinetic phenomena at polarizable surfaces, which are the focus of this chapter. For recent reviews, see Bazant et al. (2009b), Squires (2009) and Bazant and Squires (2010). The development of this subject in microfluidics began with the discovery by Ramos et al. (1999) of alternating-current electro-osmotic flow (ACEO) over microelectrodes, which Ajdari (2000) showed could be exploited for low-voltage microfluidic pumping using asymmetric arrays of inter-digitated electrodes. These breakthroughs, supported by the early experiments of Green et al. (2000a); González et al. (2000); Green et al. (2002); Brown et al. (2000); Studer et al. (2004) and others, focused attention on nonlinear AC electrokinetics in microfluidics. This work clearly demonstrated that electrokinetic phenomena can derive from non-uniform, transient charge on an electrode surface, controlled more by the applied voltage than by chemical equilibrium.

Bazant and Squires (2004) pointed out that the underlying physical mechanism of an electric field acting on its own induced charge near a polarizable surface is more general and coined the term "induced-charge electro-osmosis" (ICEO) to describe it (Squires and Bazant, 2004). Through a variety of examples, such as those in Fig. 1, they argued that ICEO flows can occur around any polarizable (metal or dielectric) surface in the presence of any (DC or low-frequency AC) electric field – i.e. not exclusively over electrodes whose voltage is directly forced to oscillate at a certain frequency, as in ACEO. The same fundamental physical process, sketched in Figure 2, thus unifies ACEO and travelling-wave electro-osmosis (TWEO)

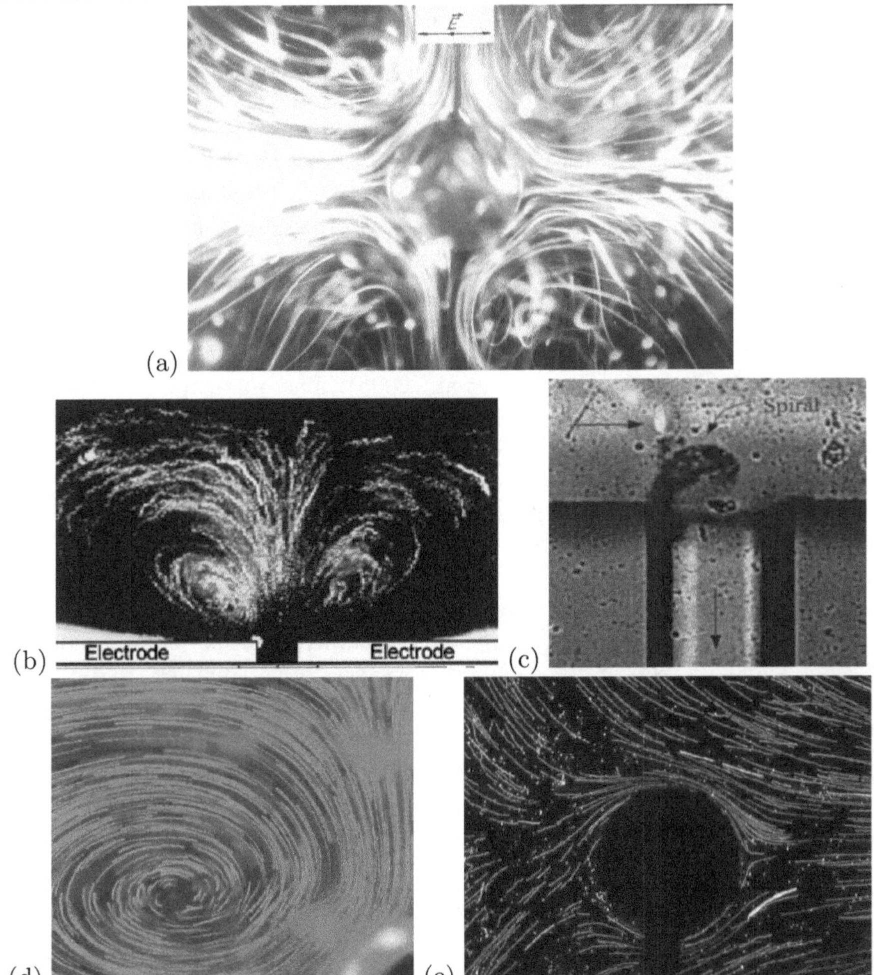

Figure 1. Experimental photographs of induced-charge electro-osmosis, imaged by streaks of tracer particles. (a) Nonlinear flow around a 500 μm spherical ionite particle driven by a weak 10 V/cm, 80 Hz background AC electric field, by V. A. Murtsovkin (courtesy of A. S. Dukhin); (b) AC electro-osmosis at a pair of titanium electrodes applying a 2 V 100 Hz AC voltage by Green et al. (2002); (c) DC electrokinetic jet at a dielectric microchannel corner by Thamida and Chang (2002); (d) one roll of quadrupolar ICEO flow around a 100 μm cylindrical gold post in a 100 V/cm 300 Hz electric field and (e) fixed potential ICEO flow around a gold post connected to one electrode supplying the background AC field by Levitan (2005). (Reproduced from Bazant (2008b) © 2008 Springer.)

(Cahill et al., 2004; Ramos et al., 2005) over micro-electrode arrays (Fig. 2d), with other seemingly unrelated phenomena, such as DC electrokinetic jets at dielectric microchannel corners (Thamida and Chang, 2002) (Fig. 2c), AC electrohydrodynamic interactions and self-assembly of dielectric colloids on electrodes (Trau et al., 1997; Yeh et al., 1997; Nadal et al., 2002; Ristenpart et al., 2003), and hydrodynamic interactions among polarizable particles (Gamayunov et al., 1986; Murtsovkin, 1996) (Fig. 2a).

The latter effect was apparently the earliest example of "ICEO" reported in the literature, from the pioneering work of V. Murtsovkin, A. S. Dukhin and collaborators in the 1980s on polarizable colloids, as reviewed by Murtsovkin (1996), long before analogous ICEO flows were observed in a microfluidic device by Levitan et al. (2005). The quadrupolar ICEO flow around an ideally polarizable sphere in a uniform electric field, and the resulting relative motion of two spheres, were first predicted by Gamayunov et al. (1986). Murtsovkin and collaborators proceeded to observe these flows around mercury drops (Murtsovkin and Mantrov, 1991) and metallic particles (Gamayunov et al., 1992). For larger particles, the flow was in the opposite direction of the theory, which was conjectured to be due to the onset of Faradaic reactions at large induced voltages, consistent with recent experiments on millimeter scale metal objects by Barinova et al. (2008).

The development of microfabrication technology has led to unprecedented control over the geometries of particles and channels, so a major focus of recent research has been on the design of polarizable structures and particles to control, enhance, optimize induced-charge electrokinetic phenomena. The original ACEO micropumps tested in experiments by Brown et al. (2000) and Studer et al. (2004) involved 2D planar arrays of interdigitated electrodes. Bazant and Ben (2006) predicted that faster flows are possible with non-planar stepped electrodes, and such "3D ACEO" designs have since been reduced to practice (Urbanski et al., 2006a,b; Huang et al., 2010). The concept of ICEO mixing by applying electric fields around fixed 3D metal microstructures (Bazant and Squires, 2004; Levitan et al., 2005) is now beginning to be reduced to practice as well (Harnett et al., 2008; Wu and Li, 2008b). ICEO flows around simple 2D metal structures also offer the chance to precisely test the standard model of electrokinetics, as recently shown by Pascall and Squires (2010). Asymmetric geometries of channels and particles also give rise to some surprising phenomena from the classical colloidal standpoint. Bazant and Squires (2004) predicted that an anisotropic particle subjected to a DC or AC field (below the frequency of double-layer charging) will generally translate and/or rotate by "induced-charge electrophoresis" (ICEP), while a fixed anisotropic object will pump the fluid by ICEO (Yariv, 2005; Squires and Bazant, 2006). These nonlinear

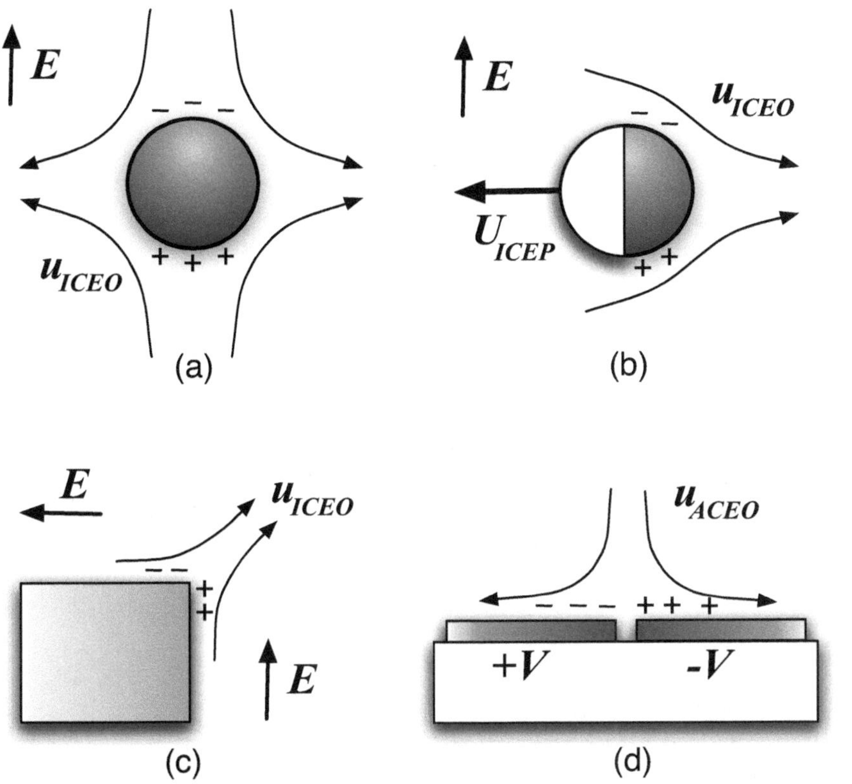

Figure 2. Examples of nonlinear electrokinetic phenomena driven by induced charge (+, −)in the diffuse part of the electrochemical double layer at ideally polarizable, blocking surfaces, subject to an applied electric field E or voltage V. (a) Induced-charge electro-osmosis (ICEO) around a metal post (Bazant and Squires, 2004; Squires and Bazant, 2004; Levitan et al., 2005) or particle (Gamayunov et al., 1986; Murtsovkin, 1996), (b) induced-charge electrophoresis (ICEP) of a metal/insulator Janus particle (Squires and Bazant, 2006; Gangwal et al., 2008), (c) a nonlinear electrokinetic jet of ICEO flow at a sharp corner in a dielectric microchannel (Thamida and Chang, 2002; Yossifon et al., 2006), and (d) AC electro-osmosis (ACEO) over a symmetric pair of microelectrodes (Ramos et al., 1999; Ajdari, 2000; Green et al., 2002). (Reproduced from Bazant et al. (2009b) © 2009 Elsevier.)

phenomena are very different from classical electrophoresis with surfaces of constant charge and are also beginning to be observed in experiments (Rose and Santiago, 2006; Gangwal et al., 2008).

In this chapter, we survey recent progress in induced-charge electrokinetics and teach the basic physical concepts, theoretical models, and experimental observations. Along the way, we also highlight various open questions for future research.

2 Background

Before discussing nonlinear induced-charge electrokinetic phenomena in electrolytes, we briefly review linear and nonlinear electrokinetic phenomena in weakly conducting liquids, as well as linear ("fixed charge") electrokinetic phenomena in electrolytes. The latter subject was mainly developed in colloid science over the past century, and there are excellent books available, e.g. by Levich (1962), Dukhin and Derjaguin (1974), Lyklema (1995) and Hunter (2001). Recent application-specific reviews are also available, such as Anderson (1989) and Delgado et al. (2007) on electrophoresis of colloids and Kirby and Hasselbrink (2004) and Tandon and Kirby (2008) on electro-osmosis in microfluidic devices.

2.1 Electrohydrodynamics in dielectric liquids

The term "electrokinetic phenomena" refers to electrically driven fluid flow or particle motion, but it is often used more narrowly, as we do here, to describe fluid or particle motion in electrolytes, consisting of large numbers of dissolved ions in a solvent, typically water. In contrast, the term "electrohydrodynamics" is often used more narrowly to refer to electrokinetic phenomena in low-conductivity dielectric liquids (Melcher and Taylor, 1969; Saville, 1997). A simple example of the latter is the electrophoretic motion of a charged particle in a non-conducting dielectric liquid in a uniform, constant, electric field, as shown in Fig. 3(a). The reader may be familiar with R. Millikan's famous oil-drop experiment, which first measured the electron charge e a century ago by showing the quantization of the charge $Q = ne$ inferred from the velocities of oil drops suspended in air between capacitor plates, based on the following simple analysis. To calculate the drop velocity $\boldsymbol{U}$, the electric force $\boldsymbol{f}_e = Q\boldsymbol{E}$ is balanced by the viscous drag force, approximated by Stokes' formula for a sphere, $\boldsymbol{f}_d = 6\pi\eta R U$, where R is the radius and η the fluid viscosity. This force balance yields the scaling of the

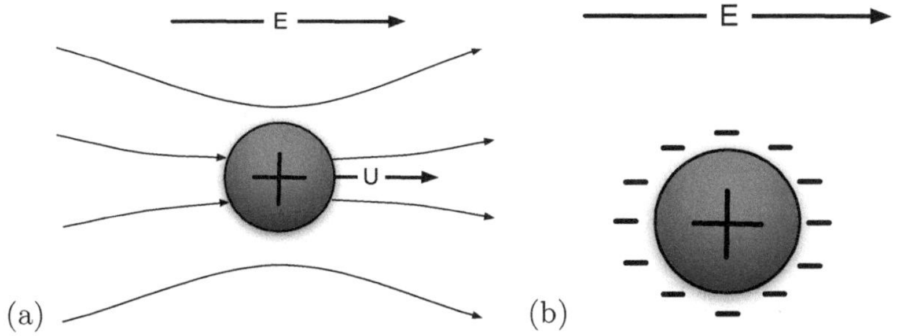

Figure 3. (a) Sketch of the electrophoretic motion of a particle of fixed charge in a non-conducting liquid in a uniform electric field, which drags fluid along with it, analogous to sedimentation under gravity; streamlines are shown in the fixed lab frame of reference. (b) The analogous situation for a charged solid particle in a (conducting) electrolyte, where a thin double layer screens the charge; since the net charge on the interface is zero, no motion results if the fluid velocity is continuous across the interface, as in the Leaky Dielectric Model for liquid drops (Saville, 1997). Instead, particle motion does occur, due to electro-osmotic flow in the double layer, which leads to an effective slip (or velocity discontinuity) over the surface, shown below in Fig. 4.

drop velocity

$$\boldsymbol{U} \propto \frac{Q}{\eta R}\boldsymbol{E} \quad \text{(electrophoresis in a dielectric liquid)} \tag{1}$$

which is proportional to the electric field and the (fixed) charge, and inversely proportional to the drop size. This is perhaps the simplest example of a "linear" electrohydrodynamic phenomenon, where $\boldsymbol{U} \propto \boldsymbol{E}$.

There can also be nonlinear electrohydrodynamic phenomena in non-conducting liquids. The most familiar example is "dielectrophoresis" (DEP) of polarizable solid particles in non-uniform electric fields (Pohl, 1978; Ramos et al., 1998). The applied field induces a dipole moment on the particle, $\boldsymbol{p} = \alpha\left(\frac{4\pi}{3}R^3\right)\varepsilon\boldsymbol{E}$, proportional to its volume and the field, where α is the Maxwell-Wagner factor which depends on electrical properties and, for an AC field, also the frequency. The induced dipole is then pulled by the field gradient toward regions of higher or lower field intensity. The electrostatic force, $\boldsymbol{f}_e = \boldsymbol{p}\cdot\nabla\boldsymbol{E}$, is again balanced by viscous drag $\boldsymbol{f}_d$ to yield a steady translational velocity which scales as

$$\boldsymbol{U} \propto \frac{\varepsilon R^2}{\eta} \nabla E^2 \quad \text{(dielectrophoresis)} \tag{2}$$

The velocity varies with the square of the applied field intensity, and thus survives in an AC field. In uniform DC or AC field, a polarizable particle can also rotate to align its induced dipole with the field axis, in response to the electrical torque, $\boldsymbol{p} \times \boldsymbol{E}$, but it cannot translate or rotate continuously.

For weakly conducting liquids, including many oils and non-aqueous solutions, the theory must also account for bulk current and charge accumulation at interfaces. This alters the analysis of electrophoresis and dielectrophoresis and also leads to some new electrohydrodynamic phenomena, directly tied to a non-zero conductivity. Taylor (1966) first described the nonlinear deformation of oil drops in electric fields, using what is now called the "Leaky Dielectric Model" (Melcher and Taylor, 1969; Saville, 1997). The applied field drives a small current which induces a charge on the fluid/fluid interface. Although, by symmetry, the drop cannot move, the interfacial induced charge is pulled by the electric field to produce counter-rotating quadrupolar vortices, both inside and outside the drop, while maintaining a continuous fluid velocity at the interface. The flow scales as

$$\boldsymbol{u} \propto \frac{\varepsilon R E^2}{\eta} \quad \text{(flow around a leaky dielectric drop)} \tag{3}$$

which generally arises in electrohydrodynamics from the balance of electric body force $\rho_e \boldsymbol{E} \propto$ and the viscous force $\eta \nabla^2 \boldsymbol{u} \propto \eta u / R^2$, using Poisson's equation for a linear dielectric response, $\rho_e = \nabla \cdot \varepsilon \boldsymbol{E} \propto \varepsilon E / R$. Outside the deformed drop, the steady flow resembles the quadrupolar ICEO flow around a solid polarizable particle, in Figs. 1(a) and 2(a). Indeed, below we will encounter the same scaling (3) for ICEO flow around a polarizable particle, but this is clearly a different phenomenon. If Taylor's liquid drop were replaced by a solid particle, the Leaky Dielectric Model would predict no fluid motion, because the tangential velocity is assumed to be continuous across the interface.

2.2 Electrokinetics in electrolytes

The situation is fundamentally different in an electrolytic solution containing large numbers of dissolved ions. Besides the high conductivity of the bulk electrolyte, ions can easily move to the interface to screen the surface charge, so that the net interfacial charge (for the surface plus its diffuse ionic screening cloud) is zero, as sketched in Figure 3(b). The length scale for this screening process, where electrostatic attraction is balanced by diffusion, is the Debye screening length,

$$\lambda_D = \sqrt{\frac{\varepsilon k_B T}{\sum_i (z_i e)^2 c_0}} \tag{4}$$

where ε is the permittivity, k_B is Boltzmann's constant, T is the absolute temperature, $z_i e$ are the ionic charges, and c_0 is the bulk neutral salt concentration. In aqueous solutions, the double layers have a tiny extent $\lambda_D = 0.5 - 100$ nm, which is typically much smaller than any geometrical scale L, such as the particle size or microchannel thickness. Such a "thin double layer" with $\lambda_D \ll L$, resembles a capacitor skin on the surface.

In electrolytes with thin double layers, the charge density (per volume) is zero everywhere, including the interface, so how can there be any electrokinetic effects? Indeed, the Leaky Dielectric Model would predict no motion, since there can be no force on an interface of zero net charge. The model also assumes continuity of the velocity field across the interface, which precludes relative motion of the two sides. Electrokinetic effects are readily observed at solid surfaces, however, so clearly an electrolyte is not a standard leaky dielectric.

The flaw in these arguments is that the interface in an electrolyte is a *double layer*, equivalent to a sheet of dipoles, which experiences a nonzero *torque* in a tangential electric field. The electrostatic torque accelerates the fluid on one side relative to the other (fluid or solid) phase, until it is balanced by an opposing viscous torque. In a quasi-steady Stokes flow, this process is instantaneous, and the tangential field produces a steady *electro-osmotic slip*, or velocity discontinuity between the fluid and the surface. This is how Helmholtz (1879) resolved the paradox of electrophoresis in electrolytes with thin double layers, many years after Reuss first observed the electrophoresis of clay particles in water in 1808.

By modeling the double layer as a thin capacitor with a voltage drop ζ from the surface to the bulk solution, Helmholtz derived a simple formula for effective slip across the double layer given by

$$\boldsymbol{u}_s = -\frac{\varepsilon\zeta}{\eta}\boldsymbol{E}_{\parallel} \quad \text{(electro-osmotic slip in an electrolyte)} \tag{5}$$

where ε is the permittivity and η the viscosity of the electrolyte, and $\boldsymbol{E}_{\parallel}$ is the tangential electric field, which is continuous across the interface. This phenomenon of "electro-osmotic flow" forms the basis for electrokinetic phenomena in electrolytes (Hunter, 2001; Lyklema, 1995). In particular, the electrophoresis of a particle with thin double layers in an infinite fluid can

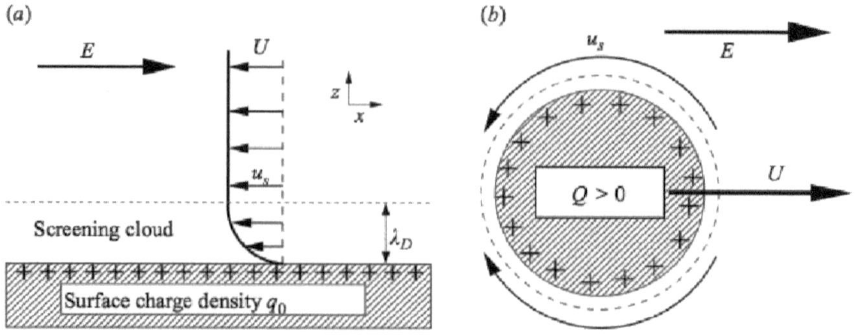

Figure 4. (a) A charged solid surface in an electrolyte attracts a "screening cloud" of excess counter-ions (of the opposite sign) to form a capacitor-like "double layer". An applied tangential electric field acts on the screening charge to drive electro-osmotic flow parallel to the surface, which builds from no-slip on the walls to an effective nonzero slip velocity outside the double layer. (b) If a particle with thin double layers is freely suspending in an electrolyte, the velocity discontinuity from electro-osmotic flow in an electric field leads to a net swimming motion termed "electrophoresis". (Reproduced from Squires and Bazant (2004) © 2004 Cambridge University Press.)

be understood as a phenomenon of "force-free" motion (akin to swimming) in the direction of the applied field due to electro-osmotic slip. Similar phoretic motion can also be driven by gradients in temperature and salt concentration, which also produce tangential gradients in osmotic pressure within the double layer (Anderson, 1989).

Smoluchowski (1905) extended the theory of electro-osmotic flow for diffuse screening charge in the double layer and showed that Eq. (5) holds more generally whenever the fluid permittivity and viscosity are constant across the double layer (using the classical continuum electrokinetic equations). As shown in Fig. 4(a), the tangential fluid velocity builds up exponentially across the diffuse part of the double layer from zero on the surface (assuming no hydrodynamic slip) to u_s outside the double layer. In Smoluchowski's theory, the zeta potential corresponds to the potential difference between the hypothetical "shear plane" at the inner edge of the diffuse layer and the neutral bulk solution, just outside the double layer, although this need not be the case in more general theories (Bazant et al., 2009b). If the zeta potential is small compared to the thermal voltage, $k_B T/e$ (= 26 mV at room temperature), then the diffuse double layer has a constant capacitance (per area), $C_D \sim \varepsilon/\lambda_D$, as expected for charged parallel plates separated by the screening length (Bazant et al., 2004). In that case, we can recast the electro-osmotic slip formula (5) as

$$\boldsymbol{u}_s \sim -\frac{\lambda_D q}{\eta} \boldsymbol{E}_{\|}, \text{ if } |e\zeta| \ll k_B T \tag{6}$$

since $C_D = q/\zeta$.

Smoluchowski also considered the motion of colloidal particles with thin double layers ($\lambda_D \ll R$) driven by electro-osmotic flow in applied electric fields, as shown in Fig. 4(b). He showed that a particle with uniform zeta potential (or surface charge) in an infinite fluid translates at a velocity,

$$\boldsymbol{U} = \frac{\varepsilon\zeta}{\eta} \boldsymbol{E}_\infty \quad \text{(electrophoresis in an electrolyte)} \tag{7}$$

where $\boldsymbol{E}_\infty$ is a uniform background electric field, applied "at infinity". It can be shown that this result is independent of the shape and size of the particle, assuming thin double layers and uniform ζ, since in the case, the fluid velocity is proportional to the electric field everywhere, $\boldsymbol{u} \propto \boldsymbol{E}$, as shown in Fig. 5(a).

For the same reason, a colloid of many such particles will experience zero hydrodynamic interactions (Morrison and Stukel, 1970; Anderson, 1989), as shown in Fig. 5(b). In other words, all the particles will move at the same velocity (7), regardless of the sizes, shapes or concentration. Of course, this can pose a problem for electrophoretic separations of like-charged particles, such as DNA molecules, which explains why DNA electrophoresis is done in a gel in order to exploit entropic (i.e. trapping), rather than purely electrokinetic, effects.

The inability to separate like-charged particles is related to other surprising features of *linear* electrokinetic phenomena involving surfaces of *constant surface charge* (or zeta potential) in electrolytes with thin double layers. A porous medium with the these physical properties exhibits the same linear relationship (7) between the mean fluid velocity $\langle \boldsymbol{u} \rangle = -\boldsymbol{U}$ and the electric field, regardless of the microstructure. This can be understood as the same effect, if the porous medium is a packed bed of particles held in place, driving flow in the opposite direction of electrophoresis.) The crucial microscopic principle behind these results is that electro-osmotic flow is irrotational, $\nabla \times \boldsymbol{u} = \boldsymbol{0}$, in the limit of thin double layers and constant zeta potential, because the fluid velocity is proportional to the electric field, $\boldsymbol{u} \propto \boldsymbol{E}$, everywhere in the bulk electrolyte. This is no longer the case if any

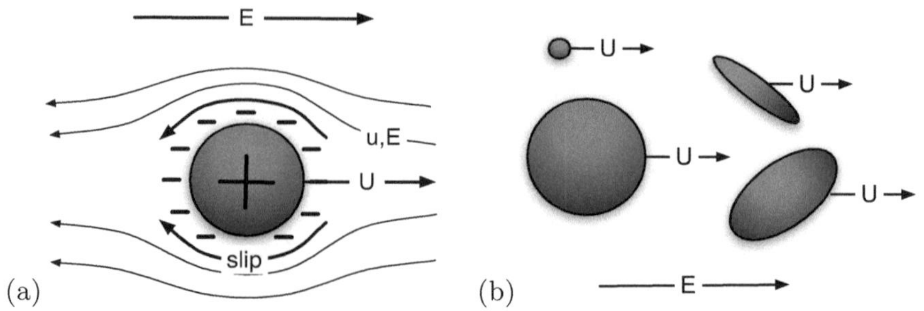

Figure 5. Electrophoresis of particles with thin double layers and constant, uniform surface charge (or zeta potential). (a) The fluid flow in the frame of the moving particle, driven by electro-osmotic slip, is proportional to the electric field, everywhere. (b) As a result, a colloid consisting of many such particles will move at the same velocity as a single particle, regardless of the sizes and shapes of the particles.

of these assumptions break down, and it is generally possible to produce vortices by electro-osmosis. In particular, we shall now discuss *nonlinear* ICEO flows involving *polarizable surfaces*, whose charge is not fixed.

3 Principles of Induced-Charge Electrokinetics

3.1 Flows around metal surfaces

The simplest example of ICEO involves a metal sphere (Gamayunov et al., 1986) or cylinder (Bazant and Squires, 2004) in an electrolyte with thin double layers, suddenly subjected to a uniform electric field, sketched in Figure 6. As a first approximation, the sphere is "ideally polarizable", meaning that its potential is held constant without any Faradaic electron-transfer reactions occurring. Conceptually, there are two steps in the dynamics: (1) electrochemical relaxation of the surface charge in response to the applied field, and (2) electro-osmotic flow around the particle, driven the induced charge.

1. Induced surface charge. When the field is turned on, electrons on the metal surface immediately drift toward one pole to induce a dipole moment, in order to make the surface equipotential (Figure 6a). In a non-conducting dielectric liquid this could be the steady state, but this is an unsteady configuration in an electrolyte. Since the field drives an

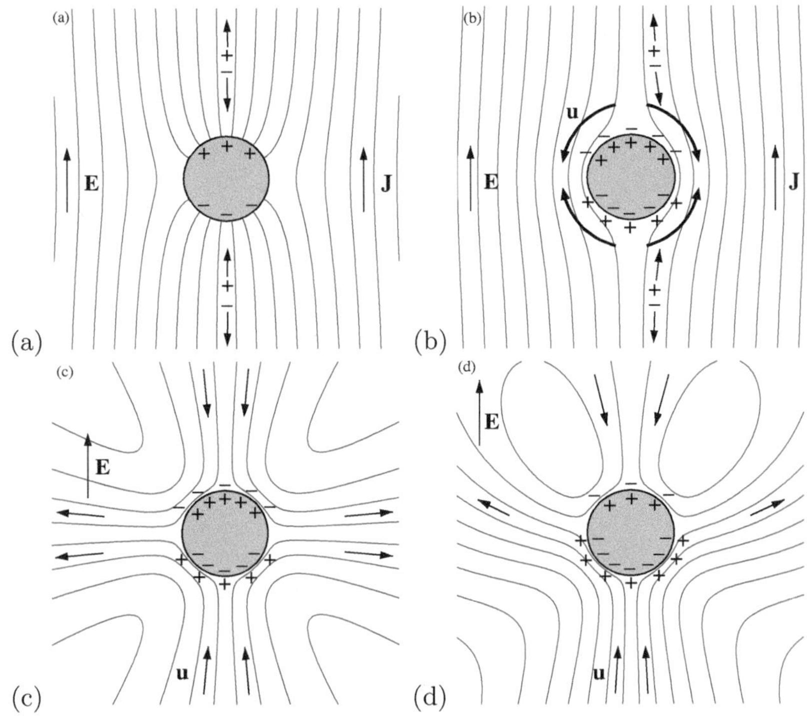

Figure 6. Physical mechanism for induced-charge electro-osmosis around an ideally polarizable metal cylinder in a suddenly applied electric field. (a) When the field is turned on, electronic charges relax to make the surface an equipotential, but the normal current drives double-layer charging. (b) After charging, the field lines are expelled by a nonuniform distribution of induced double-layer charge. (c) The tangential field acts on the induced charge to drive quadrupolar ICEO flow around a neutral cylinder. (d) If the cylinder has a nonzero total charge, then the dipolar flow of linear electrophoresis is superimposed on the quadrupole. (Reproduced from Bazant and Squires (2004) © 2004 American Physical Society).

ionic current, any normal component transports charge in or out of the diffuse layer. Neglecting surface conduction through the double layer (for thin double layers) and Faradaic reactions passing current through the particle (at low voltage), the normal current locally charges double layer, like a capacitor. This process continues until all the field lines are expelled in steady state (Figure 6b). As first noted by Simonov and Shilov (1977), the basic time scale for this process is the RC time for the equivalent circuit of the bulk resistance (through the solution around the equator) coupled to the diffuse-layer capacitances (on the surface near the poles):

$$\tau_0 = \frac{RC_D}{\sigma} = \frac{\lambda_D R}{D} \tag{8}$$

where σ and D are the bulk conductivity and diffusivity, and R is the particle radius. In more general situations, R is a length scale characterizing the distance between oppositely polarized surfaces, such as an electrode separation (Ramos et al., 1999), as reviewed by Bazant et al. (2004). In microfluidic devices, the typical double-layer charging time τ_0 ($\approx$ ms) is much larger than the Debye relaxation time $\varepsilon/\sigma = \lambda_D^2/D$, for bulk ionic screening ($\approx \mu$s) and much smaller than the diffusion time L^2/D for the relaxation of bulk concentration gradients ($\approx$ s) . For nano-channels or nano-particles, however, all of these time scales can be comparable ($\approx \mu$s).

2. Induced electro-osmotic flow. The tangential field acts on the non-uniform induced-charge (or ζ) distribution to produce quadrupolar ICEO flow, sucking fluid at the poles and ejecting it at the equator (Figure 6c). The scaling of the flow can be easily understood as follows. Capacitive charging transmits a non-uniform voltage to the double layer of order ER. If (5) holds, then ICEO flow scales as

$$\boldsymbol{u}(\boldsymbol{r}) \propto u_0 = \frac{\varepsilon R E^2}{\eta} \quad \text{(ideally polarizable surface)} \tag{9}$$

which is the same, generic electrohydrodynamic scaling arising in Taylor's flow around a leaky dielectric drop, Eq. (3), as noted above. Unlike linear electro-osmosis, ICEO flow is rotational and depends on the geometry via the size R as well as the shape of the particle (see below). In response to a DC voltage step, the flow approaches the steady state over the RC time scale τ_0. For AC field of frequency ω, the steady state flow decays above the RC frequency as $[1 + (\omega\tau_0)^2]^{-1}$ (Squires and Bazant, 2004).

From these simple physical arguments, we expect similar flows to be produced around any polarizable object in any electric field, but any *broken symmetry* will generally lead to fluid flow past the object, if it is held fixed, or motion by "induced-charge electrophoresis" (ICEP), if it is freely suspended (Gangwal et al., 2008). If the object has a nonzero total charge Q, as in the case of a charged metal colloidal particle, then the ICEO flow is superimposed on the familiar streaming flow of linear electrophoresis (Figure 6d). Whenever (5) holds, the two effects are additive, since the total charge corresponds to a constant ζ offset, relative to the background potential.

In fixed-potential ICEO, the potential of a polarizable object is controlled so as to induce total charge in phase with a (steady or oscillating) background field (Squires and Bazant, 2004). This effect is essentially an AC generalization of the "flow field effect transistor" (Schasfoort et al., 1999; van der Wouden et al., 2005), similar to the work of van der Wouden et al. (2006). The effective length R above is then set by the distance between the object and the electrodes supplying the background field. As a result, fixed-potential ICEO flow can be much faster than locally produced (e.g. quadrupolar) ICEO flow and has a different frequency response.

Other broken symmetries include irregular shapes (e.g. rods, polyhedra, etc.), non-uniform surface properties (e.g. partial dielectric or metallic coatings), and non-uniform background electric fields (Squires and Bazant, 2006). In each case, net pumping of the fluid by ICEO results if the object is held fixed, which requires a certain force and torque. Conversely, if the object is a colloidal particle, then broken symmetries cause it to translate and rotate by ICEP, as described below.

3.2 Flows around dielectric surfaces

The canonical example above assumes an ideally polarizable surface, where the double layer charges capacitively to sustain the entire voltage applied to the object, but the phenomenon of ICEO is more general and occurs at any polarizable surface, to varying degrees (Squires and Bazant, 2004). For example, if the metal object described above has a thin dielectric coating of width h_S and permittivity ε_S, then both the time scale (8) and the flow scale (9) are multiplied by a factor $\Lambda = (1+\delta)^{-1}$, where

$$\delta = \frac{C_D}{C_S} = \frac{\varepsilon}{\varepsilon_S}\frac{h_S}{\lambda_D} \tag{10}$$

is the ratio of the dielectric-layer capacitance to the diffuse-layer capacitance, which are placed in series in an equivalent circuit for the double layer. This shows that dielectric coatings thicker than the Debye length can substantially reduce ICEO flows at metal surfaces. In the limit of a purely dielectric object of characteristic length scale, $R \gg \lambda_D$, ICEO flow scales as

$$\boldsymbol{u} = \frac{\varepsilon_S \lambda_D E^2}{\eta} \quad \text{(dielectric surface)} \tag{11}$$

which is smaller than for an ideally polarizable object by a factor $\delta \gg 1$.

Although often small, ICEO flows at dielectric surfaces need not be negligible in microfluidic devices, since there can be large local electric fields at sharp geometrical features. As shown in Figure 1(c), an electric field passing around a sharp corner in a dielectric microchannel can drive a strong nonlinear electrokinetic jet of ICEO flow due to the corner field singularity (Thamida and Chang, 2002; Yossifon et al., 2006). In very simple terms, illustrated in Fig. 2(c), this phenomenon can be understood as half of the quadrupolar flow around a polarizable particle, where the jet corresponds to the outward flow at the equator in Figure 6(c).

As noted above, dielectric objects also experience electrostatic forces, leading to DEP motion of freely suspended particles. The uniform component of a background electric field induces a dipole on the object, which then feels a torque to align it with the field. A field gradient applies a force to the induced dipole. Higher-order multipoles in the background field can likewise cause forces and torques by acting on higher-order induced multipole moments on the object. In the case of colloidal dielectric particles, these forces and torques (balanced by hydrodynamic drag) produce translational velocity $u \sim \varepsilon R^2 \nabla E^2/\eta$ and rotational velocity $\Omega \sim \varepsilon E^2/\eta$ of DEP, respectively. The theory of DEP has mostly been developed for dielectric liquids, but in electrolytes ICEO flows also occur, with the very same scalings with field and particle size (Squires and Bazant, 2006). The net electrokinetic motion of polarizable particles in non-uniform fields results from a competition between DEP and ICEP, originally termed "dipolophoresis", which was first analyzed for colloidal spheres by Shilov and Simonova (1981).

4 Standard Model for thin double layers

The mathematical description of ICEO flows began with the pioneering work of Murtsovkin (1996) on metallic colloids and Ramos et al. (1999) and Ajdari (2000) on AC pumping of liquids by micro-electrode arrays. Bazant and Squires (2004) unified these theories in a simple "Standard Model", derivable from the full Poisson-Nernst-Planck (PNP) equations of ion transport and Navier Stokes equations of viscous fluid flow in the asymptotic limit of thin double layers (DL), compared to geometrical length scales. The model is based on the assumption of "linear" or "weakly nonlinear" charging dynamics (Bazant et al., 2004), which further requires that the applied voltage is small enough not to significantly perturb the bulk salt concentration, whether by double-layer salt adsorption or Faradaic charge-transfer reaction currents. By neglecting Faradaic reactions, we focus on "blocking" or "ideally polarizable" metal surfaces. In the same limit, surface conduction through the diffuse part of the double layers can also be neglected (Chu and Bazant, 2007; Khair and Squires, 2008), unless there is a large pre-existing fixed charge, upon which a small perturbation is induced, as described by Murtsovkin (1996).

With these assumptions, the problem is greatly simplified, and the electrokinetic problem decouples into one of electrochemical relaxation, similar to Leaky Dielectric Model (Saville, 1997), and another of viscous flow, driven by electro-osmotic slip. Although this model can be rigorously justified only for very small voltages, $\Psi_D \ll kT/e$, in a dilute solution (González et al., 2000; Squires and Bazant, 2004), it manages to describe many features of ICEO flows at much larger voltages. Extensions of the model for large voltages are reviewed by Bazant et al. (2009b) and discussed at the end of this chapter.

4.1 Electrochemical relaxation

The first step in the Standard Model is to solve Laplace's equation for the electrostatic potential across the bulk resistance,

$$\boldsymbol{\nabla} \cdot \boldsymbol{J} = \boldsymbol{\nabla} \cdot (\sigma \boldsymbol{E}) = 0 \quad \Rightarrow \quad \nabla^2 \phi = 0 \tag{12}$$

assuming Ohm's Law with a constant conductivity σ. For a blocking polarizable surface, which cannot pass a normal current, a capacitance-like boundary condition closes the equivalent RC circuit:

$$\frac{dq}{dt} = \hat{\boldsymbol{n}} \cdot \boldsymbol{J} \quad \Rightarrow \quad C\frac{d\Psi}{dt} = \sigma\, \hat{\boldsymbol{n}} \cdot \nabla\phi, \tag{13}$$

where $-q$ is the total surface charge (equilibrium + induced), q is the screening charge, $\Psi = \phi_0 - \phi$ is the local double-layer voltage drop from the metal surface at ϕ_0 to the bulk solution just outside the double layer at ϕ, and C is the total differential capacitance of the double layer, also assumed to be constant. A simple and important special case of (13) is the low-frequency or DC limit, $\hat{\boldsymbol{n}} \cdot \nabla\phi = 0$, where the surface behaves like an insulator since the double layer is fully charged and cannot sustain any normal current (since we neglect surface conduction and Faradaic reactions).

The potential of the surface ϕ_0 is either controlled externally, in the case of an electrode, or determined self-consistently by the condition of fixed total charge Q integrated over the surface, in the case of a freely suspended colloidal particle:

$$Q = \oint_S d\boldsymbol{r} \int_0^{\Psi(\boldsymbol{r})} C(\psi)d\psi \tag{14}$$

where we allow for nonlinear response of the double layer, via a voltage-dependent capacitance, $C(\Psi)$. Examples of such nonlinear extensions of the Standard Model are discussed below and reviewed by Bazant et al. (2009b). In the limit of linear response with a constant capacitance, valid for small voltages $|e\Psi| \ll k_B T$, the total charge is proportional to the surface-averaged double-layer voltage

$$Q \sim C \oint_S d\boldsymbol{r}\Psi(\boldsymbol{r}) \tag{15}$$

which implies

$$\phi_0 \sim \bar{\phi}_0 + \langle\phi\rangle \tag{16}$$

where $\bar{\phi}_0 = Q/CA$ is the surface potential assuming a constant capacitance over the area $A = \oint_S d\boldsymbol{r}$ and $\langle\phi\rangle = \oint_S d\boldsymbol{r}\phi(\boldsymbol{r})/A$ is the surface-averaged potential in the solution, just outside the double layer. In symmetric problems involving uncharged colloids ($Q = 0$), it is common to set $\phi_0 = \langle\phi\rangle$, but this is only valid for linear response to a small induced double-layer voltage. More generally, Equation (14) is a nonlinear integral constraint, which self-consistently determines the potential of a metal particle, ϕ_0.

4.2 Fluid flow

After solving for the electrostatic potential ϕ, the second step in the Standard Model is to solve for the fluid velocity $\boldsymbol{u}$ satisfying the unsteady Stokes Stokes equations for creeping flow,

$$\rho_m \frac{\partial \boldsymbol{u}}{\partial t} = -\boldsymbol{\nabla} p + \eta \nabla^2 \boldsymbol{u}, \quad \boldsymbol{\nabla} \cdot \boldsymbol{u} = 0, \tag{17}$$

where ρ_m is the mass density. The unsteady term is only important to describe transient flows at high frequencies, where momentum diffusion becomes comparable with mass diffusion. For steady AC response or low frequency transients, this term is generally neglected.

Since it is assumed that the bulk electrolyte remains quasi-neutral with constant conductivity, there is no body force on the fluid in (17). Instead, coupling to the electrochemical problem only arises through the Helmholtz-Smoluchowski boundary condition (5) of effective fluid slip outside the double layer,

$$\boldsymbol{u}_s = \boldsymbol{u} - \boldsymbol{U} - \boldsymbol{r} \times \boldsymbol{\Omega} = -\frac{\varepsilon \Psi_D}{\eta} \boldsymbol{E}_{\|} \tag{18}$$

where $\boldsymbol{E}_{\parallel}$ is the tangential field and Ψ_D is the voltage from the "shear plane" (where the velocity vanishes) to the bulk solution (Lyklema, 1995). The slip velocity represents a tangential velocity discontinuity between $\boldsymbol{u}$, the fluid velocity just outside the double layer, and $\boldsymbol{U}+\boldsymbol{r}\times\boldsymbol{\Omega}$, the velocity of the solid surface, where $\boldsymbol{U}$ and $\boldsymbol{\Omega}$ are the local translational and rotational velocities, which are either prescribed or determined by mechanical constraints, as described below. The assumption of uniform bulk salt concentration allows us to neglect tangential osmotic pressure gradients, which modify the slip formula with a term for diffusio-osmotic flow (Rubinstein and Zaltzman, 2001; Zaltzman and Rubinstein, 2007).

To close the model, following Green et al. (2002), it is common to assume that only a constant fraction $\Lambda \leq 1$ of the double-layer voltage falls across the diffuse screening cloud: $\Psi_D = \Lambda\Psi$. The simplest model of ICEO flow assumes $\Lambda = 1$, i.e. all of the double-layer voltage drop contributes to the induced zeta potential. In that case, the theory has no adjustable parameters, but it tends to over-estimate ICEO flows, sometimes by orders of magnitude compared to experimental data. As shown in Fig. 7, some experimental data can be fitted fairly well by allowing Λ to be an adjustable parameter. See Table 1 of Bazant et al. (2009b) for a summary of Λ values extracted from a wide range of experiments on different induced-charge electrokinetic phenomena and a discussion of the limitations of this approach.

The great simplification of the Standard Model is that the full nonlinear Poisson-Nernst-Planck/Navier-Stokes equations are replaced by two *linear* boundary-value problems, which are amenable to analytical solutions in many cases and more convenient for numerical solutions. First, the potential ϕ is obtained by solving the linear problem (12)-(13), and this allows the slip velocity profile (18) to be calculated. The fluid velocity $\boldsymbol{u}$ and pressure p are then obtained by solving another linear problem (17)-(18). This procedure avoids solving a nonlinear problem because the slip boundary condition (18) is linear in $\boldsymbol{u}$ and only nonlinear in ϕ, which can be solved separately, without knowing $\boldsymbol{u}$ in advance.

4.3 Particle motion

The total mechanical force $\boldsymbol{F}$ and torque $\boldsymbol{T}$ acting on any body are related to the electric field and fluid velocity profiles via integrals over an

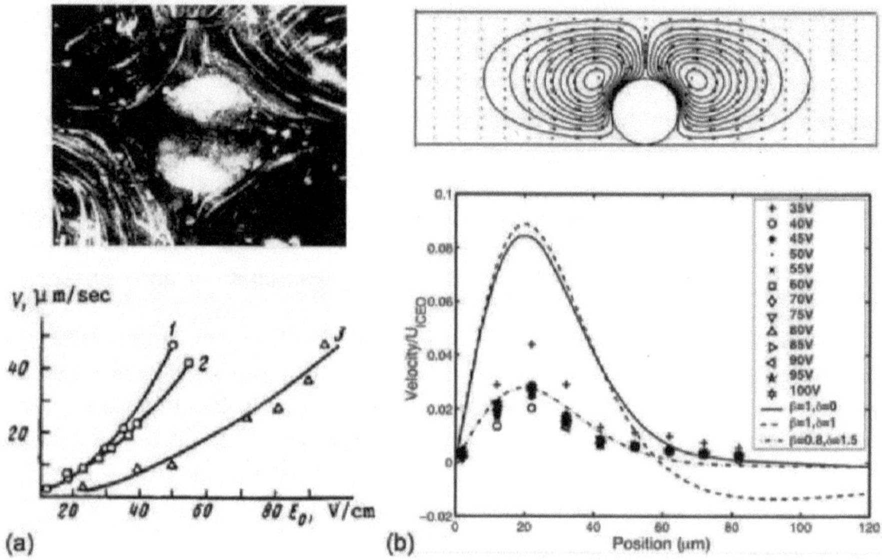

Figure 7. Early tests of the Standard Model for ICEO flows around metal objects in AC electric fields. (a) ICEO flow around a tin colloidal particle (photograph above); experimental data by Gamayunov et al. (1992) for the fluid velocity sampled at different points around the particle (top) versus the field strength, which demonstrated the quadratic scaling of Eq. (9), but not any details of the flow profile. (b) The velocity profile around a 100 μm platinum cylinder in a microchannel, simulated by the Standard Model (above) and compared with measurements by micro-particle-image velocimetry (below) by Levitan et al. (2005); a horizontal slice of the velocity profile 5 μm above the wire at different voltages shows good data collapse with the scaling (9); a reasonable fit is obtained with $\Lambda = 1/(1+1.5) = 2/5$ using (26), and a better fit (without accounting for ion adsorption) can be obtained using a constant-phase-angle impedance model. (Reproduced from Bazant (2008b) © 2008 Springer.)

enclosing surface S,

$$\boldsymbol{F} = \int_S d\boldsymbol{r}\, \hat{\boldsymbol{n}} \cdot \sigma \tag{19}$$

$$\boldsymbol{T} = \int_S d\boldsymbol{r}\, \boldsymbol{r} \times (\hat{\boldsymbol{n}} \cdot \sigma) \tag{20}$$

of the stress tensor,

$$\sigma_{ij} = -p\,\delta_{ij} + \eta\left(\frac{\partial u_j}{\partial x_i} + \frac{\partial u_i}{\partial x_j}\right) + \varepsilon\left(E_i E_j - \frac{1}{2}E^2\delta_{ij}\right) \tag{21}$$

where δ_{ij} is the unit tensor. The first term in (21) is the isotropic pressure tensor; the second is the viscous stress tensor for a Newtonian fluid; and the third is the Maxwell stress tensor for a linear dielectric medium of constant permittivity. In the case of steady state flow (or time-averaged periodic flow, described below), the Stokes equation (17) expresses local mechanical equilibrium, $\nabla \cdot \boldsymbol{T} = 0$, so the bounding surface S can be deformed arbitrarily to any convenient shape to calculate the force and torque integrals. For example, for a bounded collection of colloidal particles in an infinite fluid, it is usually best to deform S to infinity.

For fixed geometries in microfluidic devices, the translational and rotational velocities $\boldsymbol{U}$ and $\boldsymbol{\Omega}$ of solid boundaries are set to zero. The integrals (20) and (20) then give the force $\boldsymbol{F}$ and torque $\boldsymbol{T}$ exerted by the fluid on the solid, which are equal and opposite to the force and torque needed to hold the solid in place, respectively.

For a freely suspended colloidal particle, the situation is more subtle. In a quasi-steady Stokes flow, the translational and rotational velocites, $\boldsymbol{U}$ and $\boldsymbol{\Omega}$, are determined implicitly by the constraints that there be no net force, $\boldsymbol{F} = 0$, and no net torque, $\boldsymbol{T} = 0$, exerted on the particle by the fluid, since there is negligible translational and angular acceleration, respectively. This assumes that viscous dissipation is strong enough to neglect the inertial term, $\boldsymbol{u} \cdot \nabla \boldsymbol{u}$, and fast enough (compared to other relaxation times or the AC period) to neglect the unsteady term, $\partial \boldsymbol{u}/\partial t$, in the Navier-Stokes equations.

In practice, the translation and rotational dynamics of a particle can be calculated as follows (Kilic and Bazant, 2011). In the Standard Model, the linearity of the bulk Stokes flow (outside the double layers) allows us to express the fluid velocity as a superposition of two flows:

1. the electro-osmotic flow $(\boldsymbol{u}_s, p_s)$ resulting from the slip profile (18) around a fixed particle with $\boldsymbol{U} = \boldsymbol{\Omega} = \boldsymbol{0}$, which exerts a force $\boldsymbol{F}_s$ and torque $\boldsymbol{T}_s$ on the particle;

2. the purely viscous flow $(\boldsymbol{u}_v, p_v)$ with $\boldsymbol{E} = \boldsymbol{0}$ resulting from the particle's motion $\boldsymbol{U}$ and $\boldsymbol{\Omega}$, which exerts $\boldsymbol{F}_v$ and $\boldsymbol{T}_v$ on the particle.

These flows are subject to the constraints, $\boldsymbol{F} = \boldsymbol{F}_s + \boldsymbol{F}_v = \boldsymbol{0}$ and $\boldsymbol{T} = \boldsymbol{T}_s + \boldsymbol{T}_v = \boldsymbol{0}$, which implicitly determine $\boldsymbol{U}$ and $\boldsymbol{\Omega}$. In nontrivial geometries involving asymmetric particles or nearby channel walls, there is generally a coupling between translation and rotation due to viscous flow, which can be expressed as

$$\begin{pmatrix} \boldsymbol{F}_v \\ \boldsymbol{T}_v \end{pmatrix} = \boldsymbol{M}_v^{-1} \begin{pmatrix} \boldsymbol{U} \\ \boldsymbol{\Omega} \end{pmatrix} \tag{22}$$

where $\boldsymbol{F}_v$ and $\boldsymbol{T}_v$ are the force and torque exerted by the fluid on the particle, due only to its motion, and $\boldsymbol{M}_v$ is a mobility tensor, taking into account viscous dissipation in the instantaneous geometry. The inverse tensor $\boldsymbol{M}_v^{-1}$ is a generalization of the drag coefficient, e.g. $6\pi\eta R\boldsymbol{I}$ for a sphere in an infinite fluid, and can be calculated by solving Stokes equations with no slip on a moving and rotating particle and then performing the integrals (20) and (20) for the force and torque.

Armed with the particle's mobility tensor, $\boldsymbol{M}$, the particle motion can then be determined from

$$\begin{pmatrix} \boldsymbol{U} \\ \boldsymbol{\Omega} \end{pmatrix} = -\boldsymbol{M}_v \begin{pmatrix} \boldsymbol{F}_s \\ \boldsymbol{T}_s \end{pmatrix} \tag{23}$$

where $\boldsymbol{F}_s$ and $\boldsymbol{T}_s$ are the force and torque on the particle (electrostatic + viscous) in response to the slip profile around a fixed particle in the same position. In a numerical simulation, these equations can be iterated to self consistency in an implicit scheme, or $\boldsymbol{M}_v$, $\boldsymbol{F}_s$, and $\boldsymbol{T}_s$ can be calculated once using $\boldsymbol{E}$, $\boldsymbol{U}$ and $\boldsymbol{\Omega}$ from the previous time step, or by interpolation to the new time step, in an explicit scheme.

4.4 Symmetric geometries

For certain symmetric geometries, the reciprocal theorem for Stokes flow can be used to avoid having to actually solve the Stokes equations, if one only wants to calculate the motion of the solid body. For a spherical particle with an arbitrary slip distribution $\boldsymbol{u}_s(\boldsymbol{r})$ on its surface, Stone and Samuel (1996) showed that the translational velocity is equal and opposite to surface-averaged the slip velocity:

$$\boldsymbol{U} = -\frac{\oint_S d\boldsymbol{r}\, \boldsymbol{u}_s(\boldsymbol{r})}{\oint_S d\boldsymbol{r}} \tag{24}$$

while the rotational velocity is given by

$$\boldsymbol{\Omega} = -\frac{\oint_S d\boldsymbol{r}\, \hat{\boldsymbol{r}} \times \boldsymbol{u}_s(\boldsymbol{r})}{2 \int_V d\boldsymbol{r}} \tag{25}$$

where V is the volume of the particle. Squires and Bazant (2006) pointed out that the same relations also hold for cylindrical geometries, in spite of various subtleties of two-dimensional Stokes flows. (Note that we have written (25) in a different form, which clarifies this connection.) These results make it much easier to solve for the motion of inhomogeneous particles with symmetric shapes (and variable surface properties), rather than for homogeneous particles with asymmetric shapes. As in many mathematical problems, complicated boundary conditions are more tractable than complicated geometries.

An analogous result to (24) also holds for fluid pumping in a parallel-plate microchannel with arbitrary slip distributions on both surfaces. The total flow rate through the channel is the same as that of a linear shear flow driven by the surface-averaged slip on the two walls, or, equivalently, a plug flow $\boldsymbol{u}_{plug} = -\boldsymbol{U}$ driven by the overall average slip (over both walls) from (24). This property was noted by González et al. (2000) in the context of a Fourier analysis of AC electro-osmotic pumping with periodic slip, but it holds more generally for any slip profile, as shown by Squires (2008). Most theoretical studies of slip-driven microfluidic pumping have used this property to calculate the time-averaged flow rate, without having to solve for the time-averaged velocity field, but it can only be applied to flat plate geometries. For three-dimensional electrodes, as discussed below, one must solve the full Stokes flow to obtain the flow rate.

5 Double-layer models

A microscopic model of the double layer is required to predict how the capacitance C and voltage Ψ_D (or Λ) in the Standard Model depend on experimental conditions, such as the bulk salt concentration and the interfacial chemistry. Most studies of ICEO flow have adopted the classical two-part model of the double layer (Bockris and Reddy, 1970), which adds a molecular-scale "compact part" described by boundary conditions between the surface and the outer "diffuse part" described by continuum equations for mobile ions. It is typically assumed that the electrokinetic (zeta) potential Ψ_D is the same as the diffuse-layer potential drop from the edge of the compact layer to the bulk solution. This sharp partitioning of the double layer into two distinct regions is convenient for mathematical modeling, but not precisely defined. In principle, many features of the molecular compact

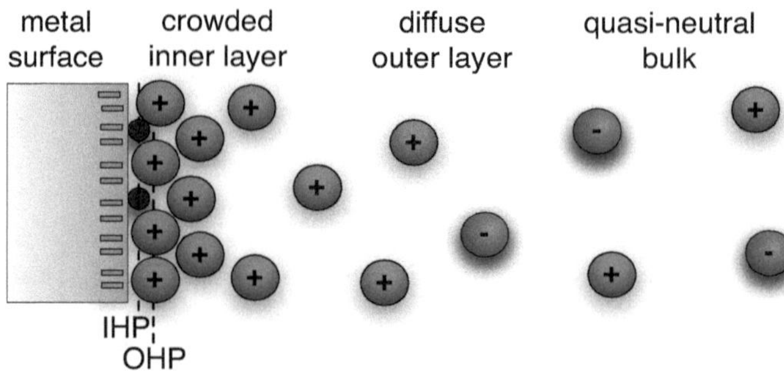

Figure 8. Sketch of the double layer near a blocking electrode at high voltage. Solvated counterions (green) are crowded in the inner region and smoothly transition across the outer diffuse region to a dilute solution with solvated anions (orange). An ion can break free from its solvation shell and adsorb on the surface (black), thereby moving from the outer Helmholtz plane (OHP) to the inner Helmholtz plane (IHP). (Reproduced from Bazant et al. (2009b) © 2009 Elsevier.)

layer can be reproduced by more realistic continuum models of the diffuse layer accounting for ion-specific effects, such as steric volume constraints, dielectric saturation, and the viscoelectric effect Bazant et al. (2009a,b).

The concept of the compact layer was introduced by Stern (1924) to account for the finite solvated ion size in the simplest possible way, by positing a distance of closest approach of solvated ions to the surface h_S, at the "outer Helmholtz plane" (Bockris and Reddy, 1970) shown in Fig. 8. After separating a layer of thickness h_S from the continuum region, it is also convenient to assign it a reduced permittivity $\varepsilon_S < \varepsilon$ to describe dielectric saturation (aligning of solvent dipoles) in the large normal field in the innermost portion of the double layer. Stern's model effectively adds capacitance $C_S = \varepsilon_C/h_S$ *in series* with the diffuse-layer capacitance C_D. The concept of a "surface capacitance" in series with the bulk solution is quite general and not limited to a Stern monolayer of algined solvent molecules. It could also describe a thin dielectric coating on the surface, such as an oxide layer, contaminant film, self-assembled polymer monolayer, etc.

The resulting simple model,

$$\frac{C}{C_D} = \Lambda = \frac{1}{1+\delta} \quad \text{(Stern layer or dielectric coating)} \tag{26}$$

has only one fitting parameter $\delta = C_D/C_S$ in (10) to describe the the physical properties of double layer (Ajdari, 2000; Green et al., 2002; Bazant et al., 2004; Levitan et al., 2005; Olesen et al., 2006). Note that the time scale (8) for double-layer charging is generally reduced by the same factor as the capacitance, $\tau/\tau_0 = C/C_D$. This model has been widely applied to ICEO flows, but it is unable to fit detailed experimental measurements in microfluidic devices without some additional modification of the boundary conditions, as shown in Fig. 7. It also fails to predict the strong decay of ICEO flow with increasing concentration and various flow reversals that can occur at large voltage and/or large frequency (Bazant et al., 2009b). It is becoming clear that the simplest version of the Standard Model (26) is incomplete, but simple extensions are being developed that improve the agreement with experiments, at least in the regime of small diffuse-layer voltages where the Standard Model has theoretical justification.

Another important role of the compact layer is to mediate the adsorption/desorption of ions, which react with ionizable sites on the surface and thus regulate the surface charge. The storage of charge by specific adsorption of ions introduces an effective "chemical" capacitance of the double layer *in parallel* with its "physical" capacitance due to purely electrostatic effects (van Hal et al., 1996). In the case of deprotonization reactions in water, the surface effectively buffers the pH of the solution, so this parallel capacitance is sometimes called the "buffer capacitance" (van der Wouden et al., 2006). We will more generally refer to it as the "adsorption capacitance" C_A.

Pascall and Squires (2010) recently showed that including the adsorption capacitance is essential to fit experimental data for ICEO flows over metal electrodes with silica coatings, as shown in Figures 9 and 10. In their version of the Standard Model, C_A is in parallel with C_D, and the pair is in series with C_S:

$$\frac{C}{C_D} = \frac{1+\beta/\delta}{1+\delta+\beta}, \quad \Lambda = \frac{1}{1+\delta+\beta}, \quad \text{(ion adsorption on dielectric)} \qquad (27)$$

where $\beta = C_A/C_S$ is a second dimensionless parameter, taken to be constant in a given experiment. This model is reasonable for their experiments where C_A represents the deprotonization of silanol groups on the silica coating, $SiOH \leftrightarrow SiO^- + H^+$ and C_S represents a dielectric layer, much thicker than the molecular scale, inserted between the surface and the electrolyte.

For bare metal surfaces, however, the situation is different, as sketched in 8. In that case, C_A describes desolvated ions adsorbed on the surface at the "inner Helmholtz plane", while C_S represents the dielectric response of the Stern solvent monolayer up to the first layer of solvated ions at the "outer Helmholtz plane" (OHP) (Bockris and Reddy, 1970). In that case, the

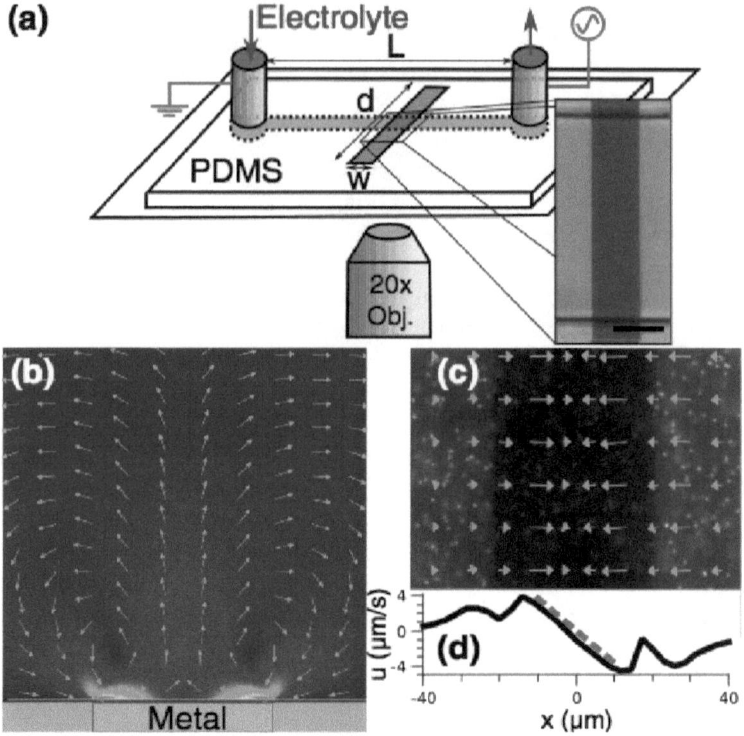

Figure 9. Experimental setup of Pascall and Squires (2010) to measure ICEO flows at high throughput with different surfaces. A planar gold strip (50μm) sits perpendicular to a PDMS microchannel (a), along which an AC field is applied, driving two counter-rotating ICEO rolls in aqueous KCl solutions, as simulated by the Standard Model (b). Micro-PIV velocity measurements just above the strip (c) recover the predicted ICEO slip velocity varying linearly with distance from the strip center (d). (Reproduced from Pascall and Squires (2010) © 2010 American Physical Society.)

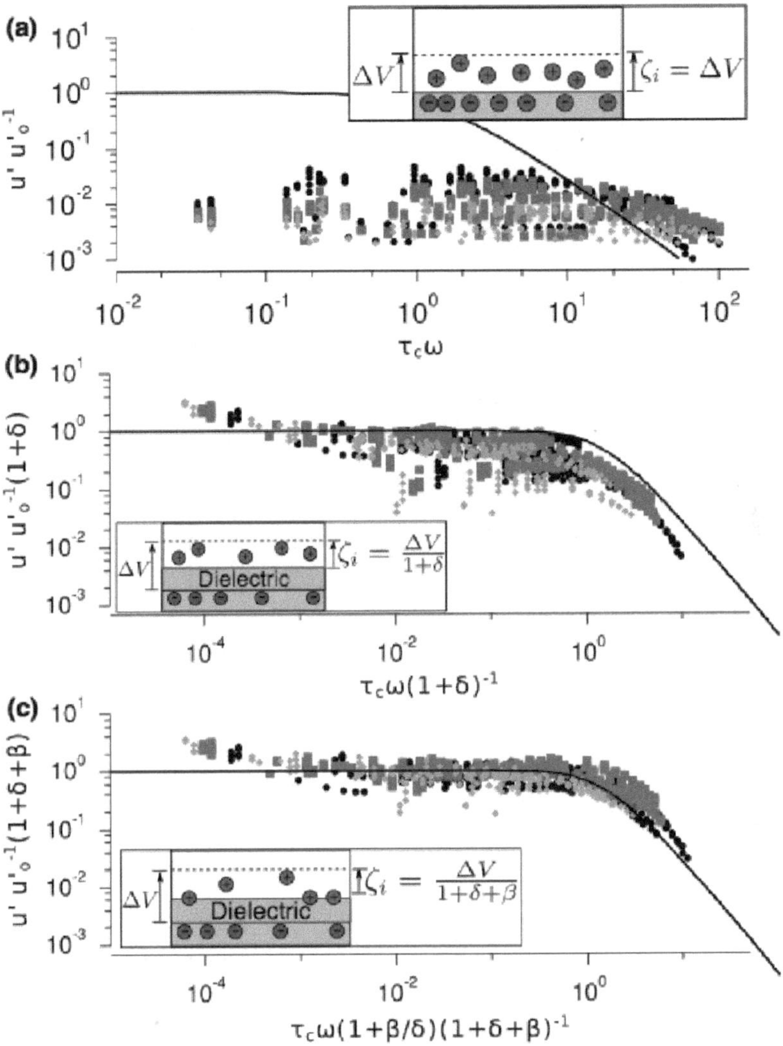

Figure 10. Experimental validation of the Standard Model of ICEO flow using the setup of Figure 9. Measurements over gold strips "controllably contaminated" with SiO_2 films for 987 conditions of varying thickness (33-100 nm) and frequency ω show poor agreement with a theory that ignores the SiO_2 (a), improved agreement when the known surface capacitance of the film is included via $\delta = C_D/C_S$ (b), and remarkable collapse when the adsorption capacitance of SiOH deprotonization is included via $\beta = C_A/C_S$ (c). (Reproduced from Pascall and Squires (2010) © 2010 American Physical Society.)

Stern layer does not separate the adsorbed ions from the surface, as shown in Fig. 8, so it is more reasonable to place the adsorption capacitance C_A in parallel with the total physical capacitance of C_D and C_S in series, leading to

$$\frac{C}{C_D} = \frac{\beta}{\delta} + \frac{1}{1+\delta}, \quad \Lambda = \frac{1}{1+\delta} \quad \text{(ion adsorption on bare metal)} \tag{28}$$

in place of (27). It appears that this simple model with constant δ and β has not yet been applied to ICEO flows, but more sophisticated models of potential-dependent ion adsorption have recently been developed by Suh and Kang (2008) and successfully fit to the ACEO pumping data of Green et al. (2002) by Suh and Kang (2009).

It is important to note that the capacitance of the double layer generally depends on the interfacial voltage, due to the *nonlinear* electrochemical response of ions in the diffuse layer, as well as adsorption and reaction kinetics in the compact layer. See Bazant et al. (2009b) for a recent review, in the present context of induced-charge electrokinetics. For example, long ago, Grahame (1947) showed that the total differential capacitance of mercury drop electrodes in aqueous solutions can be well described using the Gouy-Chapman model for the nonlinear differential capacitance of the diffuse layer in a dilute symmetric binary electrolyte,

$$C_D(\Psi_D) = \frac{\varepsilon}{\lambda_D} \cosh\left(\frac{ze\Psi_D}{2k_BT}\right) \tag{29}$$

placed in series with a (fitted) nonlinear compact-layer capacitance, which depends only on the state of charge, but not the bulk salt concentration (Bockris and Reddy, 1970). The Gouy-Chapman-Stern model assumes a constant compact layer capacitance in series with $C_D(\Psi_D)$ from (29), so that $C(\Psi) = (C_S^{-1} + C_D(\Psi_D)^{-1})^{-1}$ in the RC boundary condition (13). Such nonlinear effects have also been included in some Standard Model calculations as the first correction to the linear response theory, but the most common approximation is a constant capacitance C for the double layer, although this can only be rigorously justified for small voltages, $|ze\Psi_D| \ll k_BT$.

6 AC forcing

6.1 The complex potential

It is common to study ICEO flows under alternating current (AC) conditions at driving frequency ω. In that case, the Standard Model with constant double-layer capacitance can be placed in a simple time-dependent form, following González et al. (2000) and Levitan et al. (2005). We neglect transient vorticity diffusion within each period and focus only on deriving the time-averaged flow profile $\langle \boldsymbol{u} \rangle$ in a periodic steady state, which satisfies the steady Stokes equations,

$$\nabla p = \nabla^2 \langle \boldsymbol{u} \rangle, \ \nabla \cdot \langle \boldsymbol{u} \rangle = 0 \tag{30}$$

due to the linearity of the unsteady equations.

We begin by switching to dimensionless variables. We scale length to the geometrical scale R and time to the RC charging time $\tau = (C/C_D)\tau_0$ in Eq. (8). The dimensionless frequency is $\omega\tau$. For an applied field amplitude E, we scale the potential to ER, and velocity to Λu_0 in Eq. (9). These scalings contain all the information about the chemical and physical properties of the system using the definitions above, leaving only one parameter in the equations, the dimensionless frequency, $\omega\tau$. In the remainder of this section, we abuse notation and use the same variables to represent their dimensionless counterparts, to keep the presentation simple.

For constant double-layer capacitance, the electrochemical relaxation problem is linear, so the response to any periodic forcing is simply a linear superposition of the response to individual Fourier modes at frequency ω. To solve for a particular Fourier mode, we introduce the dimensionless *complex potential amplitude*, Φ, defined by

$$\phi(\boldsymbol{r}, t) = \mathrm{Re}\left\{\Phi(\boldsymbol{r}) e^{i\omega t}\right\} \tag{31}$$

The real part of the complex amplitude, $\mathrm{Re}\Phi$, represents the response which is in phase with the forcing, while the imaginary part, $\mathrm{Im}\Phi$ is the out-of-phase response. Both parts are harmonic functions,

$$\nabla^2 \Phi = 0 \tag{32}$$

subject to the boundary conditions $\hat{\boldsymbol{n}} \cdot \nabla\Phi = 0$ on an insulating surface and

$$i\omega(\Phi - \Phi_0) = \hat{\boldsymbol{n}} \cdot \nabla\Phi \tag{33}$$

on an ideally polarizable surface at complex potential Φ_0.

More generally, we should write

$$\Phi - \Phi_0 = Z(\omega)\, \hat{\boldsymbol{n}} \cdot \nabla\Phi \tag{34}$$

where $Z(\omega)$ is the (dimensionless) impedance of the double layer (Barsoukov and Macdonald, 2005). Our simple capacitor model corresponds to $Z = (i\omega)^{-1}$. Green et al. (2002) and Levitan et al. (2005) have also considered "constant-phase-angle impedance", $Z = (i\omega)^p$, with a fitted exponent $p < 1$. This model was found to improve the fit of their experimental data for bare metal surfaces, as shown in Fig. 7(b) (where $p = \beta$), and also helped to fit independent cell impedance measurements. It is difficult to interpret the data unambiguously, however, since the microscopic justification of constant-phase-angle impedance is controversial, and other effects, such as ion adsorption (see below), were neglected.

Once the time-independent linear boundary-value problem (32)-(33) is solved for $\Phi(\boldsymbol{r})$, we can solve the linear Stokes equations (30) for the time-averaged velocity profile, $\langle \boldsymbol{u}_s \rangle$, subject to no-slip on the insulating surfaces and a time-averaged ICEO slip boundary condition,

$$\langle \boldsymbol{u}_s \rangle = -\frac{1}{4}\nabla_{\|}|\Phi - \Phi_0|^2 \tag{35}$$

on the polarizable surfaces, which is linear in the unknown velocity, but nonlinear in the known potential. Levitan et al. (2005) derived Eq. (35) with $\Phi_0 = 0$ for a symmetric geometry with a cylindrical metal wire at potential. For the general case of an electrode or metal structure with $\Phi_0 \neq 0$, the (dimensionless) oscillating slip velocity is

$$\boldsymbol{u}_s = -\Psi_D \boldsymbol{E}_\perp = \mathrm{Re}\left\{(\Phi_0 - \Phi)e^{i\omega t}\right\} \mathrm{Re}\left\{\nabla_\perp \Phi e^{i\omega t}\right\} \tag{36}$$

Using $\mathrm{Re} z = (z + \overline{z})/2$ and averaging over a time period, we find

$$\langle \boldsymbol{u}_s \rangle = \frac{1}{4}\left[(\Phi_0 - \Phi)\nabla_\perp \overline{\Phi} + (\overline{\Phi_0} - \overline{\Phi})\nabla_\perp \Phi\right] \tag{37}$$

The desired result (35) follows for a metal surface whose potential is constant in space (but not time), so that $\nabla_\perp \Phi_0 = 0$.

6.2 Analytical example: Metal sphere in an AC field

To demonstrate the ease of applying the our general mathematical framework, we derive the formula of Murtsovkin (1996) for the frequency-dependent flow around an ideally polarizable colloidal sphere in a uniform AC field (neglecting surface conduction). Many other examples appear in the literature cited above and in recent reviews (Bazant et al., 2009b; Squires, 2009; Bazant and Squires, 2010), but this canonical example serves to illustrate the basic steps in any analysis of ICEO flow with the Standard Model.

We work with dimensionless variables in spherical coordinates (r, θ, φ) with length scaled to the sphere radius R and voltage scaled to E_∞/R, where E_∞ is the uniform field strength far from the sphere. By symmetry the solution depends only on (r, θ), and we can set the sphere potential to zero $\Phi_0 = 0$ relative to the background applied potential. The solution to Laplace's equation (32) satisfying the far field boundary condition,

$$\boldsymbol{E} = -\nabla\Phi \sim \cos\omega t\,\hat{z}, \quad r \to \infty \tag{38}$$

has the form

$$\Phi(r,\theta) = -r\cos\theta\left(1 + \frac{p}{r^3}\right) \tag{39}$$

where the first term represents the uniform background field and the second the induced dipole on the particle and its screening cloud. The (dimensionless) complex induced dipole moment,

$$p = \frac{1 - i\omega}{2 + i\omega}, \tag{40}$$

is obtained by imposing the RC boundary condition (33). Consistent with the simple physical arguments in Fig. 6, the metal particle behaves like a bare conductor at high frequency, $\lim_{\omega\to\infty} p = -1$, and like an insulator at low frequency, $\lim_{\omega\to 0} p = 1/2$, due to complete screening by the induced double layer. The former is due to electron charge separation in the particle, required to make the particle an equipotential surface with a normal electric field, while the latter is the result of ionic charge relaxation in the double layers, which fully expels the normal electric field.

The complex dipole moment (40) captures the time-dependent polarization of the particle and its screening cloud in response to the AC forcing (Dukhin and Shilov, 1980). The real part is the in-phase polarization,

$$\mathrm{Re}\,p = \frac{2 - \omega^2}{4 + \omega^2} \tag{41}$$

which transitions from 1/2 at low frequency to −1 at high frequency, and passes through zero at $\omega_c = \sqrt{2}$. At this critical frequency, only the uniform background electric field remains, as if the particle were "invisible" in phase with the AC forcing. The out-of-phase polarization

$$\mathrm{Im}\,p = -\frac{3\omega}{4 + \omega^2} \tag{42}$$

is reaches a maximum at ω_c, due to capacitive charging of the double layers on opposite sides of the particle, carried by ionic currents passing through the bulk electrolyte resistance to complete an equivalent RC circuit (Simonov and Shilov, 1977). These time-dependent polarization phenomena are also seen in more complicated geometries, as illustrated in Fig. 11 below.

The time-averaged electro-osmotic slip profile on the sphere is easily calculated from (35) as

$$\langle u_\theta(1,\theta)\rangle = \left(\frac{9}{4}\right)\frac{\sin 2\theta}{4+\omega^2} = U_s(\omega)\sin 2\theta \tag{43}$$

which drives a quadrupolar Stokes flow with angular and radial components

$$\langle u_\theta(r,\theta)\rangle = \frac{U_s(\omega)\sin 2\theta}{r^4} \tag{44}$$

$$\langle u_r(r,\theta)\rangle = \frac{U_s(\omega)(1+3\cos 2\theta)}{2}\left(\frac{1}{r^4}-\frac{1}{r^2}\right) \tag{45}$$

respectively.

As in Fig. 6, the long-ranged part of the ICEO flow, decaying as r^{-2}, is a radial flow that sucks fluid in toward the poles of the sphere and ejects fluid away from the equator. The flow has a longer range than classical electro-osmotic flow, decaying as r^{-3}, although still a shorter range than forced electrophoresis (Fig. 3(a)), decaying as r^{-1}. Gamayunov et al. (1986) first calculated this flow for $\omega = 0$ and noted how the long-range part dominates the interaction between two polarizable colloidal particles, causing them to move together (or apart) if aligned parallel (or perpendicular) to the field axis. The same anisotropic hydrodynamic interactions due to ICEO flow are also evident in the Brownian-dynamics simulations of Saintillan et al. (2006a) and the experiments of Rose and Santiago (2006) for rod-like metal particles in a uniform electric field.

6.3 Numerical example: Metal cylinder in a microchannel

Microfluidic devices involve bounded channel geometries, which make analytical progress difficult in most cases. (Exceptions include simple geometries in two dimensions, amenable to conformal mapping analysis following Yossifon et al. (2006).) Nevertheless, the formalism above, based on the Standard Model with AC forcing, is still useful to reduce the full, time-dependent, nonlinear equations to a relatively simple time-independent form, involving only the linear Laplace and Stokes equations. Aside from dealing with the non-standard boundary conditions, it then becomes straightforward to solve the problem using well known algorithms or common software packages.

An example of the numerical solution of the model using the Finite Element Method is shown in Figure 11, using the package, FEMLAB, a precursor of COMSOL Multiphysics. The simulations describe the ICEO flow around an ideally polarizable metal cylinder lying on the floor of a flat

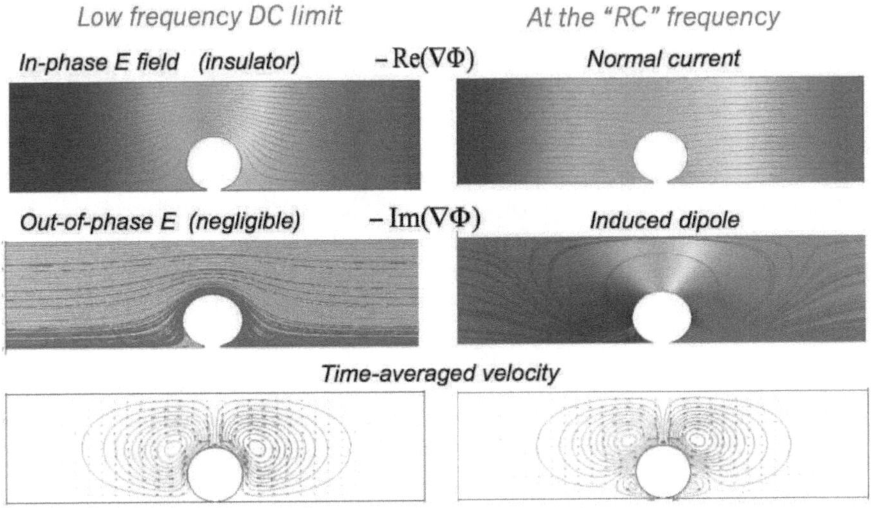

Figure 11. Finite-element numerical solution of the Standard Model for AC forcing from Section 6.1 for the experimental geometry of Levitan et al. (2005), consisting of a 100-micron diameter metal cylinder in a straight microchannel with an longitudinal applied field. Top row: electric field lines in phase with the AC forcing (gradient of the real part of the complex potential). Middle row: field lines 90° out of phase with the forcing (gradient of the imaginary part). Bottom row: streamlines of the time-averaged ICEO flow. Left column: Zero frequency (DC, steady state). Right column: Unit dimensionless frequency $\omega\tau = 1$, where the AC period equals the "RC time" of the equivalent circuit. A slice of the computed flow profile in this figure is compared to experimental data in Figure 7(b) above. (Simulations by Y. Ben from the work of Levitan et al. (2005).)

microchannel, whose height is twice the cylinder's diameter. An AC electric field is applied along the microchannel, perpendicular to the orientation of the cylinder. This is the geometry of the experiments of Levitan et al. (2005), which were the first to demonstrate ICEO flow around an electrically floating object (whose voltage is not controlled as an electrode) in a microfluidic device. This study was also able to make the first quantitative comparison of the theoretical flow profile, computed numerically, with experimental data, taken by particle-image velocimetry, as shown in Fig. 7.

The simulations in Fig. 11 nicely illustrate the physical principles in Fig. 6 and show similar frequency-dependent behavior as the analytical solution of the previous section:

- At low frequency ($\omega\tau \ll 1$), the AC period is long enough to allow complete charging of the double layer in phase with the forcing. As a result, the in-phase electric field ($\mathrm{Re}\nabla\Phi$) resembles that of an insulator in a DC field, going around the cylinder, while the out-of-phase field ($\mathrm{Im}\nabla\Phi$) is negligible. The time-averaged ICEO flow is directed from the poles of the cylinder toward the equator, but due to the nearby no-slip wall, only half of the quadroplar flow for an isolated cylinder (Fig. 6) is visible. The bounded geometry also causes the flow to recirculate in two counter-rotating vortices.

- At the transition frequency ($\omega\tau = 1$), the AC period is in resonance with the relaxation of the double layer. As a result, the in-phase electric field is almost unaffected by the presence of the cylinder, and the out-of-phase field lines show the induced dipole with two lobes emanating from the cylinder. The time-averaged flow now shows the appearance of secondary, small vortices completing a quadrupolar structure, due to incomplete charging of the double layers, which has not had enough time to proceed between the poles and the wall.

- At high frequency ($\omega\tau \gg 1$, not shown), there is not enough time for double-layer charging, and the in-phase field resembles that of a perfect conductor in a DC field, where the cylinder remains an equipotential surface, even outside the double layers. Since there is little charge relaxation, the out-of-phase field and ICEO flow are negligible.

These three frequency regimes are generally present for all ICEO flows around a polarizable surface, driven by a weak AC field in the electrolyte. More complicated microfluidic geometries can lead to a distribution of charging time scales, and thus additional frequency regimes, where ICEO flows develop in different locations for different ranges of frequencies.

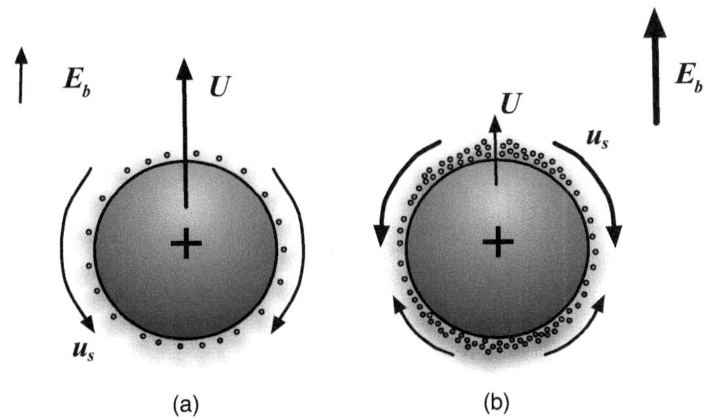

Figure 12. Field-depenent electrophoretic velocity U of an ideally polarizable, charged sphere of radius R with thin double layers in a background field E. (a) In small fields, the mobility $\mu_{ep} = U/E$ is a constant, set by the uniformly distributed double-layer charge. (b) In large fields, $E \gg kT/eR$, the dipolar induced charge overwhelms the pre-existing uniform charge and alters the mobility, $\mu_{ep}(E)$, if cations and anions do not condense at the same density and must redistribute to conserve total charge. (Reproduced from Bazant et al. (2009b) © 2009 Elsevier.)

7 Electrophoresis of ideally polarizable particles

7.1 Field-dependent mobility and aperiodic electrophoresis

A well-known prediction of the classical theory of electrophoresis is that the mobility (7) only depends on the total charge (or average zeta potential), in the limits of thin double layers, small charge, and weak fields (Hunter, 2001; Anderson, 1989). This remarkable result holds for any size or shape, even if the particle is polarizable and acquires a non-uniform charge (or zeta) profile in response to the applied field. It is not widely appreciated, however, that this follows from the assumption of linear response of the double layer with a constant capacitance, which reduces (14) to (16)

In the 1970s, S. S. Dukhin's group was perhaps the first to recognize that the electrophoretic mobility of polarizable particles must generally depend on the electric field (Dukhin and Dukhin, 2005). In a series of Russian papers, which have yet to gain widespread attention, they predicted perturbations of the electrophoretic mobility as, $\Delta\mu_{ep} \propto E^2$, and thus nonlinear electrokinetic motion, $\Delta U \propto E^3$, which they have come to call the "Stotz-Wien effect".

One general mechanism for nonlinear electrophoresis in steady DC fields, first predicted by A. S. Dukhin (1993), is a voltage-dependent double-layer capacitance. In the limit of weak applied electric fields, $E \ll kT/eR$, he showed that an ideally polarizable sphere with equilibrium zeta potential ζ_0 and radius R has a field-dependent DC electrophoretic mobility,

$$\mu_{ep}(E) = \frac{U(E)}{E} \sim \frac{\varepsilon}{\eta}\left(\zeta_0 - \frac{3}{8}\frac{C_D'(\zeta_0)}{C_D(\zeta_0)}(ER)^2 + \dots\right) \tag{46}$$

where ζ_0 is the surface averaged diffuse layer voltage, $\langle \Psi_D \rangle$. This result follows directly from the Standard Model formalism developed above, by applying perturbation methods to Eq. (14). The same equations could be used to analyze the frequency dependence and shape dependence of this effect, but apparently this has not yet been done.

Dukhin's correction (46) has mainly been applied in the context of the Gouy-Chapman model, Eq. (29), which predicts decreased mobility, $\Delta\mu_{ep} < 0$ since $dC_D/d\psi > 0$ for $\zeta_0 > 0$. It has also recently been derived as the small field limit of a general formula for thin double layers with the Gouy-Chapman model by Yariv (2008). The same formula was also derived by Bazant et al. (2009b) as the dilute limit of a still more general theory that also accounted for the significant influence of finite ion sizes at high voltage (see below). Effects of background concentration gradients during the passage of direct current, such as asymmetric concentration polarization and diffusio-osmotic flows which lead to nonlinear "electro-diffusiophoresis," were analyzed for dilute solutions with thin double layers by Rica and Bazant (2010).

The basic physics of this nonlinear effect is illustrated in Fig. 12. If the double-layer voltage varies enough to cause spatial variations in its differential capacitance, then counterions aggregate with varying density (per area) around the surface of the particle upon polarization by the applied field, and this nonlinearity breaks symmetry in polarity with respect to the mean voltage. For example, if the positively charged part of the diffuse layer (relative to the mean charge) is less dense (e.g. due to larger or less charged cations than anions), it will cover more of the surface than the negatively charged part; cations are then more likely to dominate in regions of large tangential field near the equator and thus make an enhanced contribution to the electrophoretic mobility of the particle, regardless of its true surface charge.

Dukhin and Dukhin (2005) have proposed a general means to exploit of field-dependent mobility of colloidal particles for separations by "aperiodic

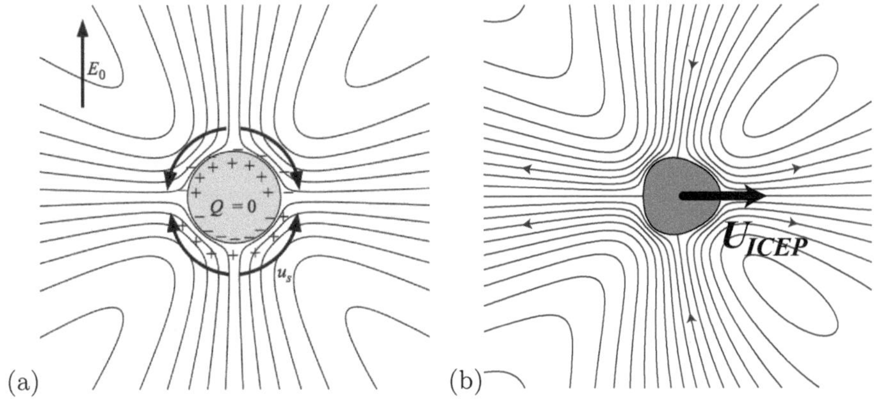

Figure 13. (a) Induced-charge electro-osmotic (ICEO) flow around a symmetric, uncharged, ideally polarizable particle (from Bazant and Squires (2004)); (b) An example of ICEO flow and the resulting induced-charge electrophoretic (ICEP) velocity for an asymmetric shape (from Ref. 4). Due to broken left-right symmetry, the unbalanced ICEO flows cause the particle to move perpendicular to the electric field, which would not be possible due to electrostatic forcing alone. (Reproduced from Squires and Bazant (2006) © 2006 Cambridge University Press.)

electrophoresis". The basic idea is to use an "unbalanced AC field", whose time-average is zero, $\langle E\rangle = 0$, but whose higher moments are nonzero. (The same concept, without reference to a particular mechanism for field-dependent mobility, apparently first appeared in a 1992 U.S. patent of Chimenti, Ser. No. 5,106,468.) If $\langle E^3\rangle \neq 0$, then Dukhin's first correction to the mobility (46) survives time averaging and leads to separation of particles with different polarizabilities. An example of such an unbalanced field is

$$E = E_1 \sin(\omega t) + E_2 \sin(2\omega t + \varphi) \tag{47}$$

where φ is a phase shift, which can be conveniently varied to control the time-averaged motion of a particle. Analyzing and exploiting this effect in practical separations would be an interesting direction for research.

7.2 Induced-charge electrophoresis

Mobility perturbations for spherical particles only hint at the rich phenomena that can arise in the electrokinetic motion of polarizable particles. Murtsovkin (1996) and co-workers were the first (and to date, perhaps the

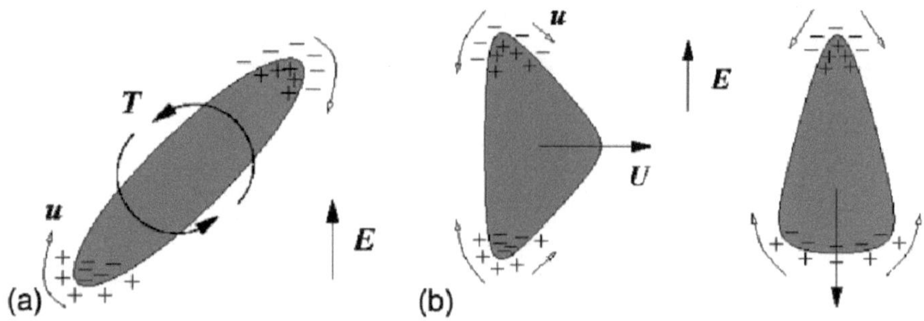

Figure 14. (a) Mechanism for ICEP torque on a rod-like, polarizable particle in a uniform electric field, which enhances dielectrophoretic (DEP) torque. (b) Possible ICEP velocities for asymmetric shapes, once their long axes have aligned with the field.

only ones) to experimentally observe the nonlinear electrokinetic motion of homogeneous particles in a uniform AC field in directions oblique to the field axis. They studied irregular quartz particles moving in water near the wall of a cuvette in surprising directions apparently set only by the particle shape. If a particle rotated enough by Brownian motion when the field was off, it could be seen to reverse direction when the field was turned back on. The velocity scaled with the square of the field amplitude and increased with the particle size. No theory was proposed for this phenomenon, in part since it was only observed near the wall and not in the bulk solution.

Bazant and Squires (2004) recently predicted that polarizable particles in the bulk can undergo essentially arbitrary translation and/or rotation by "induced-charge electrophoresis" (ICEP) in a uniform electric field, as long as they possess appropriate broken symmetries (Squires and Bazant, 2006), such as non-spherical shapes and/or non-uniform surface properties, e.g. due to coatings of varying polarizability. The former cases begin to explain Murtsovkins early observations and beg for new experiments to test a variety of specific theoretical predictions, discussed below. The latter cases, first observed by Gangwal et al. (2008), are described in the next section on heterogeneous particles.

For homogeneous particles, the canonical example is that of an uncharged, ideally polarizable particle of irregular shape in a weak, uniform DC field. In that case, the basic velocity scale U_0 is given by Eq. (9), where R is a characteristic radius scale. Using the Standard Model (with constant double-layer capacitance), Yariv (2005) general expressions for the translational and rotational velocities, respectively,

$$E_i = U_0 \sum_{jk} C_{ijk} E_i E_j \tag{48}$$

$$\Omega_i = \frac{U_0}{a} \sum_{jk} D_{ijk} E_j E_k \tag{49}$$

where C is a dimensionless tensor and D a pseudo-tensor, each expressible as surface integrals involving the bulk potential, just outside the double layer. Squires and Bazant (2006) treated a number of specific examples by solving the Standard Model equations analytically using boundary perturbation methods for nearly symmetric objects, and they developed simple principles to predict the motion of a particular shape.

The basic mechanism of ICEP for irregular particles is shown in Figure 13. As shown in (a) and described above, the ICEO flow around a symmetric particle is quadrupolar, drawing fluid in along the field axis and ejecting it radially. If the particle has broken left/right symmetry as shown in (b), then the radial flow is stronger on one side than the other, leading to ICEP motion perpendicular to the field. Such unusual motion cannot result from electrostatics alone, since there is no charge distribution which can experience an electrostatic force transverse to a uniform electric field. Similarly, breaking only fore/aft symmetry produces ICEP motion along the field axis, and combinations of these asymmetries can cause motion in an arbitrary direction.

ICEP can also contribute to the rotation of polarizable particles with elongated shapes, as illustrated in Figure 14(a). It is well known that DEP causes such particles to align with the axis of a uniform field, due to electrostatic torque on the induced dipole. At low AC frequency (or in the DC limit), if the field persists in one direction long enough for ICEO flow to occur, then ICEP causes a rotational velocity with a basic scale U_0/R that is independent of the particle size but sensitive to its shape,

$$\Omega \propto \frac{\varepsilon E^2}{\eta} \tag{50}$$

This scale happens to be the same as that of the DEP rotational velocity, so ICEP rotation is easily overlooked and mistakenly interpreted as DEP. It is possible, however, to clearly distinguish the two effects, as recently demonstrated by experiments of Rose and Santiago (2006) and simulations of Saintillan et al. (2006a) involving rod-like, metal particles in uniform AC fields. See Figure 15(a).

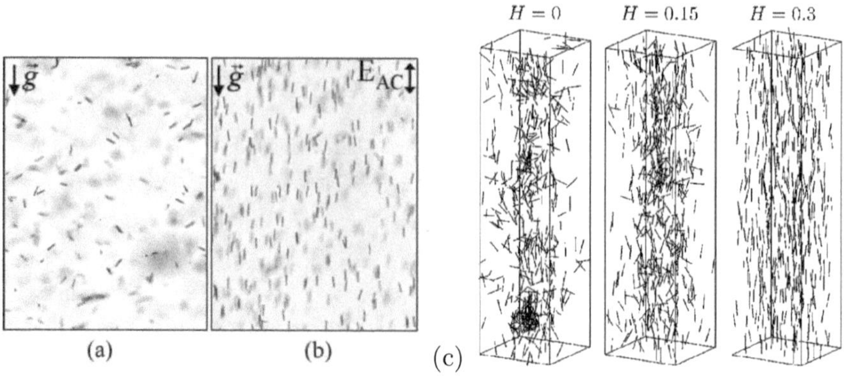

Figure 15. Experiments on cylindrical silver particles (.318μm diameter, 6μm length) sedimenting in de-ionized water by gravity alone (a) and in a 100 Hz, 100 V/cm AC field aligned with gravity (b). (Reproduced from Rose and Santiago (2006) © 2007 American Physical Society.) The experimental distribution of angles in different fields is well described by the Standard Model, taking into account both ICEP rotation and DEP electrostatic torque. (c) Brownian-dynamics simulations show the suspension of sedimenting rods can be stabilized by ICEO flows upon increasing the field strength (H). (Reproduced from Saintillan et al. (2006b) © 2006 American Institute of Physics.)

More complicated asymmetric particles can undergo essentially arbitrary ICEP motion, even in a uniform field. Even in the context of the simple model above, these effects have not yet been fully analyzed, but some general principles have been identified by Squires and Bazant (2006). A striking example is shown in Figure 14(b), which illustrates how arrow-like particles of slightly different shapes can move in perpendicular directions in a uniform field, depending on their broken symmetries: On the left, a short, fat arrow rotates to align its long axis with the field and then moves perpendicular to the field, toward its pointed end; on the right, a long, thin arrow also rotates to align its long axis, but then moves parallel to the field, toward its blunt end. Such predictions are quite recent, however, and remain to be tested experimentally.

A telltale sign of ICEP is the presence of non-uniform ICEO flow around the particle, which leads to complex hydrodynamic interactions with other particles and walls. For example, the basic quadrupolar flow in Figure 13(a) causes two symmetric particles to move toward each other along the field axis and then push apart in the normal direction, as first shown by Gamayunov et al. (1986). A finite cloud of such particles would thus become squashed into a disk-like spreading pancake perpendicular to the field axis (Squires and Bazant, 2004). Hydrodynamic interactions between particles due to ICEO flows are also able to stabilize a suspension of sedimenting rods, above a critical field strength, as shown in Fig. 15(c) from the work of Saintillan et al. (2006b).

The basic ICEO flow field can also cause particles to be repelled from insulating walls (perpendicular to the field), as noted by Zhao and Bau (2007a), or attracted toward electrodes (normal to the field), but these are only guiding principles. Broken symmetries in particle shape or wall geometry, however, can cause different motion due to combined effects of DEP and ICEP, even opposite to these principles, and the interactions of multiple particles can also be influenced strongly by walls. Such effects have not yet been fully explored in experiments or simulations.

7.3 Dipolophoresis

In the 1970s, Shilov and Estrella-Lopis first recognized that electrohydrodynamics (what we now call ICEO) can contribute to the motion of particles in low-frequency, non-uniform electric fields (Simonova et al., 2001), in addition to the classical effect of DEP, although the effect has not been studied much in theory or experiment. Shilov and Simonova (1981) analyzed the problem of an ideally polarizable sphere in a uniform field gradient and made the remarkable prediction that the particle does not move. Due to equal and opposite motions by DEP and ICEP, the sphere levitates in the field while driving a steady ICEO flow, but this is a unique case.

Squires and Bazant (2006) showed that broken symmetries in the field gradient and/or the particle shape generally cause a particle to move, due to subtle imbalances between ICEP and DEP. Both effects have the same basic scaling (2). Moreover, as shown in Figure 16, the DEP force and ICEP velocity tend to act in opposite directions, at least for the case of an ideally polarizable particle with thin double layers in a non-uniform electric field (of arbitrary complexity). Similar to the case of rotational motion discussed above, ICEP can be easily overlooked and the observed translational motion attributed solely to DEP, if it is along the field gradient. Experiments clearly separating ICEP and DEP effects are still lacking, and an opportunity exists to exploit these combined effects for manipulating polarizable colloids, once the effects are better understood.

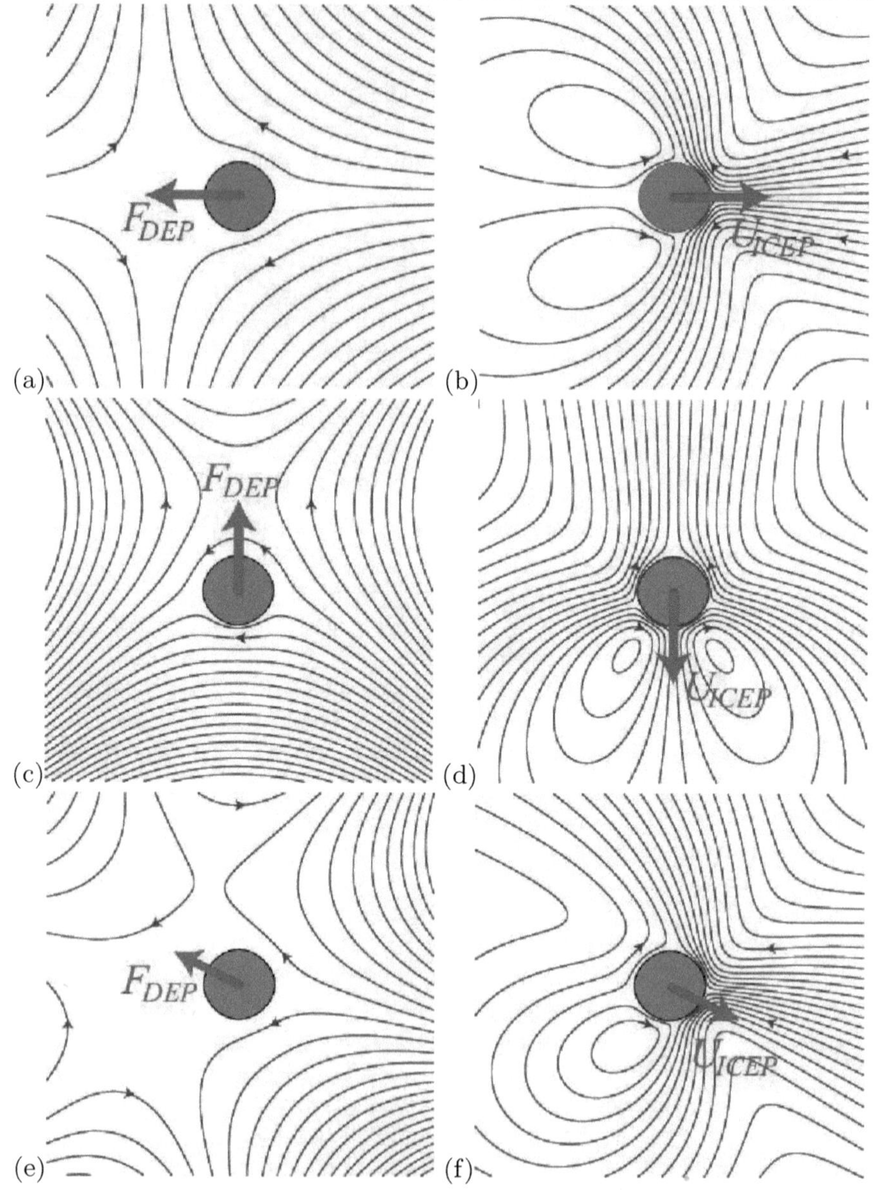

Figure 16. Analytical solutions of the Standard Model for dipolophoresis in two dimensions. Electric fields (a,c,e) and ICEO flows (b,d,f) are shown around ideally polarizable cylinders in inhomogeneous elds. Regardless of the complexity of the multipolar background electric field distribution, the DEP force and ICEO velocity are always in opposite directions, as indicated. (Adapted from Squires and Bazant (2006).)

8 Electrophoresis of heterogeneous particles

In the previous section, we considered homogeneous polarizable particles, allowing for broken symmetries in the particle shape, which lead to induced-charge electrophoretic motion. In this section, we also allow for nonuniform surface properties. In order to appreciate the nonlinear effects of ICEP due to variable polarizability, we first review the possible types of linear response of heterogeneous particles with fixed surface charge.

8.1 Linear electrophoresis

The electrokinetic motion of colloidal particles and molecules in solution in response to applied electric fields can be rather complicated, so many approximations have been made in theoretical treatments. The classical theory of electrophoresis, dating back over a century to Smoluchowski, considers homogeneous particles, which are (i) non-polarizable, (ii) spherical, (iii) uniformly charged, (iv) rigid, (v) much larger than the thickness of the electrical double layer, (vi) in an unbounded fluid, very far from any walls or other particles, and subjected to (vii) uniform and (viii) weak fields, applying not much more than the thermal voltage (kT/e=25mV) across the particle in (ix) dilute electrolytes. Under these assumptions, the particles velocity is linear in the applied electric field, $\boldsymbol{U} = \mu_{ep}\boldsymbol{E}$, where the electrophoretic mobility, $\mu_{ep} = \varepsilon\zeta/\eta$, as noted above. In Smoluchowski's theory, the zeta potential is equal to the voltage across the double layer, which is proportional to the surface charge at low voltage.

Much less attention has been paid to the electrokinetic motion of heterogeneous particles, which have non-spherical shape and/or non-uniform physical properties. By far the most theoretical work has addressed the case linear electrophoresis of non-polarizable particles with a fixed, equilibrium distribution of surface charge (Anderson, 1989). Some examples are shown in Figures 17 and 18. In that case, relaxing only assumption (ii) leads to the classical prediction that the mobility of a particle of uniform composition (uniform zeta) is independent of the shape and size of the particle. Perhaps it was this insensitivity to geometry that led to the common belief that the electrophoretic mobility measures some kind of average surface charge, until Anderson (1984) was the first to clearly point out that this is generally not the case. By carefully relaxing only assumption (iii), he predicted that a sphere of non-uniform zeta potential can move in a different direction from the field and that its mobility is not simply related to its total charge. Generalizing work of Fair and Anderson (1992) on doublet particles, Long and Ajdari (1998) showed that relaxing both (ii) and (iii) leads to even more complicated behaviour, including particles that rotate

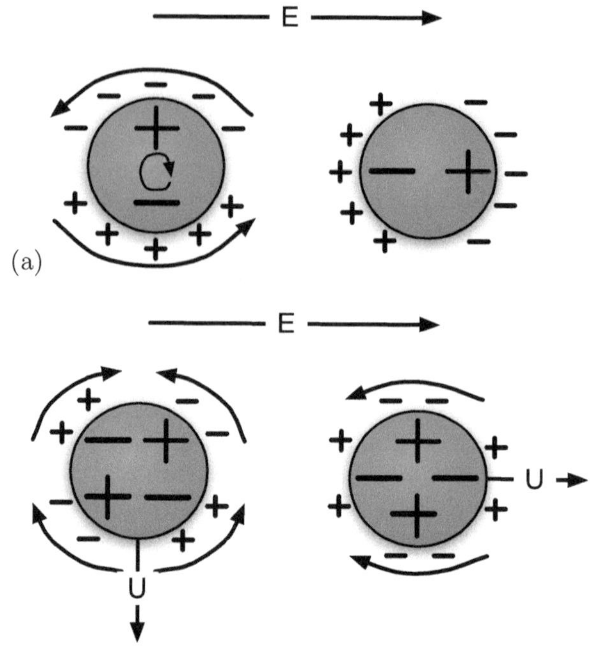

Figure 17. Linear electrophoretic motion of spherical heterogeneous particles with non-uniform fixed surface charge and thin double layers. (a) A dipolar charge distribution will rotate to align its dipole with the field, driven by both electrostatic torque and electro-osmotic flows (indicated). (b) A quadrupolar charge distribution can translate either perpendicular or parallel to the electric field, depending on its orientation.

continuously or translate perpendicular to a uniform DC field. Relaxing assumption (iv), the electrophoresis of flexible heterogeneous particles has also been studied, such as DNA molecules connected to beads (Long and Ajdari, 1996).

It is tempting to think of the electrophoretic mobility of a heterogeneous particle as a measure of its average charge, when in fact it has a nontrivial dependence on the spatial distribution of surface charge. This is clearly demonstrated by a counter-example of Long and Ajdari (1996), motivated by chain-like polyelectrolytes, such as DNA molecules. Consider a dumbbell-shaped particle consisting of two uniformly charged spheres with electrophoretic mobilities μ_1 and μ_2 and hydrodynamic drag coefficients ξ_1

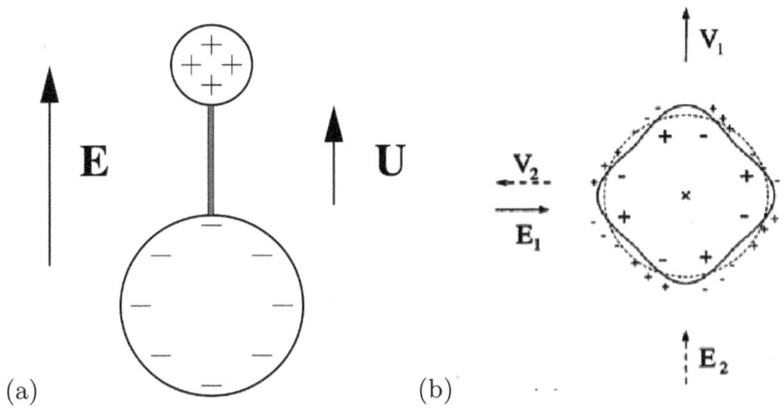

Figure 18. Examples of unusual linear electrophoretic motion of heterogeneous particles with asymmetric shapes. (a) A dumbbell consisting of two oppositely charged spheres of connected by a rigid rod rotates to align as shown and moves in the direction of the electric field (positive mobility), even though the total charge is negative, if the positive sphere is smaller. (b) A particle of zero total charge with four-fold and eight-fold perturbations in shape and surface charge, respectively, moves perpendicular to the electric field, regardless of its orientation (Adapted from Long and Ajdari (1998) © 1998 American Physical Society.)

and ξ_2 , held together by an uncharged, rigid rod. As a first approximation, the rod has negligible drag and is long enough that hydrodynamic and electrostatic interactions between the spheres can be neglected. In a uniform electric field, the dumbbell rotates to a stable configuration aligned with the field axis, as shown in Figure 18(a) and moves a velocity, $\boldsymbol{U} = \mu_{ep}\boldsymbol{E}$, where μ_{ep} is the overall mobility. In order for each particle ($i = 1, 2$) to move at the same velocity, the rod must exert a force, $\boldsymbol{F}_i = \xi_i(\boldsymbol{U} - \mu_i\boldsymbol{E})$. Force balance on the rod, $\boldsymbol{F}_1 = -\boldsymbol{F}_2$, then yields the mobility

$$\mu_{ep} = \frac{\xi_1\mu_1 + \xi_2\mu_2}{\xi_1 + \xi_2} \propto \frac{Q_1}{R_1} + \frac{Q_2}{R_2} \tag{51}$$

which is the drag-weighted average of the two mobilities. In the last step, we have used Stokes formula, $\xi_i = 6\pi\eta R_i$, and assumed that the local mobility (slip coefficient) is proportional to the surface charge density, $\mu_i \propto Q_i/4\pi R_i^2$, where Q_i is the total charge of each sphere. We see that, depending on the geometry, the mobility can have either sign, regardless of the sign the total charge, $Q_1 + Q_2$. For example, as shown in Fig. 18(a), a small sphere of charge $Q > 0$ connected to a larger sphere of charge $-2Q$ can have a positive mobility, even though its total charge is negative, as long as $R_2 > 2R_1$.

Variations in charge density and shape can lead to even more surprising transverse electrophoretic motion, which departs from the field axis. In linear electrophoresis, a spherical particle of non-uniform surface charge (or zeta potential) can move perpendicular to the field, but only for certain orientations; it can also rotate, but only transiently to align its dipole with the field axis. These behaviors are shown in Fig. 17. If both the surface charge and the shape are perturbed, however, then these restrictions do not apply, as noted by Long and Ajdari (1998). Figure 18(b) shows a cylindrical particle of zero total charge, which always moves perpendicular to the electric field, regardless of its orientation. It has four-fold shape perturbation and eight-fold surface charge perturbation, such that each bump on the surface has positive surface charge to the left and negative to the right. By constructing appropriate chiral perturbations of the shape and surface charge, it is also possible to design heterogeneous particles, which rotate continuously around a particular axis without translating, for a particular direction of the electric field.

8.2 Induced-charge electrophoresis

The preceding examples involve non-polarizable objects with fixed surface charge distributions, which do not respond to the electric field. The resulting electrophoretic motion is linear in the field amplitude and vanishes for AC fields. The electrokinetic motion of polarizable particles, however, has nonlinear field dependence due to the phenomenon of induced-charge electro-osmosis (ICEO), where the field acts on induced diffuse charge in the electrical double layer. At frequencies low enough for capacitive charging of the double layer (typically < 10 kHz), the time-averaged motion in an AC field is resembles that in a DC field. In the canonical example of an uncharged metal sphere in a uniform field, the ICEO flow is quadrupolar, drawing in fluid along the field axis and expelling it radially, but there is no net motion.

Motivated by the examples from linear electrophoresis above, Bazant and Squires (2004) pointed out that broken symmetry in ICEO flow generally causes particle motion, and coined the term, induced-charge electrophoresis (ICEP). Examples of broken symmetries include particles with irregular shapes and/or non-uniform physical characteristics, as well as non-uniform applied fields. In the latter case, ICEP occurs at the same time as dielectrophoresis (DEP), although the combined effects of ICEP and DEP on

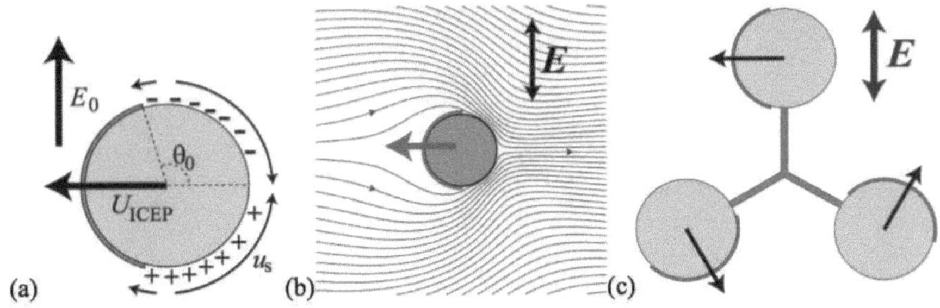

Figure 19. Induced-charge electrophoresis of Janus particles, illustrated for the case of metal partially coated with insulating thin films (adapted from Kilic and Bazant (2011)). (a) Stable orientation in a uniform field, showing induced charge and slip velocities on the metallic side, resulting in motion toward the insulating end, perpendicular to the field. (b) Streamlines of ICEO flow. (c) An ICEP pinwheel, consisting of three Janus particles connected by rigid rods, which tilts to align and then spins continuous around the field axis.

heterogeneous particles remain to be explored. Besides persisting in AC fields, ICEP also depends much more sensitively on particle shape and surface properties than does linear DC electrophoresis. Cases of non-spherical particles with uniform polarizability are discussed above, so we now focus on ICEP due to heterogenous surface polarizability.

The canonical example of Squires and Bazant (2006) is that of a Janus particle with one metallic and one insulating hemisphere, using the standard low-voltage model for electrokinetic motion of polarizable particles. In response to an applied electric field, the Janus particle rotates to align the interface between the two hemispheres with the field axis, due to both ICEP (electrohydrodynamics) and DEP (electrostatics). At the same time, for any orientation, the particle translates in the direction of its insulating end, propelled by ICEO flow on the metallic end, with a velocity

$$U = \frac{9\varepsilon\Lambda R E^2}{64\eta} \tag{52}$$

In particular, once the particle aligns in the field, it continues to move perpendicular to the electric field, with an azimuthal angle set by its initial orientation.

All the generic features of the dynamics still hold if the particles insulating end is smaller or larger than the metallic end, since it is determined by

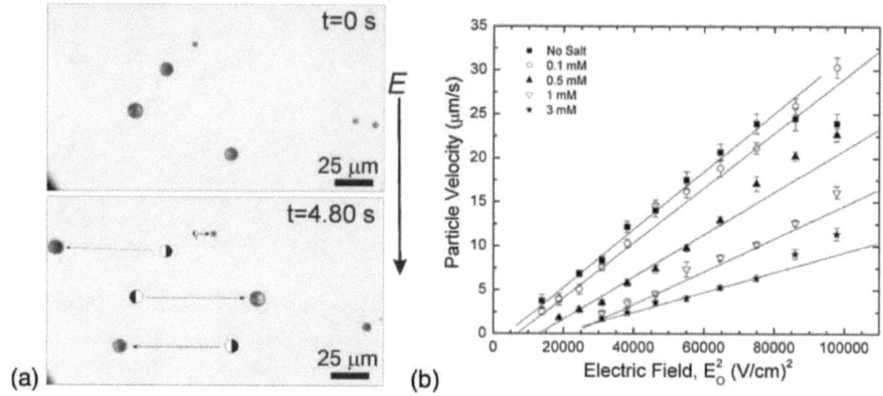

Figure 20. Experimental observation of induced-charge electrophoresis of metallo-dielectric Janus particles in a uniform 10 kHz AC field. (a) Sequence of micrographs demonstrating motion transverse to the field in the direction of the dielectric (light) end propelled by the metallic (dark) end, where the velocity increases with the particle size as in Eq. (52). (b) Velocity versus field amplitude squared at different bulk concentrations of NaCl. (Adapted from Gangwal et al. (2008).)

the broken symmetry. Motion transverse to a uniform AC field cannot have any contribution from DEP, but it is easily understood by considering the ICEO flow in Figure 19(a). After alignment in the field, part of the usual quadrupolar ICEO flow is suppressed on the insulating end. The remaining ICEO flow over the metallic end sucks in fluid along the field axis and pushes it outward from the metallic pole, as shown in Figure 19(b), which propels the particle toward the insulating pole.

This example suggests how to design particles that spin continuously in a uniform field, as noted by Squires and Bazant (2006). Since a Janus particle always translates towards its less polarizable end, a set of three Janus particles connected by rigid rods can be set into continuous motion like a pinwheel, if connected as shown in Figure 19(c). This ICEP pinwheel responds to any DC or AC electric field (of sufficiently low frequency) by tilting to align the particle plane perpendicular to the field and then spinning around the field axis until the field is turned off. Perhaps such particles could be used to sense electric fields or to apply torques to attached molecules or cells.

Transverse ICEP motion of metallo-dielectric Janus particles in a uni-

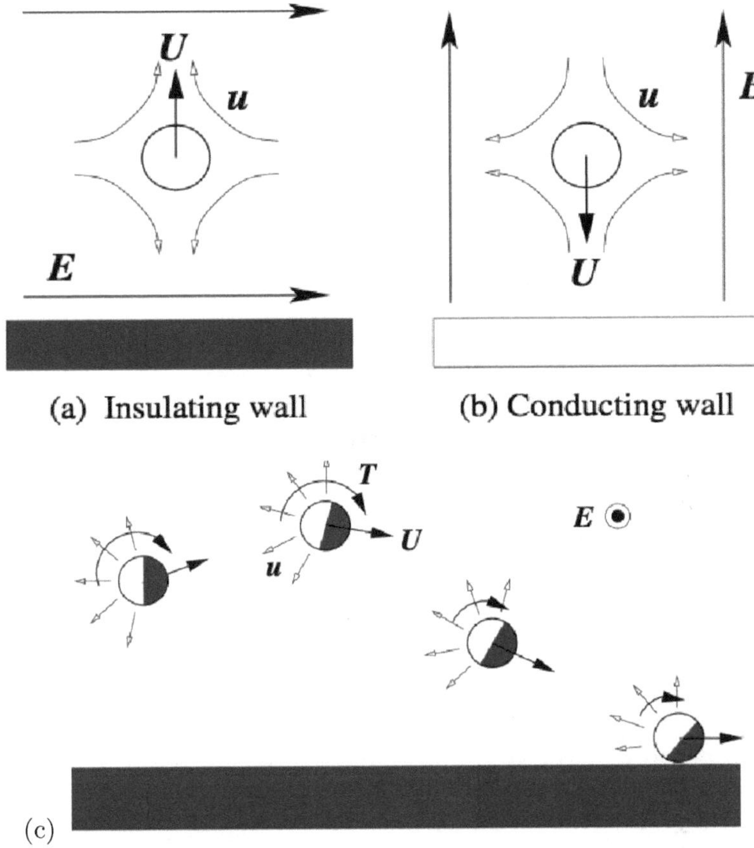

Figure 21. Wall interactions in induced-charge electrophoresis. Dominant hydrodynamic interactions between a homogeneous ideally polarizable particle and (a) an insulating wall and (b) an unscreened, ideally polarizable wall in a parallel electric field. (c) Theory by Kilic and Bazant (2011) of the interaction of a metallo-dielectric Janus particle and a insulating wall, as in the experiments of Gangwal et al. (2008). The particle rotates to align its equatorial plane with the field, with an arbitrary azimuthal angle. Asteady ICEO flow sucks in fluid along the field axis (perpendicular to the page) and ejects it radially on the metallic side, which causes the particle to rotate to face the wall, due to a (mostly) hydrodynamic torque $\boldsymbol{T}$. Near the wall, electrostatic torque can balance the viscous torque to enable motion parallel to the surface without contact, while maintaining a steady tilt angle. (Adapted from Kilic and Bazant (2011).)

form AC field has recently been observed by Gangwal et al. (2008). Consistent with theoretical predictions in Figure 19, the particles align and translate perpendicular to the field in the direction of the less polarizable (light) end, as shown in Figure 20. Larger particles move faster than smaller ones, as expected from Eq. (52), and the velocity scales like the field squared in dilute NaCl solutions. The ICEP velocity decays at higher concentrations, extrapolating to zero around 10 mM. The same concentration dependence is also observed in AC electro-osmotic flow and other nonlinear electrokinetic phenomena, which, although poorly understood, further reinforces that the motion is indeed due to ICEP.

Current research is focusing on how heterogeneous particles undergoing electrokinetic motion due to ICEP and DEP interact with walls and other particles. An interesting feature of the experiments in Gangwal et al. (2008) is that the Janus particles are attracted to nearby glass walls, and the transverse motion is also observed close to the walls, where the theory of Squires and Bazant (2006) does not strictly apply. This behavior is perhaps surprising because, according to the Standard Model, homogeneous particles should be repelled from insulating walls (and attracted to polarizable walls) by ICEO flows (Zhao and Bau, 2007a). Kilic and Bazant (2011) show that this attraction can be understood as a consequence of ICEP torque, which redirects the Janus particle toward a nearby wall and causes it to tilt while translating transverse to the field, as shown in Fig. 21

A major motivation to develop this subject is the possibility of new applications, opened by advances in microfluidics and nanotechnology. In principle, heterogenous particles of specific irregular shapes and non-uniform electrical and/or chemical properties can be designed and fabricated for specific applications. The complex electrokinetic motion of these particles could potentially be used for separation or sample concentration in chemical or biological assays, self-assembly in the fabrication of anisotropic materials, directional transport of attached cargo, electric-field sensing and applying forces and torques to molecules or cells.

9 Induced-charge electro-osmotic mixing

Bazant and Squires (2004) proposed the use of ICEO flow around metallic microstructures (posts, surface patterns, etc.) for microfluidic mixing, switching, and pumping. The potential advantages of such elements in a microfluidic system include low power, programmability, and local flow control. As shown in Fig. 22, the basic physics of ICEO flow immediately suggests a number of designs involving metal posts or surface structures placed in microchannels with applied electric fields. Theoretical work using the Stan-

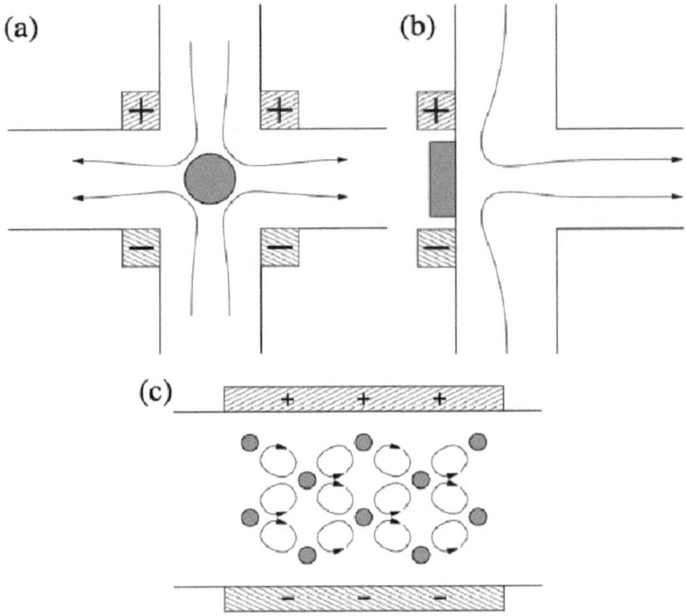

Figure 22. Simple microfluidic devices involving fixed metal posts or surface structures (shaded areas) driven by weak AC or DC fields applied at nearby microelectrodes (cross-shaded areas). Broken symmetries, such as triangular shapes, can lead to transverse flows, and sharp corners can enhance local ICEO flows in electrokinetic jets. (Reproduced from Bazant and Squires (2004) © 2004 Cambridge University Press.)

dard Model has shown that ICEO-based micro-mixing can be enhanced by broken symmetries (Squires and Bazant, 2006) or by the introduction of sharp corners in dielectric channel side walls (Yossifon et al., 2006).

In order to achieve rapid, programmable mixing of the fluid and any suspended particles, temporal modulation of the applied field can be used to produce *chaotic streamlines* (Zhao and Bau, 2007b). This concept is illustrated in Fig. 23. The basic idea is to switch between different asymmetric patterns to produce chaotic trajectories. Even if the underlying flows, $\boldsymbol{u}(\boldsymbol{r}, t)$, satisfy the linear Stokes equations and can be superimposed, the trajectories of passive tracer particles, $\boldsymbol{r}_i(t)$, generally satisfy a nonlinear

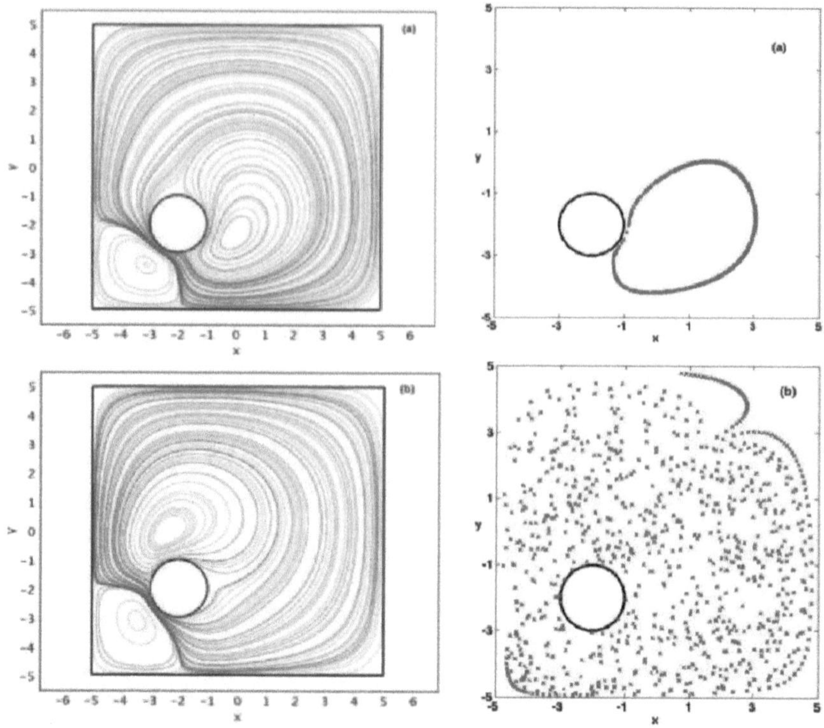

Figure 23. Simulation of an ICEO chaotic mixer with four electrodes on the side walls of a chamber (not shown) driving time-dependent ICEO flows around an off-center metal post. Left: by applying the field either north-south (top) or east-west (bottom), two different flows can be generated. Right: by alternating between these flows, chaotic advection be achieved, as evidenced by the stroboscopic plots (Poincaré section) of a particle, showing the transition from a nearly periodic loop (top) to a chaotic streamline (bottom) with increasing time. (Adapted from Zhao and Bau (2007a).)

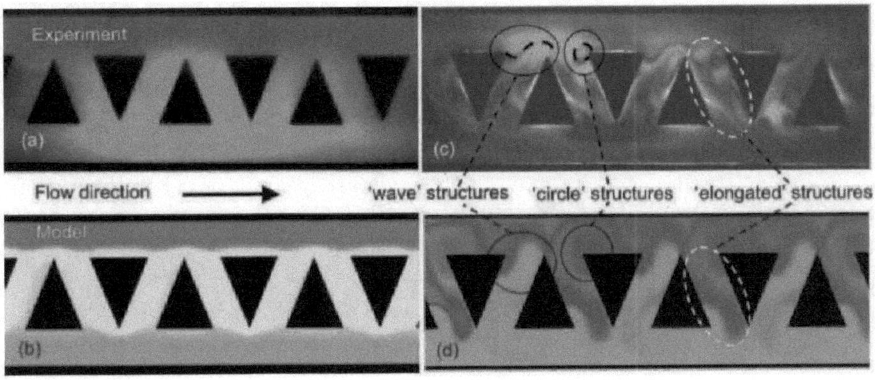

Figure 24. Experimental observation of induced-charge electro-osmotic mixing by Harnett et al. (2008), consistent with the theoretical predictions of Bazant and Squires (2004), sketched in Fig. 22c. Two colored fluid streams of 0.1 mM KCl flowing at 0.1 μl/min from left to right undergo convective mixing by an array of asymmetric metal posts in a transverse AC field (6 Vpp, 100 Hz applied by electrodes above and below, separated by the channel width 200μm). Images from experiments (a,c) and simulations of advection-diffusion in ICEO flow in the same geometry (b,d) show the distribution of red and green fluorescent beads after loading (a,b) and during mixing (c,d). (Reproduced from Harnett et al. (2008) © 2008 Royal Society of Chemistry.)

ordinary differential equation,

$$\frac{d\boldsymbol{r}_i}{dt} = \boldsymbol{u}(\boldsymbol{r}_i(t), t) \tag{53}$$

which can have chaotic solutions, suitable for mixing. The same principle of chaotic advection was originally developed for passive microfluidic mixing by pressure-driven flows in grooved channels (Stone et al., 2004).

The first microfluidic demonstration of ICEO flow around a metal cylinder by Levitan et al. (2005) showed steady vortices, but did not study mixing. In 2008, two groups reported the first experimental demonstrations of microfluidic mixing by ICEO flow around metallic microstructures, effectively reducing to practice the theoretical predictions of Ref. Bazant and Squires (2004). (i) Harnett et al. Harnett et al. (2008) integrated an array of gold-coated posts of triangular cross section in a microchannel with electrodes applying a low-frequency AC field on the side walls, as shown in Fig. 24. The post-array mixer was placed at the junction of two Y-channels, and programmable on/off mixing of two different streams of dilute electrolytes was demonstrated. Good agreement with theoretical predictions was noted, albeit with a correction factor of $\Lambda = 0.25$. (ii) Wu and Li (2008b,a) reported simulations and experiments on ICEO mixing in flow past pointed platinum "hurdles" (floating electrodes) and different geometrical designs were compared. Further design improvements could benefit from numerical optimization methods for ICEO flows developed by Gregersen et al. (2009).

ICEO mixers can be used to enhance the transport of slowly diffusing molecules to an active surface. In biomedical microfluidics, ICEO flows can improve the sensitivity biological assays by passing probe molecules, such as DNA or proteins, rapidly over a detection surface. In electrochemical systems for water purification and desalination, ICEO mixers can enhance the transport or salt and impurities to a membrane or porous electrode for rapid removal, beyond diffusion limitation.

10 AC electro-osmotic pumping

10.1 Slip-driven microfluidic pumps

There are many strategies for microfluidic pumping, as reviewed by Laser and Santiago (2004), Stone et al. (2004), Squires and Quake (2005). Pumps based on fluid body forces, due to externally applied pressure gradients, magnetic fields, electrothermal forces, etc., lose their efficiency with miniaturization, due the overwhelming viscous drag at no-slip walls. On the other other hand, the same viscous drag can be put to use in pumps that generate flow by effective fluid slip on the walls, which only get more efficient with miniaturization. This is the principle behind all electro-osmotic micropumps, whether linear or nonlinear in the applied voltage.

The basic physics of slip-driven pumping are illustrated in Figure 25. For any pump operating in the viscous regime of low- Reynolds number, the flow rate decays linearly with the back pressure P according to

$$\frac{Q}{Q_{max}} = 1 - \frac{P}{P_{max}} \tag{54}$$

where Q_{max} is the flow rate at $P = 0$ and P_{max} is the back pressure that yields $Q = 0$ and effectively stops the pump. By linearity, whenever the pump operates against a hydraulic resistance, the slip-driven flow in the forward direction is superimposed with a pressure-driven parabolic Poiseuille flow profile in the opposite direction. The situation can be modeled by an

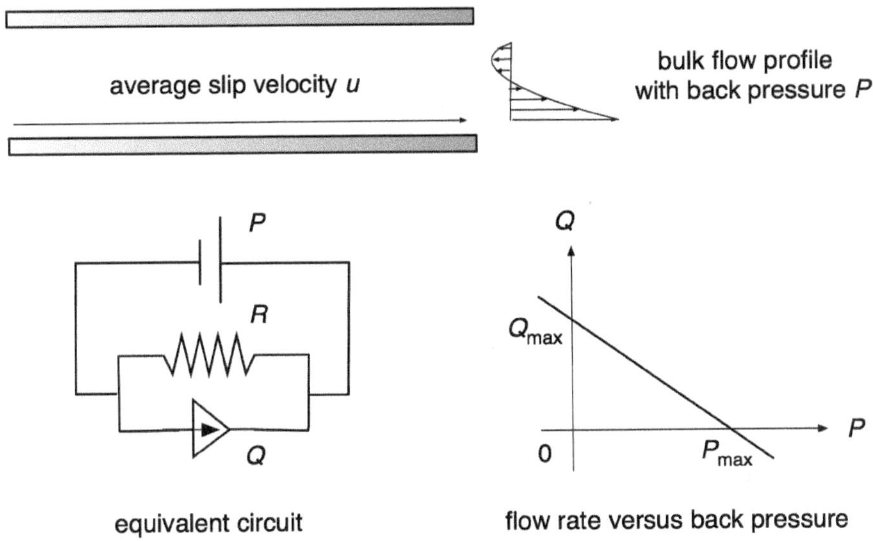

Figure 25. Basic physics of slip-driven micropumps. The pump generates a flow rate Q in a microchannel driven by fluid slip over a wall. If the pump encounters a back pressure drop P, driving a back-flow through its hydraulic resistance R, the net flow is given by the linear relation, $Q - PR$. The maximum flow rate (at $P = 0$) is $Q_{max} = Q$, while the maximum back-pressure which stops the pump ($Q = 0$) is $P_{max} = Q/R$.

equivalent electrical circuit shown in the Figure 25, where the pump consists of constant current Q_{max} in parallel with the back-flow resistance R_B.

To estimate these quantities, consider a microchannel of rectangular cross section with a wide floor of width W producing slip and a much smaller height $H \ll W$, and solve for the resulting Stokes flow. The total flow rate due to a mean slip velocity U over the bottom surface is

$$Q_{max} = \frac{\alpha HWU}{2} \tag{55}$$

where $\alpha \sim 1 - (1/2)(H/W)^2$ corrects for fringe flows in the limit $H \ll W$. The back pressure required to stop the net forward flow is given by

$$P_{max} = R_B Q_{max} = \frac{UL}{k} = \frac{6\eta\alpha UL}{H^2}\left[1 + \left(\frac{H}{W}\right)^2\right] \tag{56}$$

k is the hydrodynamic Darcy permeability and L is the length of the microchannel.

10.2 DC versus AC electro-osmotic pumps

This simple calculation shows that the hydraulic resistance to back flow, and thus P_{max}, can be increased by reducing the micro-channel height H. This strategy has been used to boost the pressure of DC electro-osmotic pumps by employing linear electro-osmotic flows in porous glass frits with submicron pores (Yao and Santiago, 2003). Electro-osmotic micropumps can achieve large head pressures, exceeding 50 atm, without any moving parts, but they require large DC voltages up to kilo-Volts. Such large voltages cause Faradaic reactions, such as water electrolysis, leading to hydrogen and oxygen gas production at the electrodes, which must be managed carefully.

The large operating voltage and the need to manage reaction products can hinder the application of DC electro-osmotic pumps in portable or implantable lab-on-a-chip devices. Moreover, it is difficult to locally manipulate the fluid within the microchannels by applying an electric field across the entire device. These drawbacks can be overcome using small AC voltages applied at microelectrodes suitably distributed inside a microfluidic system. The integration of electrodes in the channel limits the extent to which the channel height H can be reduced, but useful pressures can still be generated using small AC applied voltages, around 1 Volt (root mean square), with greatly reduced Faradaic reactions.

As described above, classical electrokinetic phenomena are linear in the applied voltage and thus cannot produce any net flow under alternating current conditions. A variety of nonlinear electrokinetic phenomena, which persist in AC fields, have been known for decades in colloid science, but the focus has been on electrophoretic mobility and particle interactions. The advent of microfluidics has stimulated interest in the use of electric fields to drive fluid flows, without any moving parts. In this context, nonlinear electrokinetics offers some unique advantages, such as the reduction of unwanted electrochemical reactions (using AC voltages) and the ability to drive fast, programmable flows at low voltages (using closely spaced micro-electrodes).

This area of research in nonlinear electrokinetics began with the discovery of Ramos et al. (1999) of steady electro-osmotic flow over a pair of micro-electrodes applying an AC voltage and dubbed the effect AC electro-osmosis. Around the same time, Ajdari (2000) predicted ACEO flow over periodic electrode arrays and showed how the effect could be used for long-range pumping. As the performance of ACEO pumps has advanced (Huang

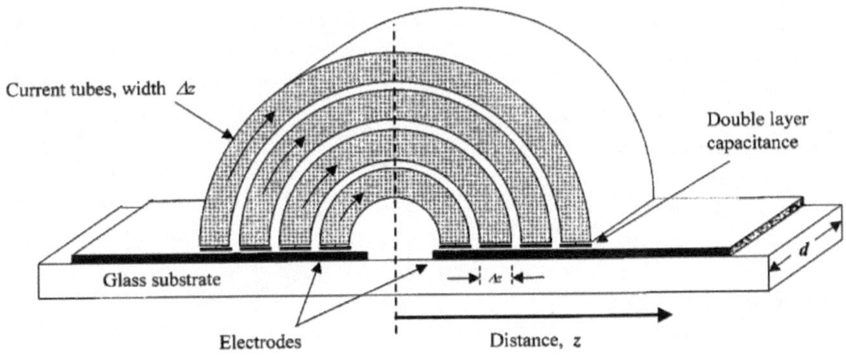

Figure 26. Equivalent RC circuit model for double-layer charging over a pair of electrodes. The inner edges of the electrodes encounter less bulk resistance (due to shorter current tubes) and thus charge more quickly than the outer edges. (Reproduced from Ramos et al. (1999) © 1999 Elsevier.)

et al., 2010), ACEO has also been exploited, in conjunction with dielectrophoresis in different geometries to manipulate particles and cells in microfluidic devices (Green et al., 2000b; Wong et al., 2004; Wu, 2006).

10.3 Flows over ideally polarizable electrodes

ACEO is a phenomenon of induced-charge electro-osmosis, where flow is generated by the action of an electric field on its own induced diffuse charge near a polarizable surface. The main difference with other examples of ICEO flows discussed above is that ACEO involves electrode surfaces, which supply both the electric field and the induced screening charge, in different regions at different times. For this reason, ACEO is inherently time-dependent (as the name implies) and tied to the dynamics of diffuse charge, as ions move to screen the electrodes.

Perhaps the easiest way to understand ACEO is to consider a pair of ideally polarizable planar electrodes applying a sudden DC voltage (which is analogous to ICEO flow around a polarizable particle in an sudden electric field). As shown in Figure 26, charge relaxation can initially be described by an equivalent RC circuit, where the diffuse layers act as capacitors, connected to current tube resistors of varying length through the bulk solution. Since the resistance is smaller (and the field larger) near the gap, the inner portions of double layers on the electrodes charge more quickly than

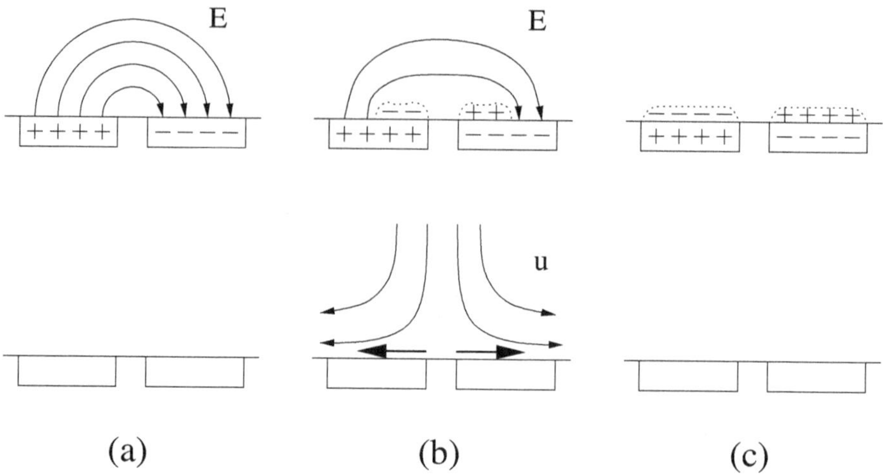

Figure 27. The basic physics of AC electro-osmosis. Electrochemical relaxation (top) and induced - charge electro-osmotic flow (bottom) in response to a suddenly applied voltage across an electrode pair. (a) At first the electric field has no tangential component on the electrodes, since they are equipotential surfaces, and thus there is no electro-osmotic flow. (b) Capacitive double-layer charging begins near the gap where the initial normal current is strongest and causes the unscreened field lines dip down and provide tangential components over the induced charge; the result is ICEO flow directed away from the electrode gap. (c) After the charging time passes, the electrodes are fully screened, leaving no electric field and thus no flow. An AC voltage can drive a steady time-averaged flow, similar to (b), if its period is comparable to the charging time. (Reproduced from Bazant (2008a) © 2008 Springer.)

the outer portions. As shown in Figure 27, this causes ICEO flow to occur, directed outward from the gap, only when the electrodes are partially screened, as the tangential field from the unscreened outer portions acts on induced charge on the inner portions. Note that the flow is independent of the sign of the applied voltage: If the polarity were reversed, then the field and induced charges would both change sign, resulting in the same ICEO flow.

Under AC forcing, the flow peaks when the oscillation period is comparable to the charging time (Fig. 27b). ACEO flow decays at higher frequencies, since there is not enough time for charge relaxation (Fig. 27a). It also decays at lower frequencies, since there is enough time to completely screen the bulk electric field (Fig. 27c).

The theory of ACEO is mostly based on the Standard Model using the complex potential for AC forcing, following González et al. (2000) as described in Section 6. In this regime, the basic scaling of time-averaged ACEO flow is

$$\langle u \rangle \sim \frac{\Lambda(\omega/\omega_c)^2}{[1+(\omega/\omega_c)^2]^2}\frac{\varepsilon V^2}{\eta L} \tag{57}$$

where V is the applied voltage amplitude and L is electrode spacing (roughly from center to center). The basic velocity scale for ACEO is the same as the electroviscous scale u_0 for ICEO flow with the characteristic field, $E \sim V/L$, and induction length $R = L$. The frequency-dependent prefactor is a Lorentzian spectrum peaking at the critical frequency,

$$\omega_c \sim \frac{D}{\lambda_D L}\frac{C_D}{C} \tag{58}$$

which is the inverse of the RC time scale τ defined above.

10.4 Fluid pumping by micro-electrode arrays

Some useful general principles have been developed to guide the design of ACEO pumps. The flows discovered by Ramos et al. (1999) over small numbers of electrodes can be used for local fluid mixing or particle trapping at stagnation points, but the flow decays quickly away from the electrode surfaces. A symmetric, periodic array of many inter-digitated electrodes (of alternating polarity at each moment in time) similarly produces an array of counter-rotating convection rolls, but no net pumping of the fluid in one direction. Instead, long-range pumping over an electrode array requires *broken symmetry within each spatial period* to rectify the AC forcing.

There are several ways to design ACEO pumps by breaking symmetry in a periodic electrode array. Ajdari (2000) originally suggested modulating

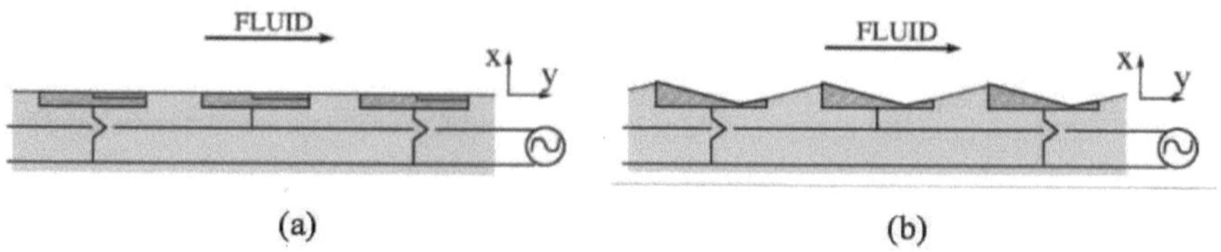

Figure 28. Sketches of local broken symmetries in a periodic electrode array which lead to global time-averaged ACEO pumping: (a) non-uniform surface coatings; (b) non-uniform surface height. (Reproduced from Ajdari (2000) © 2000 American Physical Society.)

either the electrode capacitance via a dielectric coating (Figure 28a) or the surface height (Figure 28b) with half the spatial period of the array, so that the one side of each electrode drives stronger ACEO flow compared to the other side and thus wins to produce net pumping over the array. In the first implementation of an ACEO pump, Brown et al. (2000) opted instead to break symmetry by using planar electrodes of different widths and gaps, and, until recently, this design was the only one studied experimentally (Studer et al., 2004) or theoretically (Olesen et al., 2006). It has been shown to generate velocities over 100 microns/sec, although it also exhibits poorly understood flow reversals (see below).

The performance of ACEO pumps can be greatly enhanced by designing appropriate non-planar electrode geometries. As recently predicted by Bazant and Ben (2006), various 3D ACEO designs exhibit dramatically increased flow rate without flow reversal, due to a special geometry in which the non-uniform slip profile on the electrodes all contributes to flow in the same direction. The basic idea is to create a fluid conveyor belt with electrodes each having steps of two different heights, as shown in Figure 30: On each electrode, the region of desired forward flow is raised up, while the region of reverse flow is recessed below, so as to recirculate in a vortex aiding the forward flow (rather than fighting it, as in planar designs). This can be accomplished with electrodes having electroplated metal steps, as shown in Figure 29, although other designs are possible, such as flat electrode steps deposited on a grooved surface (without the vertical metal surfaces). Simulations predict that 3D ACEO pumps are faster than planar pumps by more than an order of magnitude, at the same voltage and minimum feature size, and thus can achieve mm/sec velocities with only a few volts. This suggests using 3D ACEO pumps to drive flows in battery-powered, portable or

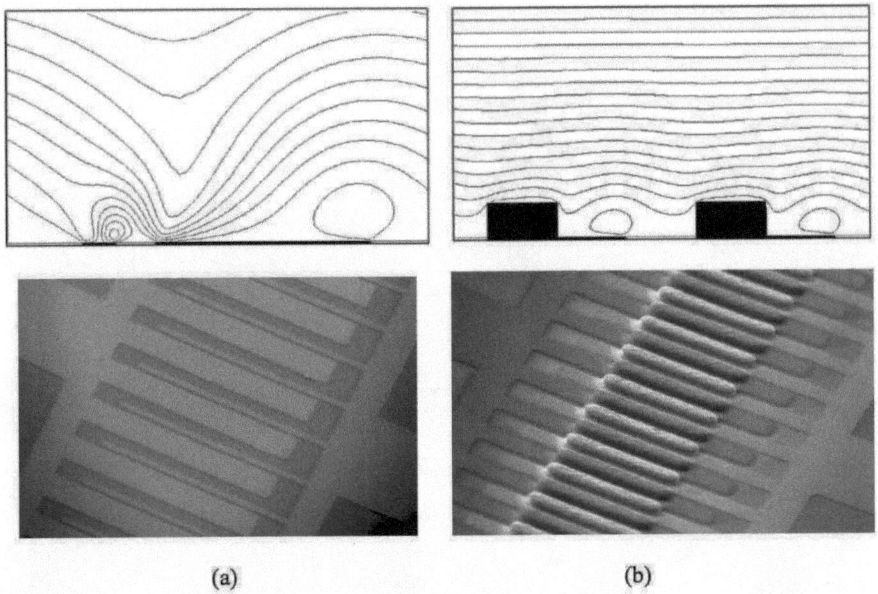

Figure 29. Top (a,b): Simulations of ACEO microfluidic pumps, showing the time-averaged flow over a pair of micro-electrodes (dark regions) in one spatial period of an interdigitated-electrode array. (a) A nearly optimal planar design with different electrode sizes and gaps; the smaller electrode has the largest local slip velocity, but the larger electrode wins in overall pumping from left to right. (b) A more efficient 3D ACEO design with stepped electrodes having a symmetric footprint and the same minimum feature size; the reverse slip now recirculates in a vortex to create a fluid conveyor belt for the raised pumping flow from left to right. (Reproduced from Bazant and Ben (2006) © 2006 Royal Society of Chemistry.). Bottom (a,b): Scanning electron microscopy images of each design fabricated in gold on glass with minimum feature size (gap) of 5 microns. (Courtesy of J. P. Urbanski and J. A. Levitan, using methods of Urbanski et al. (2006b)).

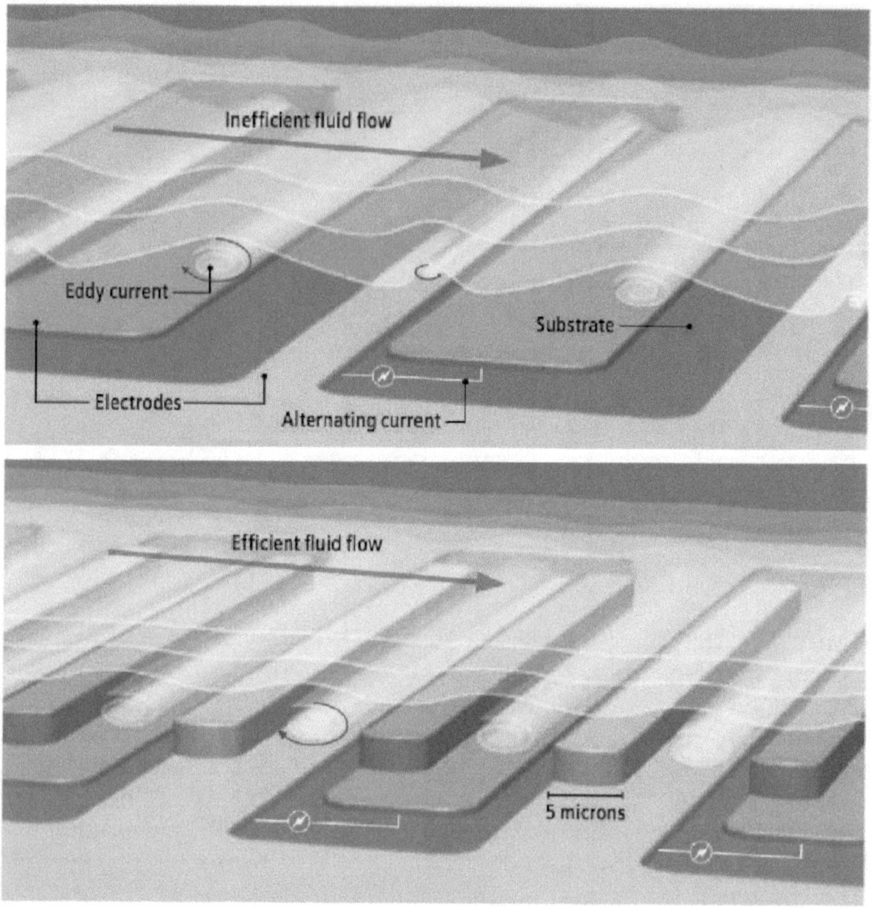

Figure 30. Artist's rendering of the flows in two ACEO micropump designs, based on the simulations in Fig. 29 below. Top: the original planar interdigitated electrode array of Brown et al. (2000). Bottom: the 3D design of Bazant and Ben (2006), which produces a more efficient "fluid conveyor belt", as verified in the experiments of Huang et al. (2010) in Fig. 31 below. (Reproduced from Choi (2007) © 2007 George Retseck.)

implantable microfluidic devices.

Huang et al. (2010) recently reported the state-of-the-art in 3D ACEO micropumps and demonstrated the first integration of ACEO (or ICEO) flow control in a portable biomedical lab-on-a-chip device (Fig. 31). Their design is based on (i) theoretically optimal electrode shapes for ultrafast flows predicted by Burch and Bazant (2008) and (ii) long, serpentine microchannels to dramatically boost the head pressure, by an order of magnitude over previous devices. The latter effect takes advantage of the scaling of the head pressure with length, $P_{max} \sim L$, in Eq. (56). With 1.06 Volt (rms) applied at 1-10 kHz, the pump achieved pressures over 1% atm and mean velocities over 1 mm/sec in water, sufficient to drive flows for an on-chip DNA micro-array assay. The current (mA) and power consumption (mW) are easily provided by a small Li-ion battery, so this work opens new possibilities for portable or implantable microfluidic systems. As with other ICEO phenomena, however, ACEO pumps require dilute electrolytes, which may be a fundamental limitation (Bazant et al., 2009b).

Simple scaling arguments show how to design serpentine ACEO pumps with desired characteristics. The flow rate or pressure can be increased by connecting multiple pumps in parallel or in series, respectively. For example, since our prototype pump consists of only one thin layer (25 mm (h) channels), its pressure can be increased by a factor of ten, exceeding 10% atm simply by stacking ten layers (for a total thickness below 1 mm). Regardless of the channel layout, for a given device volume, there is always trade-off between maximum flow rate and maximum pressure. We have already noted that to maximize pressure, the channel height, H, should be reduced as much as possible, given the electrode sizes and fabrication methods, so this should be viewed as a constant when designing the channel layout. To tune the flow rate, we can vary the channel width W.

Material and fabrication constraints limit the total device cross-sectional area per channel A, which includes the surrounding walls and substrate thickness, and is thus larger than the internal channel cross-sectional area HW. The fabrication method thus sets the ratio $\beta = HW/A$. For a given volume Ω, the total length of the channel can be estimated as $L = \Omega/A$, ignoring any corner effects in regions of the channel without a pumping surface. Using Eq. (56), we find

$$P_{max}Q_{max} = \beta\gamma\frac{U^2\Omega}{H^2} \tag{59}$$

where $\gamma < 1$ is a constant, reflecting the hydraulic resistance of corners and connections, compared to the pumping regions. For fixed velocity, volume, and channel height, we see that the maximum pressure is inversely propor-

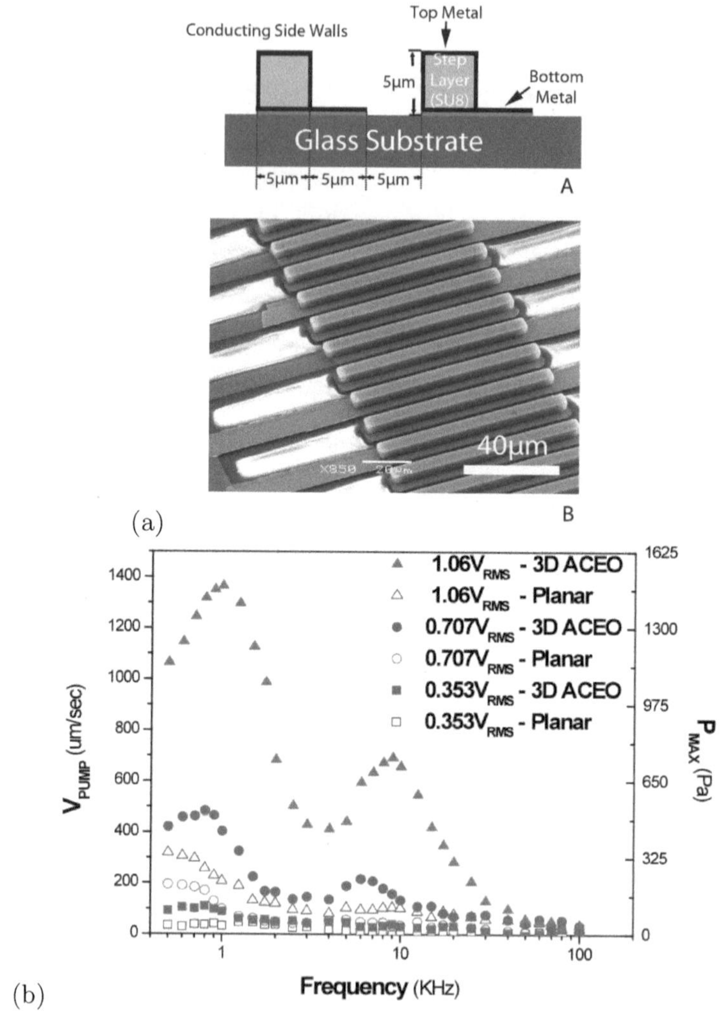

Figure 31. State-of-the-art ACEO micropumps by Huang et al. (2010), using theoretical predicted optimal electrode shapes to create a "fluid conveyor belt" . (a) Fabrication schematic and SEM image of a 3D stepped electrode array, close to the predicted optimal geometry. (b) Experimental demonstration of ultra-fast (> 1 mm/sec) mean velocity over the pump for water in a microfluidic loop with 1.06 Volt rms (3 Vpp), outperforming the standard planar pump of Brown et al. (2000) shown below in Fig. 34(b). The head pressure (> 1% atm) is increased by an order of magnitude using long serpentine channels to hinder reverse pressure-driven flow. (Reproduced from Huang et al. (2010) © 2010 Royal Society of Chemistry.)

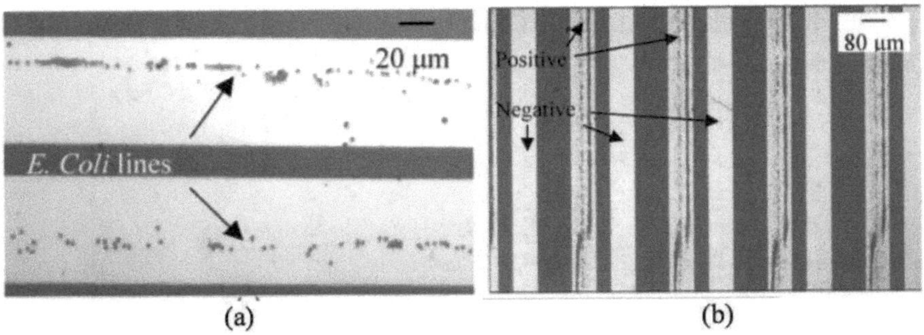

Figure 32. (a) Collection of E. Coli bacteria in tap water along the stagnation lines of ACEO flow on Au microelectrodes at low frequency (100 Hz) and moderate voltage (1 V). (b) Preferential particle trapping by asymmetric polarization ACEO on electrodes with positive DC bias at higher voltage (>3 V). (Reproduced from Wu (2006) © 2006 IEEE.)

tional to the maximum flow rate. This formula also determines the required volume for the pump, given target specifications of flow rate and pressure for a given application.

Fluid pumping over electrode arrays can also be achieved by applying a traveling wave of voltage. At low frequency, a similar induced-charge electro-osmotic mechanism, which peaks at the RC frequency ω_c, is responsible for the flow (Cahill et al., 2004). At high frequency (or with a thick dielectric coating on the electrodes), the classical Erlich-Melcher effect used to pump dielectric liquids, which peaks at the Debye frequency, D/λ_D^2, can also be observed (Ramos et al., 2005). Although traveling-wave ACEO seems to produce slower net flow than standing-wave ACEO with planar electrodes, the possibility of designing suitable non-planar electrodes has not yet been considered.

ACEO flows can also be used to manipulate colloidal particles and biological cells in microfluidic devices (Green et al., 2000b; Wong et al., 2004; Wu, 2006). The basic strategy is to use ACEO flow to draw particles to stagnation points on the electrodes, where they are trapped. By increasing the voltage, the ACEO flow can be reversed, and particles are observed to move away from the stagnation lines, overcoming any remaining trapping force. In this way, it is possible to write and erase suspended particles, bacteria, or micro-algae on microelectrodes, as shown in figure 7(a). This effect can be enhanced by added a DC bias voltage to the low-frequency AC voltage (50-100 Hz) between adjacent electrodes, as proposed by Wu (2006). Particles are observed to collect only on the positively biased electrode, as shown in Figure 32(b). It has been suggested that opposing ACEO flows are produced by the competition between Faradaic charging on one electrode (positive bias) and capacitive charging on the other (negative bias), but a quantitative theory of these experiments remains to be developed. Perhaps the effect is related to bulk electrohydrodynamic flows (Sides, 2001) and related AC colloidal aggregation phenomena on electrodes, recently reviewed by Prieve et al. (2010)

11 Beyond the Standard Model

In spite of many successes, the Standard Model has some serious shortcomings, recently reviewed and analyzed by Bazant et al. (2009b). It generally over-predicts fluid velocities compared to experiments, sometimes by orders of magnitude. It also fails to capture key experimental trends, such as the decay of ICEO flow with increasing salt concentration, flow reversals at high voltage and/or high frequency, and ion-specificities. The reasons for these discrepancies are not yet fully understood.

Bazant and Squires (2010) have reviewed various recent theoretical advances, which extend the Standard Model in the following ways: (i) thin-double-layer approximations for large induced voltages based on the classical Poisson-Nernst-Planck (PNP) equations of ion transport and Navier-Stokes (NS) equations of fluid flow, (ii) thick-double-layer approximations for the PNP/NS equations at low voltages, (iii) modified boundary conditions for electrochemical processes, and (iv) modified PNP/NS equations for large voltages and/or concentrated solutions. The reader is referred to the review articles for details, and here we only mention two interesting new effects of the latter type, crowding and overscreening, which become important in highly charged double layers.

As shown in Fig. 33, when the surface potential relative to the bulk solution greatly exceeds the thermal voltage $k_B T/e$, there is inevitably *crowding* of solvated ions in the inner portion of the double layer, and this pushes apart the diffuse screening cloud away from the surface, thus effectively separating the two plates of the double layer capacitor. Since capacitance is inversely proportional to the plate separation, the crowding of ions at high voltage causes the differential capacitance to decrease at large voltages, as the square root of the voltage, once a condensed layer of uniform charge density forms. In contrast, the classical Gouy-Chapman model of dilute-solution theory predicts the opposite voltage dependence, an exponential growth of capacitance with voltage, given by Eq. (29), since nothing stops

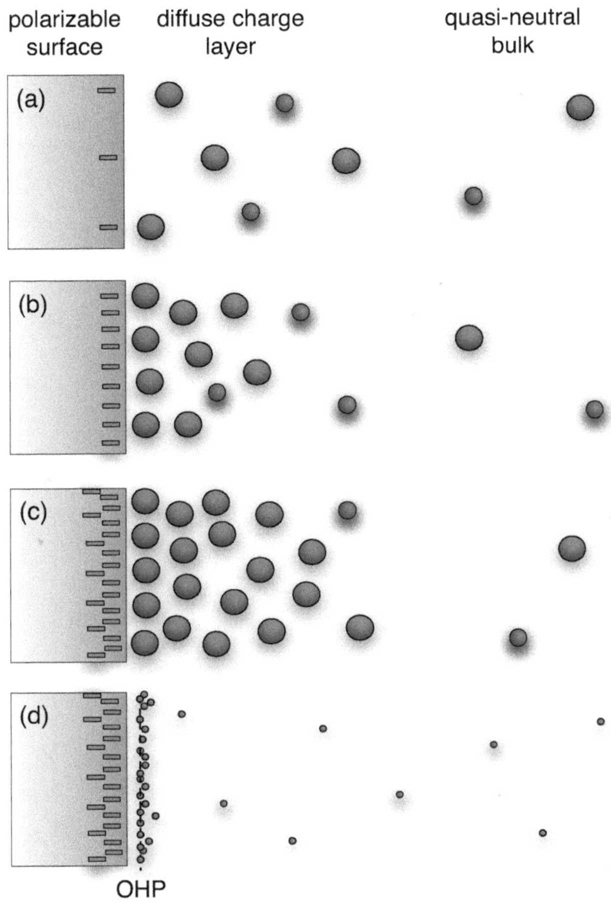

Figure 33. Solvated counterions (larger green spheres) and co-ions (smaller orange spheres) near a polarizable surface. (a) At small induced voltages, $\Psi_D \ll \Psi_c$, the neutral bulk is only slightly perturbed with a diffuse-charge layer of excess counterions at the scale of λ_D. (b) At moderate voltages, $\Psi_D \approx \Psi_c$, the diffuse layer contracts, as described by Poisson-Boltzmann (PB) theory. (c) At large voltages, $\Psi_D \gg \Psi_c$, the counterions inevitably become crowded, causing expansion of the diffuse layer compared to the predictions of the classical Gouy-Chapman-Stern model, sketched in (d), which is based Poisson-Boltzman theory for point-like ions with a minimum distance of approach, the "outer Helmholtz plane" (OHP), to model solvation of the surface. (Reproduced from Bazant et al. (2009b) © 2009 Elsevier.)

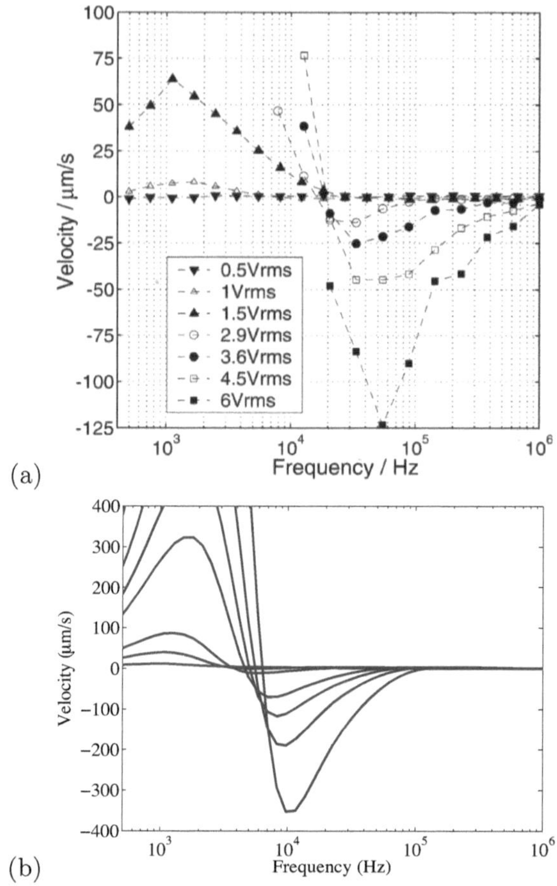

Figure 34. Crowding of finite-sized ions (Fig. 8) and high-frequency flow reversal of planar ACEO pumps. (a) Experimentally observed velocity pumping of 0.1mM KCl by the ACEO pump of Brown et al. (2000) around a microfluidic loop versus AC frequency at different peak-to-peak voltages (reproduced from Studer et al. (2004) © 2004 Royal Society of Chemistry). (b) Simulations of the same flow using a modified electrokinetic equations with an effective hydrated ion size $a = 4.4$ nm for a lattice gas in the mean-field local-density approximation; similar results are obtained using a solvated ion diameter $a \approx 1$ nm for hard spheres with dielectric saturation in water (reproduced from Storey et al. (2008) © 2008 American Physical Society).

point-like ions from piling up at the surface (or Outer Helmholtz Plane). Using a simple mean-field theory of excluded volume effects for finite-sized ions, a more general capacitance formula can be derived (Kilic et al., 2007; Bazant et al., 2009b)

$$C_D^\nu = \frac{\frac{\varepsilon}{\lambda_D}\sinh(\frac{ze\Psi_D}{k_BT})}{[1+2\nu\sinh^2\left(\frac{ze\Psi_D}{2k_BT}\right)]\sqrt{\frac{2}{\nu}[1+2\nu\sinh^2\left(\frac{ze\Psi_D}{2k_BT}\right)]}} \tag{60}$$

where $\nu = 2c_0/c_{max}$ is the bulk volume fraction of ions (in a binary electrolyte). Chapman's formula (29) is recovered in the dilute solution limit, $\nu \to 0$, or at low voltages. With this convenient analytical expression, Storey et al. (2008) were able to predict the flow reversal of ACEO pumps at high frequency and high voltage, in reasonable agreement with the experiments of Studer et al. (2004), as shown in Fig. 34. At high voltage and low frequency (or in the DC limit), Faradaic charge-transer reactions consume normal current and can discharge the double layer, like a short circuit. At high frequency, however, the polarity of the double layer changes too quickly and solvated ions are squashed near the surface and then quickly removed, prior to the onset of Faradaic reactions.

Another important effect related to ion crowding is the breakdown of the mean-field approximation, due to the importance of discrete Coulomb forces between individual ions (Levin, 2002). It is ubiquitous in electrokinetics to use the mean, continuum charge density in Poisson's equation, $\rho = \nabla\cdot\varepsilon\nabla\phi$, to determine the "mean electric field" $-\nabla\phi$, which exerts forces on ions to cause electromigration and electro-osmotic flows, described by the Nernst-Planck and Navier-Stokes equations, respectively. Although Poisson's equation holds instantaneously for the atomic-level charge density, the statistical averaging leading to the mean-field approximation is not always justified. For point-like ions, electrostatic correlations become important at length scales smaller than the Bjerrum length, $\ell_b = e^2/4\pi\varepsilon k_BT$ ($\approx$ 7Åin water at room temperature), where the Coulomb energy between a pair of ions exceeds the thermal energy k_BT. At this scale, especially for multivalent ions, Coulomb correlations lead to new phenomena, such as *overscreening* of a charged surface. An excess number of counterions is attacted to the surface, leading to an apparent "charge inversion" that is corrected by charge-density oscillations in subsequent layers. For a dense mixtures of finite-size ions (such as ionic liquids, molten salts, or highly concentrated or confined electrolytes), the ion size becomes the relevant scale for electrostatic correlations, and a subtle interplay with steric constraints leads to a layered structure of alternating charge in the double layer, which reduces the capacitance at high voltage. The availability of a simple continuum theory for these effects from Bazant et al. (2011) may enable their importance to be assessed in induced-charge electrokinetics in a general modeling framework for non-equilibrium thermodynamics of concentration solutions, as suggested by Bazant et al. (2009b).

12 Conclusion

Induced-charge electrokinetic phenomena comprise an active, growing, interdisciplinary field of research, which spans colloid science, microfluidics, and electrochemical systems. ICEO flows always occur to some degree at any interface between an electrolyte and a polarizable surface, subject an applied voltage or electric field. With the advent of microfabrication techniques, ICEO flows can now be probed and exploited with high precision in microfluidic devices or designer colloidal particles. Beyond these engineering opportunities, induced-charge electrokinetic phenomena raise profound scientific questions about the structure and dynamics of highly-charged double layers.

This material is based upon work supported by the U. S. National Science Foundation under Grant No. 0707641. Any opinions, findings, and conclusions or recommendations expressed in this material are those of the author and do not necessarily reflect the views of the National Science Foundation.

Bibliography

A. Ajdari. AC pumping of liquids. *Phys. Rev. E*, 61:R45–R48, 2000.

J. L. Anderson. Effect of non-uniform zeta potential on particle movement in electric fields. *J. Colloid Interface Science*, 105:45–54, 1984.

J. L. Anderson. Colloid transport by interfacial forces. *Annu. Rev. Fluid Mech.*, 21:61–99, 1989.

N. O. Barinova, N. A. Mishchuk, and T. A. Nesmeyanova. Electroosmosis at spherical and cylindrical metal surfaces. *Colloid Journal*, 70:695–702, 2008.

E. Barsoukov and J. R. Macdonald, editors. *Impedance spectroscopy: Theory, experiment and applications*. J. Wiley & Sons, 2005.

M. Z. Bazant. AC electro-osmotic flow. In D. Li, editor, *Encyclopedia of Microfluidics and Nanofluidics*, volume Part 1, pages 8–14. Springer, 2008a.

M. Z. Bazant. Nonlinear electrokinetic phenomena. In D. Li, editor, *Encyclopedia of Microfluidics and Nanofluidics*, volume Part 14, pages 1461–1470. Springer, 2008b.

M. Z. Bazant and Y. Ben. Theoretical prediction of fast 3d AC electro-osmotic pumps. *Lab on a Chip*, 6:1455–1461, 2006.

M. Z. Bazant and T. M. Squires. Induced-charge electrokinetic phenomena. *Current Opinion in Colloid and Interface Science*, 15:203–213, 2010.

M. Z. Bazant and T. M. Squires. Induced-charge electro-kinetic phenomena: Theory and microfluidic applications. *Phys. Rev. Lett.*, 92:066101, 2004.

M. Z. Bazant, K. Thornton, and A. Ajdari. Diffuse charge dynamics in electrochemical systems. *Phys. Rev. E*, 70:021506, 2004.

M. Z. Bazant, M. S. Kilic, B. D. Storey, and A. Ajdari. Nonlinear electrokinetics at large voltages. *New Journal of Physics*, 11:075016, 2009a.

M. Z. Bazant, M. S. Kilic, B.D. Storey, and A. Ajdari. Towards an understanding of nonlinear electrokinetics at large voltages in concentrated solutions. *Advances in Colloid and Interface Science*, 152:48–88, 2009b.

M. Z. Bazant, B. D. Storey, and A. A. Kornyshev. Double layer in ionic liquids: Overscreening versus crowding. *Phys. Rev. Lett.*, 106:046102, 2011.

J. O.'M Bockris and A. K. N. Reddy. *Modern Electrochemistry.* Plenum, New York, 1970.

A. B. D. Brown, C. G. Smith, and A. R. Rennie. Pumping of water with ac electric fields applied to asymmetric pairs of microelectrodes. *Phys. Rev. E*, 63:016305, 2000.

D. N. Burch and M. Z. Bazant. Design principle for improved three-dimensional ac electro-osmotic pumps. *Phys. Rev. E*, 77:055303(R), 2008.

B. P. Cahill, L. J. Heyderman, J. Gobrecht, and A. Stemmer. Electro-osmotic streaming on application of traveling-wave electric fields. *Physical Review E*, 70:036305, 2004.

C. Q. Choi. Big lab on a tiny chip. *Scientific American*, October:100–103, 2007.

K. T. Chu and M. Z. Bazant. Surface conservation laws at microscopically diffuse interfaces. *J. Colloid Interface Science*, 315:319–329, 2007.

A.V. Delgado, F. González-Caballero, R.J. Hunter, L.K. Koopal, and J. Lyklema. Measurement and interpretation of electrokinetic phenomena. *J. Colloid Interface Sci.*, 309:194–224, 2007.

A. S. Dukhin. Biospecific mechanism of double layer formation and peculiarities of cell electrophoresis. *Colloids and Surfaces A*, 73:29–48, 1993.

A. S. Dukhin and S. S. Dukhin. Aperiodic capillary electrophoresis method using alternating current electric field for separation of macromolecules. *Electrophoresis*, 26:2149–2153, 2005.

S. S. Dukhin and B. V. Derjaguin. *Surface and Colloid Science*, volume Vol. 7, chapter Ch. 2. Academic Press, New York, 1974.

S. S. Dukhin and V. N. Shilov. Kinetic aspects of electrochemistry of disperse systems. part II. induced dipole moment and the non-equilibrium double layer of a colloid particle. *Adv. Colloid Interface Sci.*, 13:153–195, 1980.

M. C. Fair and J. L. Anderson. Electrophoresis of heterogeneous colloids – doublets of dissimilar particles. *Langmuir*, 8:2850–2854, 1992.

N. I. Gamayunov, V. A. Murtsovkin, and A. S. Dukhin. Pair interaction of particles in electric field. 1. features of hydrodynamic interaction of polarized particles. *Colloid J. USSR*, 48:197–203, 1986.

N. I. Gamayunov, G. I. Mantrov, and V. A. Murtsovkin. Study of flows induced in the vicinity of conducting particles by an external electric field. *Colloid J.*, 54:20–23, 1992.

S. Gangwal, O. J. Cayre, M. Z. Bazant, and O. D. Velev. Induced-charge electrophoresis of metallo-dielectric particles. *Phys. Rev. Lett.*, 100: 058302, 2008.

A. González, A. Ramos, N. G. Green, A. Castellanos, and H. Morgan. Fluid flow induced by non-uniform ac electric fields in electrolytes on microelectrodes. II. a linear double-layer analysis. *Phys. Rev. E*, 61: 4019, 2000.

D. C. Grahame. The electrical double layer and the theory of electrocapillarity. *Chem. Rev.*, 41:441–501, 1947.

N. G. Green, A. Ramos, A. González, H. Morgan, and A. Castellanos. Fluid flow induced by nonuniform ac electric fields in electrolytes on microelectrodes. I. experimental measurements. *Phys. Rev. E*, 61:4011–4018, 2000a.

N. G. Green, A. Ramos, and H. Morgan. AC electrokinetics: a survey of submicrometre particle dynamics. *J. Appl. Phys. D*, 33:632–641, 2000b.

N. G. Green, A. Ramos, A. González, A. Castellanos, and H. Morgan. Fluid flow induced by nonuniform ac electric fields in electrolytes on microelectrodes. III. observation of streamlines and numerical simulation. *Phys. Rev. E*, 66:026305, 2002.

M. M. Gregersen, F. Okkels, M. Z. Bazant, and H. Bruus. Topology and shape optimization of induced-charge electro-osmotic micropumps. *New Journal of Physics*, 11:075016, 2009.

C. K. Harnett, J. Templeton, K. A. Dunphy-Guzman, Y. M. Senousya, and M. P. Kanouff. Model based design of a microfluidic mixer driven by induced charge electroosmosis. *Lab on a Chip*, 8:565–572, 2008.

H. Helmholtz. Studien über electrische grenzschichten. *Ann. Phys. Chem.*, 7 (ser. 3):337–382, 1879.

C. C. Huang, M. Z. Bazant, and T. Thorsen. Ultrafast high-pressure ac electro-osmotic pumps for portable biomedical microfluidics. *Lab on a Chip*, 10:80–85, 2010.

R. J. Hunter. *Foundations of Colloid Science.* Oxford University Press, Oxford, 2001.

A. S. Khair and T. M. Squires. Fundamental aspects of concentration polarization arising from nonuniform electrokinetic transport. *Phys. Fluids*, 20:087102, 2008.

M. S. Kilic and M. Z. Bazant. Induced-charge electrophoresis near a wall. *Electrophoresis*, in press, 2011. arXiv:0712.0453v1 [cond-mat.mtrl-sci].

M. S. Kilic, M. Z. Bazant, and A. Ajdari. Steric effects in the dynamics of electrolytes at large applied voltages: I double-layer charging. *Phys. Rev. E*, 75:021502, 2007.

B. J. Kirby and Ernest H. Hasselbrink. Zeta potential of microfluidic substrates: 1. theory, experimental techniques, and effects on separations. *Electrophoresis*, 25:187–202, 2004.

D. J. Laser and J. G. Santiago. A review of micropumps. *J. Micromech Microeng*, 14:R35–64, 2004.

V. G. Levich. *Physico-chemical Hydrodynamics.* Prentice-Hall, London, 1962.

Y. Levin. Electrostatic correlations: from plasma to biology. *Rep. Prog. Phys.*, 65:1577–1632, 2002.

J. A. Levitan. *Experimental study of induced-charge electro-osmosis.* PhD thesis, MIT, 2005.

J. A. Levitan, S. Devasenathipathy, V. Studer, Y. Ben, T. Thorsen, T. M. Squires, and M. Z. Bazant. Experimental observation of induced-charge electro-osmosis around a metal wire in a microchannel. *Colloids and Surfaces A*, 267:122–132, 2005.

D. Long and A. Ajdari. Electrophoretic mobility of composite objects in free solution: Application to dna separation. *Electrophoresis*, 17:1161–1166, 1996.

D. Long and A. Ajdari. Symmetry properties of the electrophoretic motion of patterned colloidal particles. *Phys. Rev. Lett.*, 81:1529–1532, 1998.

J. Lyklema. *Fundamentals of Interface and Colloid Science. Volume II: Solid-Liquid Interfaces.* Academic Press Limited, San Diego, CA, 1995.

J. R. Melcher and G. I. Taylor. Electrohydrodynamics: a review of the role of interfacial shear stresses. *Annu. Rev. Fluid Mech.*, 1:11146, 1969.

F. A. Morrison and J. J. Stukel. Electrophoresis of an insulating sphere normal to a conducting plane. *J. Colloid and Interface Science*, 33:88, 1970.

V. A. Murtsovkin. Nonlinear flows near polarized disperse particles. *Colloid Journal*, 58:341–349, 1996.

V. A. Murtsovkin and G. I. Mantrov. Steady flows in the neighborhood of a drop of mercury with the application of a variable external electric field. *Colloid J. USSR*, 53:240–244, 1991.

F. Nadal, F. Argoul, P. Hanusse, B. Pouligny, and A. Ajdari. Electrically induced interactions between colloidal particles in the vicinity of a conducting plane. *Phys. Rev. E*, 65, 2002.

L. H. Olesen, H. Bruus, and A. Ajdari. AC electrokinetic micropumps: the effect of geometrical confinement faradaic current injection and nonlinear surface capacitance. *Phys. Rev. E*, 73:056313, 2006.

A. J. Pascall and T. M. Squires. Induced charge electroosmosis over controllably-contaminated electrodes. *Phys. Rev. Lett*, 104:088301, 2010.

H. A. Pohl. *Dielectrophoresis: the behaviour of neutral matter in nonuniform electric fields*. Cambridge University Press, 1978.

D. C. Prieve, P. J. Sides, and C. L. Wirth. 2-d assembly of colloidal particles on a planar electrode. *Current Opinion in Colloid and Interface Science*, 15:160–174, 2010.

A. Ramos, H. Morgan, N. G. Green, and A. Castellanos. AC electrokinetics: a review of forces in microelectrode structures. *J. Phys. D*, 31:2338–2353, 1998.

A. Ramos, H. Morgan, N. G. Green, and A. Castellanos. AC electric-field-induced fluid flow in microelectrodes. *J. Colloid Interface Sci.*, 217: 420–422, 1999.

A. Ramos, H. Morgan, N. G. Green, A. Gonzalez, and A. Castellanos. Pumping of liquids with traveling-wave electro-osmosis. *Journal of Applied Physics*, 97:084906, 2005.

R. Rica and M. Z. Bazant. Electrodiffusiophoresis: Particle motion in electrolytes under direct current. *Physics of Fluids*, 22:112109, 2010.

W. D. Ristenpart, I. A. Aksay, and D. A. Saville. Electrically guided assembly of planar superlattices in binary colloidal suspensions. *Phys. Rev. Lett.*, 90:128303, 2003.

K. A. Rose and J. G. Santiago. Rotational electrophoresis of striped metallic microrods. *Physical Review E*, 75:197–203, 2006.

I. Rubinstein and B. Zaltzman. Electro-osmotic slip of the second kind and instability in concentration polarization at electrodialysis membranes. *Math. Mod. Meth. Appl. Sci.*, 11:263–300, 2001.

D. Saintillan, E. Darve, and E. S. G. Shaqfeh. Hydrodynamic interactions in the induced-charge electrophoresis of colloidal rod dispersions. *Journal of Fluid Mechanics*, 563:223–259, 2006a.

D. Saintillan, E. S. G. Shaqfeh, and E. Darve. Stabilization of a suspension of sedimenting rods by induced-charge electrophoresis. *Physics of Fluids*, 18:121701, 2006b.

D. A. Saville. Electrohydrodynamics: The taylor-melcher leaky dielectric model. *Annu. Rev. Fluid Mech.*, 29:27–64, 1997.

R. B. M. Schasfoort, S. Schlautmann, J. Hendrikse, and A. van den Berg. Field-effect flow control for microfabricated fluidic networks. *Science*, 286:942, 1999.

R. B. Schoch, J. Y. Han, and P. Renaud. Transport phenomena in nanofluidics. *Rev. Mod. Phys.*, 80(3):839–883, 2008.

V. N. Shilov and T. S. Simonova. Polarization of electric double layer of disperse particles and dipolophoresis in a steady (dc) field. *Colloid J. USSR*, 43:90–96, 1981.

P. J. Sides. Electrohydrodynamic particle aggregation on an electrode driven by an alternating electric field normal to it. *Langmuir*, 17:5791–5800, 2001.

I. N. Simonov and V. N. Shilov. Theory of low-frequency dielectric-dispersion of a suspension of ideally polarizable spherical particles. *Colloid J. USSR*, 39:775–780, 1977.

T. A. Simonova, V. N. Shilov, and O. A. Shramko. Low-frequency dielectrophoresis and the polarization interaction of uncharged spherical particles with an induced deby atmosphere of arbitrary thickness. *Colloid J.*, 63:108–115, 2001.

M Smoluchowski. Zur theorie der elektrischen kataphorese und der oberflchenleitung. *Phys. Z.*, 6:529, 1905.

T. M. Squires. Electrokinetics over inhomogeneously slipping surfaces. *Phys. Fluids*, 20:092105, 2008.

T. M. Squires. Induced-charge electrokinetics: fundamental challenges and opportunities. *Lab on a Chip*, 9:2477, 2009. doi: DOI: 10.1039/b906909g.

T. M. Squires and M. Z. Bazant. Induced-charge electro-osmosis. *J. Fluid Mech.*, 509:217–252, 2004.

T. M. Squires and M. Z. Bazant. Breaking symmetries in induced-charge electro-osmosis and electrophoresis. *J. Fluid Mech.*, 560:65–101, 2006.

T. M. Squires and S. R. Quake. Microfluidics: fluid physics on the nanoliter scale. *Rev. Mod. Phys.*, 77:977–1026, 2005.

O. Stern. Zur theorie der electrolytischen doppelschicht. *Z. Elektrochem.*, 30:508–516, 1924.

H. A. Stone and A. D. T. Samuel. Propulsion of microorganisms by surface distortions. *Phys. Rev. Lett.*, 77:41024104, 1996.

H.A. Stone, A.D. Stroock, and A. Ajdari. Engineering flows in small devices: Microfluidics toward a lab-on-a-chip. *Annu. Rev. Fluid Mech.*, 36:381411, 2004.

B. D. Storey, Lee R. Edwards, M. S. Kilic, and M. Z. Bazant. Steric effects on ac electro-osmosis in dilute electrolytes. *Phys. Rev. E*, 77:036317, 2008.

V. Studer, A. Pépin, Y. Chen, and A. Ajdari. An integrated ac electrokinetic pump in a microfluidic loop for fast tunable flow control. *Analyst*, 129: 944–949, 2004.

Y. K. Suh and S. Kang. Asymptotic analysis of ion transport in a nonlinear regime around polarized electrodes under ac. *Phys. Rev. E*, 77:011502, 2008.

Y. K. Suh and S. Kang. Numerical prediction of ac electro-osmotic ows around polarized electrodes. *Phys. Rev. E*, 79:046309, 2009.

V. Tandon and B. J. Kirby. Zeta potential and electroosmotic mobility in microfluidic devices fabricated from hydrophobic polymers: 2. slip and interfacial water structure. *Electrophoresis*, 29:1102–1114, 2008.

G. I. Taylor. Studies in electrohydrodynamics i. the circulation produced in a drop by an electric field. *Proc. Roy. Soc. A*, 291:15966, 1966.

S. K. Thamida and H.-C. Chang. Nonlinear electrokinetic ejection and entrainment due to polarization at nearly insulated wedges. *Phys. Fluids*, 14:4315, 2002.

M. Trau, D. A. Saville, and I. A. Aksay. Assembly of colloidal crystals at electrode interfaces. *Langmuir*, 13:6375, 1997.

J. P. Urbanski, J. A. Levitan, M. Z. Bazant, and Todd Thorsen. Fast AC electro-osmotic pumps with non-planar electrodes. *Applied Physics Letters*, 89:143508, 2006a.

J. P. Urbanski, J. A. Levitan, D. N. Burch, Todd Thorsen, and M. Z. Bazant. The effect of step height on the performance of AC electro-osmotic microfluidic pumps. *Journal of Colloid and Interface Science*, 309:332–341, 2006b.

E. J. van der Wouden, D. C. Hermes, J. G. E. Gardeniers, and A. van den Berg. Directional flow induced by synchronized longitudinal and zeta-potential controlling ac-electrical fields. *Lab on a Chip*, 6:1300 – 1305, 2006.

E.J. van der Wouden, T. Heuser, R.E. Oosterbroek D.C. Hermes, J.G.E. Gardeniers, and A. van den Berg. Field-effect control of electro-osmotic flow in microfluidic networks. *Colloids and Surfaces A: Physicochemical and Engineering Aspects*, 267:110–116, 2005.

R. E. G. van Hal, J. C. T. Eijkel, and P. Bergveld. A general model to describe the electrostatic potential at electrolyte oxide interfaces. *Adv. Colloid Interface Sci.*, 69:31–62, 1996.

P. K. Wong, T. H. Wang, J. H. Deval, and C. M. Ho. Electrokinetics in microdevices for biotechnology applications. *IEEE/AMSE Transactions on Mechatronics*, 9:366–376, 2004.

J. Wu. Biased AC electro-osmosis for on-chip bioparticle processing. *IEEE Transactions on Nanotechnology*, 5:84–88, 2006.

Z. Wu and D. Li. Micromixing using induced-charge electrokinetic flow. *Electrochimica Acta*, 53(19):5827–5835, AUG 1 2008a.

Z. Wu and D. Li. Mixing and flow regulating by induced-charge electrokinetic flow in a microchannel with a pair of conducting triangle hurdles. *Microfluidics and Nanofluidics*, 5(1):65–76, JUL 2008b.

S. H. Yao and J. G. Santiago. Porous glass electroosmotic pumps: theory. *Journal of Colloid and Interface Science*, 268(1):133–142, DEC 1 2003. ISSN 0021-9797.

E. Yariv. Induced-charge electrophoresis of nonspherical particles. *Phys. Fluids*, 17:051702, 2005.

E. Yariv. Nonlinear electrophoresis of ideally polarizable particles. *Europhysics Letters*, 82(5):54004, JUN 2008. ISSN 0295-5075.

S. Yeh, M. Seul, and B. Shraiman. Assembly of ordered colloidal aggregrates by electric-field-induced fluid flow. *Nature (London)*, 386:57, 1997.

G. Yossifon, I. Frankel, and T. Miloh. On electro-osmotic flows through microchannel junctions. *Phys. Fluids*, 18:117108, 2006.

B. Zaltzman and I. Rubinstein. Electro-osmotic slip and electroconvective instability. *Journal of Fluid Mechanics*, 579:173–226, 2007.

H. Zhao and H. H. Bau. On the effect of induced electro-osmosis on a cylindrical particle next to a surface. *Langmuir*, 23:4053–4063, 2007a.

H. Zhao and H. H. Bau. Microfluidic chaotic stirrer utilizing induced-charge electro-osmosis. *Phys. Rev. E*, 75:066217, 2007b.

FSC
www.fsc.org
MIX
Papier | Fördert
gute Waldnutzung
FSC® C083411

Kolibri 360 GmbH
Ferdinand-Jühlke-Straße 7
99095 Erfurt, Deutschland
produktsicherheit@kolibri360.de